GOING THE DISTANCE?

THE SAFE TRANSPORT OF SPENT NUCLEAR FUEL AND HIGH-LEVEL RADIOACTIVE WASTE IN THE UNITED STATES

Committee on Transportation of Radioactive Waste

Nuclear and Radiation Studies Board
Transportation Research Board

NATIONAL RESEARCH COUNCIL
OF THE NATIONAL ACADEMIES

THE NATIONAL ACADEMIES PRESS
Washington, D.C.
www.nap.edu

THE NATIONAL ACADEMIES PRESS 500 Fifth Street, N.W. Washington, DC 20001

NOTICE: The project that is the subject of this report was approved by the Governing Board of the National Research Council, whose members are drawn from the councils of the National Academy of Sciences, the National Academy of Engineering, and the Institute of Medicine. The members of the committee responsible for the report were chosen for their special competences and with regard for appropriate balance.

Support for this study was provided by the U.S. Department of Energy (Grant No. DE-FG28-02RW12177), U.S. Department of Transportation (Order No. DTRS56-02-P-70041 and Contract No. DTPH56-05-C-0002), Nuclear Regulatory Commission (Cooperative Agreement No. NRC-02-02-014), Electric Power Research Institute (Cooperative Agreement No. EP-P9056/C4563), National Cooperative Highway Research Program, and the National Academy of Sciences. All opinions, findings, conclusions, or recommendations expressed herein are those of the authors and do not necessarily reflect the views of these agencies and organizations.

International Standard Book Number 0-309-10004-6 (Book)
International Standard Book Number 0-309-65316-9 (PDF)
Library of Congress Control Number 200697629

Additional copies of this report are available from:

National Academies Press
500 Fifth Street, N.W.
Lockbox 285
Washington, DC 20055
800-624-6242
202-334-3313 (in the Washington Metropolitan Area)
http://www.nap.edu

Cover design by Van Nguyen.

Printed in the United States of America

THE NATIONAL ACADEMIES

Advisers to the Nation on Science, Engineering, and Medicine

The **National Academy of Sciences** is a private, nonprofit, self-perpetuating society of distinguished scholars engaged in scientific and engineering research, dedicated to the furtherance of science and technology and to their use for the general welfare. Upon the authority of the charter granted to it by the Congress in 1863, the Academy has a mandate that requires it to advise the federal government on scientific and technical matters. Dr. Ralph J. Cicerone is president of the National Academy of Sciences.

The **National Academy of Engineering** was established in 1964, under the charter of the National Academy of Sciences, as a parallel organization of outstanding engineers. It is autonomous in its administration and in the selection of its members, sharing with the National Academy of Sciences the responsibility for advising the federal government. The National Academy of Engineering also sponsors engineering programs aimed at meeting national needs, encourages education and research, and recognizes the superior achievements of engineers. Dr. Wm. A. Wulf is president of the National Academy of Engineering.

The **Institute of Medicine** was established in 1970 by the National Academy of Sciences to secure the services of eminent members of appropriate professions in the examination of policy matters pertaining to the health of the public. The Institute acts under the responsibility given to the National Academy of Sciences by its congressional charter to be an adviser to the federal government and, upon its own initiative, to identify issues of medical care, research, and education. Dr. Harvey Fineberg is president of the Institute of Medicine.

The **National Research Council** was organized by the National Academy of Sciences in 1916 to associate the broad community of science and technology with the Academy's purposes of furthering knowledge and advising the federal government. Functioning in accordance with general policies determined by the Academy, the Council has become the principal operating agency of both the National Academy of Sciences and the National Academy of Engineering in providing services to the government, the public, and the scientific and engineering communities. The Council is administered jointly by both Academies and the Institute of Medicine. Dr. Ralph J. Cicerone and Dr. Wm. A. Wulf are chairman and vice chairman, respectively, of the National Research Council.

www.national-academies.org

COMMITTEE ON TRANSPORTATION OF RADIOACTIVE WASTE

NEAL F. LANE, *Chair*, Rice University, Houston, Texas
THOMAS B. DEEN, *Vice Chair*, National Research Council (retired), Stevensville, Maryland
JULIAN AGYEMAN, Tufts University, Medford, Massachusetts
LISA M. BENDIXEN, ICF Consulting, Lexington, Massachusetts
DENNIS C. BLEY, Buttonwood Consulting, Inc., Oakton, Virginia
HANK C. JENKINS-SMITH, Texas A&M University, College Station
MELVIN F. KANNINEN, MFK Consulting Services, San Antonio, Texas
ERNEST J. MONIZ, Massachusetts Institute of Technology, Cambridge
JOHN W. POSTON, SR., Texas A&M University, College Station
LACY E. SUITER, Federal Emergency Management Agency (retired), Alexandria, Virginia
JOSEPH M. SUSSMAN, Massachusetts Institute of Technology, Cambridge
ELIZABETH TEN EYCK, ETE Consulting, Inc., Great Falls, Virginia
SETH TULER, Social & Environmental Research Institute, Greenfield, Massachusetts
DETLOF VON WINTERFELDT, University of Southern California, Los Angeles
THOMAS R. WARNE, Tom Warne & Associates, LLC, South Jordan, Utah
CLIVE N. YOUNG, United Kingdom Department for Transport, London

Board on Radioactive Waste Management Liaison[1]

NORINE E. NOONAN, College of Charleston, South Carolina

Staff

KEVIN D. CROWLEY, Study Director, Nuclear and Radiation Studies Board
JOSEPH MORRIS, Senior Program Officer, Transportation Research Board
DARLA J. THOMPSON, Research Associate
LAURA D. LLANOS, Senior Program Assistant
LEE FINEWOOD, Intern
BRANDON JONES, Intern

[1]Through February 2005, when the Board on Radioactive Waste Management was reorganized into the Nuclear and Radiation Studies Board.

NUCLEAR AND RADIATION STUDIES BOARD

RICHARD A. MESERVE, *Chair*, Carnegie Institution, Washington, D.C.
S. JAMES ADELSTEIN, *Vice Chair*, Harvard Medical School, Boston, Massachusetts
HAROLD L. BECK, Environmental Measurements Laboratory (retired), New York City, New York
JOEL S. BEDFORD, Colorado State University, Fort Collins
ROBERT M. BERNERO, U.S. Nuclear Regulatory Commission (retired), Gaithersburg, Maryland
SUE B. CLARK, Washington State University, Pullman
JAMES E. CLEAVER, University of California, San Francisco
ALLEN G. CROFF, Oak Ridge National Laboratory (retired), Tennessee
DAVID E. DANIEL, University of Texas at Dallas
SARAH C. DARBY, Clinical Trial Service Unit (CTSU), Oxford, United Kingdom
SHARON L. DUNWOODY, University of Wisconsin, Madison
RODNEY C. EWING, University of Michigan, Ann Arbor
ROGER L. HAGENGRUBER, University of New Mexico, Albuquerque
DANIEL KREWSKI, University of Ottawa, Ontario, Canada
KLAUS KÜHN, Technische Universität Clausthal, Germany
SUSAN M. LANGHORST, Washington University, St. Louis, Missouri
NIKOLAY P. LAVEROV, Russian Academy of Sciences, Moscow
MILTON LEVENSON, Bechtel International (retired), Menlo Park, California
C. CLIFTON LING, Memorial Hospital, New York City, New York
PAUL A. LOCKE, Johns Hopkins University, Baltimore, Maryland
WARREN F. MILLER, University of New Mexico, Albuquerque
ANDREW M. SESSLER, Lawrence Berkeley National Laboratory, Berkeley, California
ATSUYUKI SUZUKI, Nuclear Safety Commission of Japan, Tokyo
JOHN C. VILLFORTH, Food and Drug Law Institute (retired), Derwood, Maryland
PAUL L. ZIEMER, Purdue University (retired), West Lafayette, Indiana

Staff

KEVIN D. CROWLEY, Senior Board Director
EVAN DOUPLE, Scholar
RICK JOSTES, Senior Program Officer
MICAH D. LOWENTHAL, Senior Program Officer
BARBARA PASTINA, Senior Program Officer
JOHN R. WILEY, Senior Program Officer

TRANSPORTATION RESEARCH BOARD EXECUTIVE COMMITTEE

Acknowledgments

The committee conducted an extensive information-gathering effort during this study and attempted to reach out broadly to a wide range of organizations and individuals concerned with transportation of spent fuel and high-level radioactive waste. The successful completion of this report would not have been possible without the cooperation and assistance of a very large number of organizations and individuals. The committee would especially like to thank the following organizations and individuals for providing logistical support, advice, and/or information for this study:

- The project sponsors, whose financial support made this study possible: Department of Energy (DOE), Department of Transportation (DOT), Electric Power Research Institute (EPRI), National Academy of Sciences (NAS), National Cooperative Highway Research Program (NCHRP), and the Nuclear Regulatory Commission (USNRC). The committee would especially like to acknowledge the following individuals from these organizations for their support of the committee's information-gathering efforts:
  - –DOE: Gary Lanthrum, Robin Sweeney, Jeffrey Williams, Judith Holm, Corine Macaluso, Alex Thrower, James Wade, Russ Dyer, Bill Boyle, Barry Miles, Ned Larson, Jozette Booth, Paul Detwiler, and Sue Loudner
  - –DOT: Dick Hannon, Richard Boyle, Michael Conroy, James Simmons, Kevin Blackwell, and Ryan Paquet

–EPRI: John Kessler
–NAS: E. William Colglazier
–NCHRP: Crawford Jencks
–USNRC: Bill Brach, Earl Easton, Jack Guttman, Mike Mayfield, Francis (Skip) Young, Wayne Hodges, and Bernie White

- The State of Nevada (Robert Loux) and its consultants (Robert Halstead, Jim Hall, Marvin Resnikoff, and James Ballard)
- Nevada counties: Clark County (Irene Navis, Fred Dilger, and Engelbrecht von Tiesenhausen), Eureka County (Abby Johnson), Lincoln County (Kevin Phillips), and Nye County (Les Bradshaw)
- Nuclear industry: Nuclear Energy Institute (Steve Kraft and the late John Vincent); Exelon Nuclear Power Corp. (Jeffrey Benjamin, Annie Caputo, Adam Levin, and Bob Rybak); Progress Energy (Steven Edwards); Energy Resources International (Eileen Supko); Private Fuel Storage (John Parkyn); and Ron Pope (consultant)
- States and state associations: New Mexico (Anne DeLain Clark and W. Scott Field) and Utah (Bill Sinclair); the Western Governors' Association (Doug Larson); Western Interstate Energy Board (Bill Mackie); Council of State Governments—Midwest Office (Lisa Sattler and Thor Strong); Northeast Council of State Governments (Edward Wilds); and Southern States Energy Board (Chris Wells)
- Emergency management organizations: Department of Homeland Security (Eric Tolbert, Federal Emergency Management Agency); Argonne National Laboratory (Gordon Veerman); City of Chicago Fire Department (Gene Ryan); Linn County, Iowa Emergency Management Agency (Ned Wright); and Illinois Emergency Management Agency (Tim Runyon)
- Transportation organizations: Association of American Railroads (Bob Fronczak); Burlington Northern Santa Fe Railroad (Patrick Brady); Tri-State Motor Transit (David Bennett); Transportation Technology Center (Jon Hannafious); and Union Pacific Railroad (Sandi Covi)
- Universities and research organizations: Doug Ammerman and Ken Sorensen (Sandia National Laboratories); Bob O'Connor (National Science Foundation); Amy Goodin (University of New Mexico); Dennis Mileti (University of Colorado); Michael Lindell (Texas A&M University); James Flynn (Decision Research); Christiana Drew (University of Washington); Elaine Vaughan (University of California, Irvine); and Christopher Becker (University of Michigan)
- Other nongovernmental organizations: National Environmental Justice Advisory Council (Veronica Eady); Nevada Nuclear Waste Task Force (Judy Treichel); Citizen Alert (Michele Boyd and Peggy Maze); Southwest Research and Information Center (Don Hancock); International Institute for Indigenous Resource Management (Mervyn Tano); and the Keystone Center (Janesse Brewer)

• The many other individuals who attended and offered personal comments at the committee's meetings or provided information and materials for this report

A subgroup of committee members visited Germany and the United Kingdom during the study to become better acquainted with European transportation practices and experiences (see Appendix B). The committee would especially like to acknowledge and thank the following organizations for providing logistical support and information for this trip:

• German Federal Institute for Materials Research and Testing (BAM); especially Manfred Hennecke and Thomas Böllinghaus, who arranged the committee's visit to BAM to witness a package test
• AREVA; especially Frederic Bailly and Vincent Roland, who arranged a briefing for the committee subgroup on French transportation operations
• British Nuclear Fuels, Ltd. (BNFL); especially Colin Boardman and Jonathan Carter, who arranged and accompanied the committee subgroup on its tour of BNFL facilities
• Roger Evans (London Assembly), Linda Hayes (Cricklewood Against Nuclear Trains), Steve Robinson (Environment Business Management), and Patrick van den Bulck (Campaign for Nuclear Disarmament) for sharing their perspectives with the committee about BNFL's transport program
• Christopher Pecover (U.K. Department of Transport) for accompanying the committee on its visit to the United Kingdom

This report has been reviewed in draft form by individuals chosen for their diverse perspectives and technical expertise, in accordance with procedures approved by the National Research Council Report Review Committee. The purpose of this independent review is to provide candid and critical comments that will assist the institution in making the published report as sound as possible and to ensure that the report meets institutional standards for objectivity, evidence, and responsiveness to the study charge. The content of the review comments and draft manuscript remains confidential to protect the integrity of the deliberative process. We wish to thank the following individuals for their participation in the review of this report:

Mark Abkowitz, Vanderbilt University
John Ahearne, Sigma Xi and Duke University
Robert Bernero, Consultant
Sherwood Chu, Consultant
Emil Frankel, Parsons Brinckerhoff

B. John Garrick, Consultant
Richard Garwin, IBM Thomas J. Watson Research Center (emeritus)
David Kocher, SENES Oak Ridge Center for Risk Assessment
Dick Meserve, Carnegie Institution of Washington
Kathleen Meyer, Keystone Scientific, Inc.
Bill Nix, Stanford University (emeritus)
Eugene Rosa, Washington State University
Frank Saccomanno, University of Waterloo
Gordon Veerman, Argonne National Laboratory
Chelsea White III, Georgia Institute of Technology

Although the reviewers listed above have provided many constructive comments and suggestions, they were not asked to endorse the conclusions or recommendations, nor did they see the final draft of the report before its release. The review of this report was overseen by Chris Whipple, ENVIRON International Corporation, and Lester Hoel, University of Virginia. Appointed by the National Research Council, they were responsible for making certain that an independent examination of this report was carried out in accordance with institutional procedures and that all review comments were carefully considered. Responsibility for the final content of this report rests entirely with the authoring committee and the National Research Council.

Finally, this report could not have been completed without the help of many National Research Council staff. The committee especially wants to acknowledge Kevin Crowley (Nuclear and Radiation Studies Board) and Joe Morris and Steve Godwin (Transportation Research Board) for their help in organizing the committee's information-gathering meetings and developing the committee's final report; Darla Thompson, Lee Finewood, and Brandon Jones (Nuclear and Radiation Studies Board) for assistance with research and writing; Laura Llanos (Nuclear and Radiation Studies Board) for handling committee logistics; and Steve Mautner and Mary Kalamaras (National Academies Press) for helping the committee better understand the audience for this report.

Contents

Summary[1]

This study was initiated by the National Academies to meet what it perceived to be a national need for an independent, objective, and authoritative analysis of spent nuclear fuel and high-level radioactive waste transportation in the United States in light of

- The federal government's program to construct and operate a repository at Yucca Mountain, Nevada, to permanently dispose of the nation's spent fuel and high-level waste; and
- The commercial nuclear industry's program to construct and operate a facility in Utah for the temporary (interim) storage of spent fuel from some of its power reactors.

As this study was being completed, the federal government initiated another program to assess the feasibility of developing an integrated facility to receive, store, and recycle commercial spent fuel. If executed, these programs could involve the shipment of large quantities[2] of spent fuel and

[1]This summary provides a brief overview of the findings and recommendations in this report. A more detailed summary of the findings and recommendations is presented in the next chapter.

[2]This report identifies two general types of transportation programs, *small-quantity shipping programs* and *large-quantity shipping programs*. While there is no precise quantity demarcation between these two program types, the former involve shipment on the order of tens of metric tons of spent fuel or high-level waste, while the latter involve shipment on the order of hundreds to thousands of metric tons.

high-level waste over the nation's railways and roadways for periods of several decades.

The Committee on Transportation of Radioactive Waste (hereafter referred to as the committee; see Appendix A) was appointed by the National Academies to carry out this study. The original objectives of this study were to examine the risks and identify key current and future technical and societal concerns for the transport of spent nuclear fuel and high-level radioactive waste in the United States. After this study was under way, its scope was expanded to include a congressionally requested and U.S. Department of Transportation (DOT) sponsored examination of the procedures used by the U.S. Department of Energy (DOE) for selecting routes for transporting research reactor spent fuel between its facilities in the United States. This report provides background information on spent fuel and high-level waste transportation (Chapter 1) and the committee's findings and recommendations with respect to the original study charge (Chapters 2, 3, and 5) and the expanded study charge (Chapter 4).

This report does not examine the security[3] risks for spent fuel and high-level waste transportation because of an inability to access classified and otherwise restricted information. It also does not examine the viability of temporary storage or permanent disposal sites for spent fuel and high-level waste, or the risk trade-offs involved in transporting spent fuel and high-level waste to centralized interim storage or permanent disposal sites versus leaving them in place at current storage sites. The committee operated within the constraints of federal policy decisions that have set the nation on a clear path to transport spent fuel and high-level waste for permanent disposal. The intent of this study is to improve understanding of the issues associated with the transportation of spent fuel and high-level waste regardless of their ultimate destination.

A brief summary of the committee's findings and recommendations is presented below. Readers are strongly encouraged to examine the full text of the findings and recommendations in the referenced sections of this report for additional details.

- The committee could identify no fundamental technical barriers to the safe[4] transport of spent fuel and high-level radioactive waste in the United States. However, there are a number of social and institutional

---

[3]*Security* refers to measures taken to protect spent fuel and high-level waste against sabotage, attacks, and theft while it is in transport.

[4]*Safety* refers to measures taken to protect spent fuel and high-level waste during transport operations from failure, damage, human error, and other inadvertent acts.

challenges to the successful[5] initial implementation of large-quantity shipping programs that will require expeditious resolution (see Section 3.2 and Section 5.1). The challenges of sustained implementation should not be underestimated.

- Malevolent acts against spent fuel and high-level waste shipments are a major technical and societal concern, but the committee was unable to perform an in-depth examination of transportation security because of information constraints. The committee recommends that an independent examination of the security of spent fuel and high-level waste transportation be carried out prior to the commencement of large-quantity shipments to a federal repository or to interim storage (see Section 5.1).
- Transportation packages[6] play a crucial role in transportation safety by providing a robust barrier to the release of radiation and radioactive material. Current international standards and U.S. regulations are adequate to ensure package containment effectiveness over a wide range of transport conditions. However, recently published work suggests that there may be a very small number of extreme accident conditions involving very long duration fires that could compromise package containment effectiveness. The committee recommends that the U.S. Nuclear Regulatory Commission (USNRC) undertake additional analyses of very long duration fire scenarios that bound expected real-world accident conditions as outlined in Section 2.5. Based on the results of these investigations, the USNRC should implement operational controls and restrictions on spent fuel and high-level waste shipments as necessary to reduce the chances that such conditions might be encountered in service.
- The committee strongly endorses the use of full-scale testing to determine how packages will perform under both regulatory and credible extraregulatory[7] conditions. The committee recommends that full-scale package testing continue to be used as part of integrated analytical, computer simulation, scale-model, and testing programs to validate package performance. Full-scale testing of packages to deliberately cause their destruction should not be required as part of this integrated analysis or for compliance demonstrations (Section 2.5).
- This report provides quantitative health and safety risk comparisons for both normal transport conditions and accidents (Section 3.3). The

[5]The committee defines *success* in terms of the program's ability, under existing statutes, regulations, agreements, and budgets, to transport spent fuel and high-level waste in a safe, secure, timely, and publicly acceptable manner.

[6]Containers used for the transport of spent fuel or high-level waste, whether loaded or empty.

[7]That is, under conditions that exceed those embodied in current regulatory requirements.

radiological health and safety risks associated with the transportation of spent fuel and high-level waste are well understood and are generally low, with the possible exception of risks from releases in extreme accidents involving very long duration, fully engulfing fires. The likelihood of such extreme accidents appears to be very small, however, and their occurrence and consequences can be reduced further through relatively simple operational controls and restrictions. The committee recommends that transportation planners and managers undertake detailed surveys of transportation routes to identify and mitigate the potential hazards that could lead to or exacerbate extreme accidents involving very long duration, fully engulfing fires (see Section 3.4).

- The social risks[8] of spent fuel and high-level waste transportation pose important challenges to the successful initial implementation of programs for transporting spent fuel and high-level waste in the United States. Transportation implementers should take early and proactive steps to establish formal mechanisms for gathering high-quality and diverse advice about social risks and their management on an ongoing basis using the steps recommended in Section 3.4.

The committee also provides several recommendations for improving current and future programs to transport spent fuel and high-level waste to a federal repository or to interim storage. Although these recommendations are directed at DOE and DOT, they would also apply to programs for shipping commercial spent fuel to private storage.

- DOE's procedures for selecting routes within the United States for shipments of foreign research reactor spent fuel appear on the whole to be adequate and reasonable. DOT routing regulations are a satisfactory means of ensuring safe transportation, provided that the shipper actively and systematically consults with the states and tribes along potential routes and that states follow the route designation procedures prescribed by the department. The committee recommends that DOT take steps to ensure that states that designate routes for shipment of spent nuclear fuel rigorously comply with regulatory requirements that such designations be supported by sound risk assessments (see Section 4.4).
- The committee strongly endorses DOE's decisions to ship spent fuel and high-level waste to the federal repository by mostly rail using dedicated trains. The committee recommends that DOE fully implement

[8]Social risks arise from both *social processes*, which influence peoples' interactions and shape their communities, and *perceptions*, which influence peoples' behaviors, whether or not such perceptions are an accurate picture of reality.

these decisions by completing construction of the Nevada rail spur and making other necessary arrangements before commencing the large-quantity shipment of spent fuel and high-level waste to the federal repository. DOE should also examine the feasibility of further reducing its needs for cross-country truck shipments of spent fuel (see Sections 5.2.1 and 5.2.3).

- DOE should identify and make public its suite of preferred highway and rail routes for transporting spent fuel and high-level waste to a federal repository as soon as practicable to support state, tribal, and local planning, especially for emergency responder preparedness. DOE should follow the practices of its foreign research reactor spent fuel transport program (see Chapter 4) of involving states and tribes in these route selections (see Section 5.2.2).
- DOE should negotiate with commercial spent fuel owners to ship older fuel first to a federal repository or to federal interim storage. Should these negotiations prove to be ineffective, Congress should consider legislative remedies. Within the context of its current contracts with commercial spent fuel owners, DOE should initiate transport to the federal repository through a pilot program involving relatively short, logistically simple movements of older fuel from closed reactors to demonstrate its ability to carry out its responsibilities in a safe and operationally effective manner (see Section 5.2.4).
- DOE should begin immediately to execute its emergency responder preparedness responsibilities defined in Section 180(c) of the Nuclear Waste Policy Act using the innovative steps recommended in Section 5.2.5.
- DOE, the Department of Homeland Security, DOT, and USNRC should promptly complete the job of developing, applying, and disclosing consistent, reasonable, and understandable criteria for protecting sensitive information about spent fuel and high-level waste shipments. They should also commit to the open sharing of information that does not require such protection and should facilitate timely access to such information, for example, by posting it on readily accessible Web sites (see Section 5.2.6).
- The Secretary of Energy and the U.S. Congress should examine options for changing the organizational structure of DOE's program for transporting spent fuel and high-level waste to a federal repository to increase its chances for success. The following three alternative organizational structures, which are representative of progressively greater organizational change, should be examined: (1) a quasi-independent DOE office reporting directly to upper-level DOE management; (2) a quasi-government corporation; or (3) a fully private organization operated by the commercial nuclear industry (see Section 5.3).

Summary of Findings and Recommendations

The findings and recommendations from this report are reproduced in full in this summary. Additional background and supporting information can be found in the body of the report.

S.1 TRANSPORTATION SAFETY AND SECURITY

Based on its examination of spent fuel and high-level waste transportation in the United States, the committee developed a principal finding on transportation safety and a finding and recommendation on transportation security. These are described in greater detail in Chapter 5, Section 5.1.

PRINCIPAL FINDING ON TRANSPORTATION SAFETY: The committee could identify no fundamental technical barriers to the safe[1] transport of spent nuclear fuel and high-level radioactive waste in the United States. Transport by highway (for small-quantity shipments[2]) and by rail (for large-quantity shipments) is, from a technical viewpoint, a low-

[1]As noted in Chapter 1, *safety* refers to measures taken to protect spent fuel and high-level waste from failure, damage, human error, and other inadvertent acts during transport operations.

[2]This report identifies two general types of transportation programs, *small-quantity shipping programs* and *large-quantity shipping programs*. While there is no precise quantity demarcation between these two program types, the former involve the shipment on the order of tens of metric tons of spent fuel or high-level waste, while the latter involve the shipment on the order of hundreds to thousands of metric tons.

radiological-risk activity with manageable safety, health, and environmental consequences when conducted with strict adherence to existing regulations. However, there are a number of social and institutional challenges to the successful[3] initial implementation of large-quantity shipping programs that will require expeditious resolution as described in this report. Moreover, the challenges of sustained implementation should not be underestimated.

The wording of this finding is carefully and narrowly constructed and is focused on the technical aspects of transportation programs and the conduct of transportation operations. It is predicated on the assumption that these technical tasks are being carried out with a high degree of care and in strict adherence to regulations. The finding is also based on an assessment of past and present transportation programs and would apply to future programs only to the extent that they continue to exercise appropriate care and adhere to applicable regulations.

PRINCIPAL FINDING ON TRANSPORTATION SECURITY: Malevolent acts against spent fuel and high-level waste shipments are a major technical and societal concern, especially following the September 11, 2001, terrorist attacks on the United States. The committee judges that some of its recommendations for improving transportation safety might also enhance transportation security. The Nuclear Regulatory Commission is undertaking a series of security studies, but the committee was unable to perform an in-depth technical examination of transportation security because of information constraints.

RECOMMENDATION: An independent examination of the security of spent fuel and high-level waste transportation should be carried out prior to the commencement of large-quantity shipments to a federal repository or to interim storage. This examination should provide an integrated evaluation of the threat environment, the response of packages to credible malevolent acts, and operational security requirements for protecting spent fuel and high-level waste while in transport. This examination should be carried out by a technically knowledgeable group that is independent of the government and free from institutional and financial conflicts of interest. This group should be given full access to the necessary classified documents and Safeguards Information to carry out this task. The findings and recommen-

[3]The committee defines "success" in terms of the program's ability, under existing statutes, regulations, agreements, and budgets, to transport spent fuel and high-level waste in a safe, secure, timely, and publicly acceptable manner.

dations from this examination should be made available to the public to the fullest extent possible.

The committee was unable to perform this examination because much of the needed information is either classified or otherwise restricted. There appears to be sufficient information available, however, to undertake a substantive review of spent fuel and high-level waste transportation security by a cleared group if it is given unrestricted access to the relevant literature. The cooperation of several federal agencies would be required to obtain the information necessary to carry out this study.

S.2 TRANSPORTATION RISK

This report provides an examination of two types of transportation risks: *health and safety risks* and *social risks*. These risks arise both during *normal* transport operations and from *accidents* involving packages loaded with spent fuel or high-level waste. The health and safety risks arise from the potential exposure of transportation workers, as well as other people who travel, work, or live near transportation routes, to radiation that may be emitted or released from these packages. Social risks arise from social processes and human perceptions and can have both direct socioeconomic impacts and perception-based impacts. The health and safety risks and social risks are collectively referred to as *societal risks* in this report.

This report also provides comparisons between health and safety risks for transporting spent fuel and high-level waste and certain other risks that confront members of society. Comparisons are provided for both normal and severe accident conditions in Chapter 3 (see especially Figures 3.3 and 3.4). The committee's objective in presenting these comparisons is to inform readers' understanding about the risks of spent fuel and high-level waste transportation, not to persuade readers that such risks are—or are not—acceptable. Acceptability is a normative judgment; there is no basis in science for judging the acceptability of transportation risks.

FINDING: There are two potential sources of radiological exposures from transporting spent fuel and high-level waste: (1) radiation shine[4] from spent fuel and high-level waste transport packages under normal transport conditions; and (2) potential increases in radiation shine and release of radioactive materials from transport packages under accident conditions that are severe enough to compromise fuel element and package integrity. The ra-

[4]Radiation emitted from a transportation package containing spent fuel or high-level waste.

diological risks associated with the transportation of spent fuel and high-level waste are well understood and are generally low, with the possible exception of risks from releases in extreme accidents involving very long duration, fully engulfing fires. While the likelihood of such extreme accidents appears to be very small, their occurrence cannot be ruled out based on historical accident data for other types of hazardous material shipments. However, the likelihood of occurrence and consequences can be reduced further through relatively simple operational controls and restrictions and route-specific analyses to identify and mitigate hazards that could lead to such accidents.

RECOMMENDATION: Transportation planners and managers should undertake detailed surveys of transportation routes to identify potential hazards that could lead to or exacerbate extreme accidents involving very long duration, fully engulfing fires. Planners and managers should also take steps to avoid or mitigate such hazards before the commencement of shipments or shipping campaigns.

The finding that "radiological risks . . . are well understood and are generally low" is based on a large set of data and studies described in Chapters 2 and 3. These include the following:

- Rigorous international standards and U.S. regulations for the design, construction, testing, and maintenance of spent fuel packages;
- More than four decades of worldwide experience in transporting spent fuel without a significant release of radioactive materials during an accident;[5] the broad sharing of information on experiences and best practices by transportation planners, implementers, and regulators through organizations such as the International Atomic Energy Agency promotes the continued maintenance of this safety record;
- Full-scale crash testing of transport packages under severe accident conditions;
- A series of increasingly sophisticated analytical studies of spent fuel transport package performance; and
- Reconstructions of the mechanical and thermal loading conditions from severe accidents that did not involve spent fuel transport to asses how spent fuel packages would have performed under such conditions.

The finding that spent fuel transportation risks are "generally low" at present does not necessarily mean that such risks will continue to be low in

[5]Minor releases have been reported in obsolete packages that are no longer approved for use for transporting spent fuel (see Table 3.3).

the future. Future risks depend on a number of factors (e.g., the care taken in fabricating transport packages and executing transportation operations). Ongoing vigilance by regulators and shippers will be essential for maintaining low-risk programs in the future, especially during the scale-up and operation of large-quantity shipping programs.

FINDING: The *social risks* for spent fuel and high-level waste transportation pose important challenges to the successful implementation of programs for transporting spent fuel and high-level waste in the United States. Such risks, which can result in lower property values along transportation routes, reductions in tourism, and increased anxiety, have received substantially less attention than health and safety risks, and some are difficult to characterize. Current research and practice suggest that transportation planners and managers can take early proactive steps to characterize, communicate, and manage the social risks that arise from their operations. Such steps may have additional benefits: they may increase the openness and transparency of transportation planning and programs; build community capacity to mitigate these risks; and possibly increase trust and confidence in transportation programs.

RECOMMENDATION: Transportation implementers should take early and proactive steps to establish formal mechanisms for gathering high-quality and diverse advice about social risks and their management on an ongoing basis. The committee makes two recommendations for the establishment of such mechanisms for the Department of Energy's program to transport spent fuel and high-level waste to a federal repository at Yucca Mountain: (1) expand the membership and scope of an existing advisory group (Transportation External Coordination Working Group; see Chapter 5) to obtain outside advice on social risk, including impacts and management; and (2) establish a *transportation risk advisory group* that is explicitly designed to provide advice on characterizing, communicating, and mitigating the social, security, and health and safety risks that arise from the transportation of spent fuel and high-level waste to a federal repository or interim storage. This group should be comprised of risk experts and practitioners drawn from the relevant technical and social science disciplines and should be convened under the Federal Advisory Committee Act or a similar arrangement to enhance the openness of its operations. Its members should receive security clearances to facilitate access to appropriate transportation security information. The existing federal Nuclear Waste Technical Review Board, which will cease operations no later than one year after the Department of Energy begins disposal of spent fuel or high-level waste in a repository, could be broadened to serve this function.

This finding and recommendation spring from several factors: social risk is a poorly understood phenomenon; expert opinion frequently differs; the Department of Energy (DOE) does not, to the committee's knowledge, have any precedent to guide its understanding and management of social risks; and most transportation program staff are not likely to be well acquainted with either theory or practice on this issue. Consequently, the committee concluded that broad input and advice on social risks will be essential to the establishment and ultimate success of large-quantity shipping programs to transport spent fuel and high-level waste to a federal repository or interim storage.

The recommendation outlines pragmatic steps that transportation implementers can take immediately and at relatively low cost to better understand and, working with affected communities, manage the social risks from their programs. The recommendation is focused primarily on DOE, but it would also apply to any program for shipping large quantities of spent fuel to a private interim storage site (e.g., Private Fuel Storage in Utah).

S.3 CURRENT CONCERNS ABOUT TRANSPORTATION OF SPENT FUEL AND HIGH-LEVEL WASTE

The report examines two current concerns about transportation of spent fuel and high-level waste in the United States: (1) the performance of packages used to transport spent fuel and high-level waste under both normal and extreme mechanical forces and thermal loading conditions; and (2) the procedures used by DOE to select highway and rail routes for shipping research reactor spent fuel between DOE facilities in the United States.[6]

S.3.1 Package Performance

Package performance—the ability of a transportation package to maintain a high level of containment effectiveness in long-term routine use and under extreme mechanical forces and thermal loading conditions—is a crucial issue for transportation safety and key to understanding and quantifying transportation risks. Packages used to transport spent fuel and high-level waste are robust structures typically constructed of steel, lead, depleted uranium, and(or) concrete to provide structural strength and durability as well as radiation shielding. The International Atomic Energy Agency (IAEA) has established safety standards for such packages, and these standards are

[6]This assessment was requested by the Department of Transportation at the direction of Congress after the study was under way. The study schedule was extended to allow additional time for information gathering, deliberation, and expansion of this report to address the added study charge.

reflected in regulations (10 CFR Part 71) issued by the U.S. Nuclear Regulatory Commission (USNRC). Packages used in the United States must meet testing requirements in 10 CFR Part 71 involving a free-drop test onto an essentially unyielding surface, a puncture test, an immersion test, and a thermal test, all with less than the specified loss of containment effectiveness. The USNRC permits quantitative analysis, scale-model, and full-scale testing of packages or package components and comparisons with existing approved package designs to be used to demonstrate such compliance. Testing of full-scale packages is not a requirement of the regulations. Some participants at the committee's information-gathering meetings asserted that full-scale testing should be required and that some tests to intentionally destroy the packages should be carried out, presumably to establish their ultimate strength.

There is a good deal of quantitative information available on the performance of transportation packages under extreme loading conditions. This information is derived from analytical, computer modeling, scale-model, and full-scale testing studies carried out in the United States and abroad over the past three decades. Additional work is now under way at the USNRC to examine the effects on package performance of severe accidents and very long duration, fully engulfing fires.

FINDING: Transportation packages play a crucial role in the safety of spent fuel and high-level waste shipments by providing a robust barrier to the release of radiation and radioactive material under both normal transport and accident conditions. International Atomic Energy Agency package performance standards and associated Nuclear Regulatory Commission regulations are adequate to ensure package containment effectiveness over a wide range of transport conditions, including most credible accident conditions. However, recently published work suggests that extreme accident scenarios involving very long duration, fully engulfing fires might produce thermal loading conditions sufficient to compromise containment effectiveness. The consequences of such thermal loading conditions for containment effectiveness are the subject of ongoing investigations by the Nuclear Regulatory Commission and other parties, and this work is improving the understanding of package performance. Nonetheless, additional analyses and experimentation are needed to demonstrate a bounding-level understanding of package performance in response to very long duration, fully engulfing fires for a representative set of package designs.

RECOMMENDATION: The Nuclear Regulatory Commission should build on recent progress in understanding package performance in very long duration fires. To this end, the agency should undertake additional analyses of very long duration fire scenarios that bound expected real-

world accident conditions for a representative set of package designs that are likely to be used in future large-quantity shipping programs. The objectives of these analyses should be to

- Understand the performance of package barriers (spent fuel cladding and package seals);
- Estimate the potential quantities and consequences of any releases of radioactive material; and
- Examine the need for regulatory changes (e.g., package testing requirements) or operational changes (e.g., restrictions on trains carrying spent fuel) either to help prevent accidents that could lead to such fire conditions or to mitigate their consequences.

Strong consideration should also be given to performing well-instrumented tests for improving and validating the computer models used for carrying out these analyses, perhaps as part of the full-scale test planned by the Nuclear Regulatory Commission for its package performance study.

Based on the results of these investigations, the Commission should implement operational controls and restrictions on spent fuel and high-level waste shipments as necessary to reduce the chances that such fire conditions might be encountered in service. Such effective steps might include, for example, additional operational restrictions on trains carrying spent fuel and high-level waste to prevent co-location with trains carrying flammable materials in tunnels, in rail yards, and on sidings.

FINDING: The committee strongly endorses the use of full-scale testing to determine how packages will perform under both regulatory and credible extraregulatory conditions. Package testing in the United States and many other countries is carried out using good engineering practices that combine state-of-the-art structural analyses and physical tests to demonstrate containment effectiveness. Full-scale testing is a very effective tool both for guiding and validating analytical engineering models of package performance and for demonstrating the compliance of package designs with performance requirements. However, deliberate full-scale testing of packages to destruction through the application of forces that substantially exceed credible accident conditions would be marginally informative and is not justified given the considerable costs for package acquisitions that such testing would require.

RECOMMENDATION: Full-scale package testing should continue to be used as part of integrated analytical, computer simulation, scale-model, and testing programs to validate package performance. Deliberate full-scale

testing of packages to destruction should not be required as part of this integrated analysis or for compliance demonstrations.

S.3.2 Route Selection for Research Reactor Spent Fuel Transport

DOE transports and stores spent fuel from foreign and U.S. research reactors at two facilities: the Savannah River Site in South Carolina and the Idaho National Laboratory. Foreign research reactor spent fuel is transported by ship to the Charleston Naval Weapons Station in South Carolina. From there it is transported by rail or highway to the Savannah River Site in South Carolina. Some of these shipments continue onward to the Idaho National Laboratory using one of three highway routes established by DOE in consultation with states and tribes. A 2001 shipment of foreign research reactor spent fuel across one of these routes prompted a congressional mandate to the Department of Transportation (DOT) for a National Academies examination of the procedures used by DOE to select routes for highway and rail shipments of this fuel. That examination was carried out as part of this study because it is a good example of a current concern for transporting spent fuel. The committee's examination of DOE's routing practices resulted in the following findings and recommendations.

FINDING: The Department of Energy's procedures for selecting routes within the United States for shipments of foreign research reactor spent fuel appear on the whole to be adequate and reasonable. These procedures are risk informed; they make use of standard risk assessment methodologies in identifying a suite of potential routes and then make final route selections by taking into account security, state and tribal preferences, and information from states and tribes on local transport conditions. The Department of Energy's procedures reflect the agency's position (which is consistent with Department of Transportation regulations) that the states are competent and responsible for selecting highway routes. For rail route selection, the Department of Energy's practice of negotiating routes with carriers in consultation with states is analogous to its interaction with states on highway routing.

RECOMMENDATION: The Department of Energy should continue to ensure the systematic, effective involvement of states and tribal governments in its decisions involving routing and scheduling of foreign and DOE research reactor spent fuel shipments.

FINDING: Highway routes for shipment of spent nuclear fuel are dictated by DOT regulations (49 CFR Part 397). The regulations specify that ship-

ments normally must travel by the fastest route using highways designated by the states or the federal government. They do not require the carrier or shipper to evaluate risks of portions of routes that meet this criterion. These regulations are a satisfactory means of ensuring safe transportation, provided that the shipper actively and systematically consults with the states and tribes along potential routes and that states follow the route designation procedures prescribed by the DOT.

RECOMMENDATION: DOT should ensure that states that designate routes for shipment of spent nuclear fuel rigorously comply with its regulatory requirement that such designations be supported by sound risk assessments. DOT and DOE should ensure that all potentially affected states are aware of and prepared to fulfill their responsibilities regarding highway route designations.

S.4 FUTURE CONCERNS FOR TRANSPORTATION OF SPENT FUEL AND HIGH-LEVEL WASTE

The examination of future concerns focused on five operational issues and one organizational issue related to the federal program for transporting spent fuel and high-level waste to a repository at Yucca Mountain, Nevada (Appendix C). Although these recommendations are focused on DOE's program for transporting spent fuel and high-level waste to a federal repository or interim storage facility, some of them also apply to any large-quantity shipping program, whether federally or privately operated.

S.4.1 Mode for Transporting Spent Fuel and High-Level Waste to a Federal Repository

DOE has decided that it will ship spent fuel and high-level waste to a federal repository using the "mostly rail" option defined in its final Environmental Impact Statement (EIS) for Yucca Mountain. DOE estimates that this option will require 9600 rail shipments and 1100 highway shipments to transport the legally mandated limit of 70,000 metric tons of spent fuel and high-level waste to the repository. To implement this option, DOE must construct a 319-mile (~513-kilometer) rail spur in Nevada and may have to make other infrastructure improvements to provide rail access at commercial nuclear sites. DOE must also procure a fleet of rail packages and railcars. If DOE fails to complete these tasks prior to the opening of the federal repository, legal and contractual obligations to accept spent fuel may create pressure for it to initiate a large-scale interim shipping program using trucks.

FINDING: Transport of spent fuel and high-level waste by rail has clear safety, operational, and policy advantages over highway transport for large-quantity shipping programs. The committee strongly endorses DOE's selection of the "mostly rail" option for the Yucca Mountain transportation program for the following reasons:

- It reduces the total number of shipments to the federal repository by roughly a factor of five, which reduces the potential for routine radiological exposures, conventional traffic accidents, and severe accidents (Table 3.8).
- Rail shipments have a greater physical separation from other vehicular traffic and reduced interactions with people along transportation routes, which also contributes to safety.
- Operational logistics are simpler and more efficient.
- There is a clear public preference for this option.

The committee does not endorse the development of an extended truck transportation program to ship spent fuel cross-country or within Nevada should DOE fail to complete construction of the Nevada rail spur or procure the necessary rail equipment by the time the federal repository is opened.

RECOMMENDATION: DOE should fully implement its mostly rail decision by completing construction of the Nevada rail spur, obtaining the needed rail packages and conveyances, and working with commercial spent fuel owners to ensure that facilities are available at plants to support this option. These steps should be completed before DOE commences the large-quantity shipment of spent fuel and high-level waste to a federal repository to avoid the need to procure infrastructure and construct facilities to support an extended truck transportation program. DOE should also examine the feasibility of further reducing its needs for cross-country truck shipments of spent fuel through the expanded use of intermodal transportation (i.e., combining heavy-haul truck, legal-weight truck, and barge) to allow the shipment of rail packages from plants that do not have direct rail access.

S.4.2 Route Selection for Transportation to a Federal Repository

DOE's program to transport spent fuel and high-level waste to a federal repository will involve shipments from more than 70 sites in 31 states, most passing through or near one or more major U.S. metropolitan centers. DOE did not designate routes for these shipments in its final EIS for Yucca Mountain but instead plans to make selections about five years before the federal repository opens. Once DOE selects routes, it may be required to undertake selected route infrastructure improvements before any shipments can be made. It also must decide on safety and security procedures, arrangements for state inspections of shipments, communications and tracking, emergency responder training, and handling of en route contingencies. Most of these operations would be carried out by contractors.

FINDING: DOE has not made public a specific plan for selecting rail and highway routes for transporting spent fuel and high-level waste to a federal repository. DOE also has not determined the role of its program management contractors in selecting routes or specific plans for collaborating with affected states, tribes, and other parties.

RECOMMENDATION: DOE should identify and make public its suite of preferred highway and rail routes for transporting spent fuel and high-level waste to a federal repository as soon as practicable to support state, tribal, and local planning, especially for emergency responder preparedness. DOE should follow the practices of its foreign research reactor spent fuel transport program of involving states and tribes in these route selections to obtain access to their familiarity with accident rates, traffic and road conditions, and emergency responder preparedness within their jurisdictions. Involvement by states and tribes may improve the public acceptability of route selections and may reduce conflicts that can lead to program delays.

S.4.3 Use of Dedicated Trains for Transport to a Federal Repository

There has been a long-running controversy in the United States about whether rail shipments of spent fuel and high-level waste should be carried out using *dedicated trains*, which would carry only spent fuel or high-level waste, or *general trains*, which would carry other freight in addition to spent fuel or high-level waste. DOE's final EIS for Yucca Mountain did not provide a detailed analysis of the benefits of dedicated trains to support a decision on this issue. However, a decision by DOE to use dedicated trains was announced in July 2005.

FINDING: Studies carried out to date on transporting spent fuel by dedicated versus general trains have failed to show a clear radiological risk-based advantage for either option. However, the committee finds that there are clear operational, safety, security, communications, planning, programmatic, and public preference advantages that favor dedicated trains. The committee strongly endorses DOE's decision to transport spent fuel and high-level waste to a federal repository using dedicated trains.

RECOMMENDATION: DOE should fully implement its dedicated train decision before commencing the large-quantity shipment of spent fuel and high-level waste to a federal repository to avoid the need for a stopgap shipping program using general trains.

S.4.4 Acceptance Order for Commercial Spent Fuel Transport to a Federal Repository

The Nuclear Waste Policy Act (NWPA) specifies that DOE must accept spent fuel based on the amount and order in which it was discharged from owners' reactors. Each time a commercial nuclear plant discharges fuel from its reactor, the owner receives an allocation in the "acceptance queue" to ship an equivalent amount of spent fuel to a federal repository. DOE will accept commercial spent fuel for shipment to the federal repository starting at the beginning of the queue and will work its way through the queue during the planned 24-year life of the transportation program. The NWPA allows owners to make available to DOE for shipment to the federal repository any spent fuel from any of their sites for each of their allocations in the acceptance queue. There are two exceptions to this requirement: (1) DOE may accord priority for acceptance of spent nuclear fuel from reactors that have been permanently shut down; and (2) with the approval of DOE, owners of spent fuel can exchange positions in the acceptance queue.

The order for accepting commercial spent fuel that is mandated by the NWPA could require DOE to initiate its transportation program with movements of spent fuel from multiple, geographically dispersed sites. Further, it gives DOE limited control over the age and radiological content of the fuel that is provided by owners for transport. Shipping older fuel first would give DOE a better ability to optimize routing, scheduling, and emergency responder planning and training, especially during the early phases of the program.

FINDING: The order for accepting commercial spent fuel that is mandated by the Nuclear Waste Policy Act (NWPA) was not designed with the transportation program in mind. In fact, the acceptance order prescribed by the NWPA could require DOE to initiate its transportation program with long cross-country movements of younger (i.e., radiologically and thermally hotter) spent fuel from multiple commercial sites. There are clear transportation operations and safety advantages to be gained from shipping older (i.e., radiologically and thermally cooler) spent fuel first and for initiating the transportation program with relatively short, logistically simple movements to gain experience and build operator and public confidence.

RECOMMENDATION: DOE should negotiate with commercial spent fuel owners to ship older fuel first to a federal repository or federal interim storage, except in cases (if any) where spent fuel storage risks at specific plants dictate the need for more immediate shipments of younger fuel. Should these negotiations prove to be ineffective, Congress should consider legislative remedies. Within the context of its current contracts with commercial spent fuel owners, DOE should initiate transport through a pilot program involving relatively short, logistically simple movements of older fuel from closed reactors to demonstrate the ability to carry out its responsibilities in a safe and operationally effective manner. DOE should use the lessons learned from this pilot activity to initiate its full-scale transportation program from operating reactors.

S.4.5 Emergency Response Planning and Training

The transportation of spent nuclear fuel to a federal repository would utilize the same state and local emergency response capabilities that are in place to deal with existing materials transport accidents and incidents (see Appendix C). However, DOE has special responsibilities under the Nuclear Waste Policy Act for providing technical assistance and funding to states and tribal nations for training on both routine transportation procedures and emergency response. DOE will not begin providing such support until it identifies the routes for shipping spent fuel and high-level waste to Yucca Mountain.

FINDING: Emergency responder preparedness is an essential element of safe and effective programs for transporting spent fuel and high-level waste. Emergency responder preparedness has so far received limited attention from DOE, states, and tribes for the planned transportation program to the federal repository. DOE has the opportunity to be innovative in carrying out its responsibilities for emergency responder preparedness. Emergency responders are among the most trusted members of their communities. Well-trained responders can become important emissaries for DOE's transportation program in local communities and can enhance community preparedness to respond to other kinds of emergencies.

RECOMMENDATION: DOE should begin immediately to execute its emergency responder preparedness responsibilities defined in Section 180(c) of the Nuclear Waste Policy Act. In carrying out these responsibilities, DOE should proceed to (1) establish a cadre of professionals from the emergency responder community who have training and comprehension of emergency response to spent fuel and high-level waste transportation accidents and

incidents; (2) work with the Department of Homeland Security to provide consolidated "all-hazards" training materials and programs for first responders that build on the existing national emergency response platform; (3) include trained emergency responders on the escort teams that accompany spent fuel and high-level waste shipments; and (4) use emergency responder preparedness programs as an outreach mechanism to communicate broadly about plans and programs for transporting spent fuel and high-level waste to a federal repository with communities along planned shipping routes.

These recommended innovations are also potentially applicable to the transportation program operated by Private Fuel Storage. The committee judges that there would be significant benefits to that program in terms of capacity and public confidence building through early and innovative actions to support emergency responder preparedness.

S.4.6 Information Sharing and Openness

Some participants at the committee's information-gathering meetings expressed concerns that federal agencies, in reaction to the September 11, 2001, terrorist attacks, are withholding information that could help the public evaluate the safety and security of spent fuel and high-level waste shipments. The committee itself encountered information restrictions in its efforts to obtain information on the number of past spent fuel shipments in the United States.

FINDING: There is a conflict between the open sharing of information on spent fuel and high-level waste shipments and the security of transportation programs. This conflict is impeding effective risk communication and may reduce public acceptance and confidence. Post–September 11, 2001, efforts by transportation planners, managers, and regulators to further restrict information about spent fuel shipments make it difficult for the public to assess the safety and security of transportation operations.

RECOMMENDATION: The Department of Energy, Department of Homeland Security, Department of Transportation, and Nuclear Regulatory Commission should promptly complete the job of developing, applying, and disclosing consistent, reasonable, and understandable criteria for protecting sensitive information about spent fuel and high-level waste transportation. They should also commit to the open sharing of information that does not require such protection and should facilitate timely access to such information: for example, by posting it on readily accessible Web sites.

The public has a general right, subject to legitimate privacy and national security restrictions, to obtain information about government programs that affect their communities. Some general information is appropriate to share before shipments commence: This includes the reasons for making the shipments; information about the materials to be shipped; likely shipping modes; and general shipping time frames. Appropriate post-shipment information includes more details on the shipments, including specific modes and routes used for the shipments; the timing of shipments and quantities of materials shipped; accidents and incidents during the shipments; and any resulting response actions.

S.4.7 Organizational Structure of the Federal Transportation Program

The program for transporting spent fuel and high-level waste to a federal repository is embedded within the DOE's Office of Civilian Radioactive Waste Management. This agency is responsible for licensing, constructing, and operating the planned repository at Yucca Mountain, Nevada. Certain characteristics of the Yucca Mountain transportation program will make it exceptionally challenging to carry out successfully. The transportation program

- Will last for more than two decades;
- Is decentralized and involves a large number of parties in both government and the private sector over which DOE has limited control;
- Must operate with a high degree of consistency and reliability;
- Has limited flexibility over schedules because of spent fuel acceptance requirements; and
- Has limited budgetary control within DOE and is subject to the annual congressional appropriations process.

The transportation program is unusual in another sense: The committee knows of no other federal government-run program that has a requirement to take ownership of private-sector waste for the purposes of transport and disposal. Such programs are usually private-sector responsibilities.

FINDING: Successful execution of DOE's program to transport spent fuel and high-level waste to a federal repository will be difficult given the organizational structure in which it is embedded, despite the high quality of many current program staff. As currently structured, the program has limited flexibility over commercial spent fuel acceptance order (Section 5.2.4); it also has limited control over its budget and is subject to the annual federal appropriations process, both of which affect the program's ability to plan for, procure, and construct the needed transportation infrastruc-

ture. Moreover, the current program may have difficulty supporting what appears to be an expanding future mission to transport commercial spent nuclear fuel for interim storage or reprocessing. In the committee's judgment, changing the organizational structure of this program will improve its chances for success.

RECOMMENDATION: The Secretary of Energy and the U.S. Congress should examine options for changing the organizational structure of the Department of Energy's program for transporting spent fuel and high-level waste to a federal repository. The following three alternative organizational structures, which are representative of progressively greater organizational change, should be specifically examined: (1) a quasi-independent DOE office reporting directly to upper-level DOE management; (2) a quasi-government corporation; or (3) a fully private organization operated by the commercial nuclear industry. The latter two options would require changes to the Nuclear Waste Policy Act. The primary objectives in modifying the structure should be to give the transportation program greater planning authority; greater budgetary flexibility to make the multiyear commitments necessary to plan for, procure, and construct the necessary transportation infrastructure; and greater flexibility to support an expanding future mission to transport spent fuel and high-level waste for interim storage or reprocessing. Whatever structure is selected, the organization should place a strong emphasis on operational safety and reliability and should be responsive to social concerns.

The committee strongly encourages the program to seek expert advice to learn about and incorporate best industry practices for designing and operating this transportation system using an integrated systems approach.

1
Introduction

The National Academies' Board on Radioactive Waste Management[1] and Transportation Research Board initiated this study to address what they perceived to be a national need for an independent, objective, and authoritative analysis of spent nuclear fuel and high-level radioactive waste[2] transportation in the United States. The objectives of this study (Sidebar 1.1) were to identify key current and future technical and societal concerns about the transportation of spent fuel and high-level waste in the United States and technical and policy options for addressing those concerns and managing transportation risks.

This study also examined the selection of highway and rail routes for shipping research reactor spent fuel between U.S. Department of Energy (DOE) facilities in the United States (Sidebar 1.2). This additional examination was requested by the U.S. Department of Transportation (DOT) at the direction of Congress[3] after the study was under way. With the consent of the National Academies and the original study sponsors, the schedule for the study was extended to allow additional time for information gathering,

[1]The Board on Radioactive Waste Management was merged with another National Academies board in early 2005 to form the Nuclear and Radiation Studies Board.

[2]Also referred to as *spent fuel* and *high-level waste* in this report.

[3]Consolidated Appropriations Resolution, 2003, P.L. 108-7, February 20, 2003, Division I, Section 334.

deliberation, and expansion of this report to address this added study charge.

The "current concerns" referred to in the original study charge (Sidebar 1.1) are associated with currently operating transportation programs and with current planning efforts for future transportation programs. The congressionally mandated charge (Sidebar 1.2) to examine the routing of research reactor spent fuel is a good example of a concern about a currently operating program. The "future concerns" referred to in the original charge are associated primarily with plans to transport spent fuel and high-level waste for interim storage or permanent disposal (see Section 1.3.2). As described in some detail in Chapter 5, these future concerns relate primarily to the difficulties in scaling-up transportation systems from the relatively small, centralized programs that presently exist for moving small quantities of spent fuel to the more complex decentralized programs that will likely be

SIDEBAR 1.1 Transportation of Radioactive Waste Study Task

The principal task of this study is to develop a high-level synthesis of key technical and societal issues for spent fuel/high-level waste transport and to identify technical and policy options for addressing these issues and managing transportation risk. The principal focus of this study is on the transportation of spent fuel and high-level waste in the United States, but the study will draw on international experiences as well as experiences with transporting other waste types. The study addresses and provides findings and recommendations on the following four questions:

1. What are the principal risks for transporting (including container handling, modal transfers, and conveyance) radioactive waste, and how do they compare with other societal risks? To what extent have these risks been addressed by previous analyses?
2. At present, what are the principal technical and societal concerns for transporting radioactive waste? To what extent have these concerns been addressed, and what additional work is needed?
3. What are likely to be the key principal technical and societal concerns for radioactive waste transportation in the future, especially over the next two decades?
4. What options are available to address these concerns, for example, options involving changes to planned transportation routes, modes, procedures, or other limitations/restrictions; or options for improving the communication of transportation risks to decision makers and the public?

SIDEBAR 1.2 Transportation Routing Study Task

The principal task of this study will be to assess the manner in which the Department of Energy and its contractors:

(1) Select potential highway and rail routes for the shipment of spent nuclear fuel from foreign and domestic research nuclear reactors.
(2) Select specific land routes for such shipments.
(3) Conduct assessments, if any, of the risks associated with such shipments.

The following factors will be considered in conducting the assessments in point (3):

(i) Proximity of routes to major population centers and the risks associated with shipments of spent nuclear fuel from research nuclear reactors through densely populated areas.
(ii) Current traffic and accident data with respect to the routes under consideration.
(iii) Quality of the roads comprising the routes under consideration.
(iv) Emergency response capabilities along the routes under consideration.
(v) Proximity of the routes under consideration to places or venues (including sports stadiums, convention centers, concert halls and theaters, and other venues) where large numbers of people gather.

The assessment should identify deficiencies, if any, in current procedures for selecting routes that have important potential health or safety consequences. In making recommendations to address these deficiencies, a clear distinction should be made between technical and policy considerations. Recommendations should be directed at competent regulating authorities or the United States Congress.

required to move large quantities of spent fuel and high-level waste to interim storage or permanent disposal.[4]

1.1 STUDY PROCESS

Most National Research Council studies are undertaken in response to requests from federal agencies, the White House, or Congress. In contrast, the transportation study described in Sidebar 1.1 was self-initiated by the

[4]This report refers to these two types of programs as *small-quantity shipping programs* and *large-quantity shipping programs*. While there is no precise quantity demarcation between these two program types, the former involves shipment on the order of tens of metric tons of spent fuel or high-level waste, while the latter involves shipment on the order of hundreds to thousands of metric tons.

National Research Council. Three federal agencies and two not-for-profit private organizations recognized the importance of this study and provided the necessary financial support to enable the National Research Council to carry it out. These organizations are DOE, DOT, U.S. Nuclear Regulatory Commission (USNRC), Electric Power Research Institute (EPRI), and National Cooperative Highway Research Program. The National Academies also contributed funding to this effort.

The study was carried out using established National Research Council procedures to ensure its objectivity and freedom from inappropriate influences of sponsors and other outside organizations. A committee of 16 experts was provisionally appointed by the chair of the National Research Council to carry out the study. These appointments were finalized after a careful screening for conflicts of interest and consideration of public comments on committee balance and bias. The committee had diverse expertise and perspectives, including experts with experience in a variety of transportation sectors and related technical disciplines. The biographical sketches of the committee members (Appendix A) illustrate their collective range of technical and policy expertise.

The committee was responsible for designing and executing this study. It made an effort to reach out broadly to obtain information and perspectives, and it benefited greatly from the willingness of a large number of individuals and organizations to share information and viewpoints. The committee held six information-gathering meetings in different regions of the country to address its original charge. The committee chose the locations of its meetings to enhance attendance and participation of interested individuals and organizations. An additional information-gathering meeting was added to address the congressionally mandated study charge (Sidebar 1.2). A list of presentations received at these meetings is provided in Appendix B.

The committee also visited Yucca Mountain and some of the potential highway and rail routes within Nevada. Subgroups of the committee visited a spent fuel storage facility at an operating nuclear power plant (Exelon Nuclear Corporation's Dresden Plant in Chicago); the Transportation Technology Center in Pueblo, Colorado, to learn about rail transportation research and development programs; and Germany and the United Kingdom to learn about European transportation programs.

Open-microphone sessions were scheduled at each of the committee's information-gathering meetings so that any interested individual could speak directly to the committee. The committee also gathered a large amount of written material, ranging from peer-reviewed scientific articles to advocacy papers, for use in its deliberations.

The committee established an electronic notification list so that interested parties could be informed of upcoming meetings. Additionally, the committee established a Web site (*http://www.national-academies.org/transportationofradwaste*) for the study, where copies of the meeting agendas and electronic copies of the meeting presentations were posted. This site allowed visitors to provide feedback to the committee. The National Academies Press carried out an informal survey of selected government and nongovernmental organizations to obtain feedback to guide the future development of information products from this study.

The committee's final report was subjected to National Research Council peer review before being approved for unlimited public release. The report was reviewed by 15 people selected by the Report Review Committee of the National Research Council to provide a diversity of disciplinary expertise and viewpoints. The reviewers were asked to comment on whether the report addressed the study charges (Sidebars 1.1 and 1.2) in a fair and objective manner and whether the findings and recommendations were supported by fact and analysis. The committee was required by the National Research Council to make appropriate revisions to its report to address those comments.

The report's findings and recommendations were not provided to the study sponsors or to the public until the review process had been completed and the report was approved for release. This is standard National Research Council procedure to ensure that no outside organization is able to influence the outcome of its studies inappropriately.

The committee heard a wide range of opinions about spent fuel and high-level waste transportation during its information-gathering meetings. It was presented with a range of views about the safety and security[5] of spent fuel transportation in the United States and—although it was not within the purview of the study—about the desirability of a federal repository at Yucca Mountain, Nevada, or a centralized interim storage facility in Utah. It quickly became clear to the committee that there are many individuals and groups with strongly held "pro" and "anti" positions on issues related to nuclear technology, and that some of these positions are expressed in terms of support for or opposition to the transport of spent fuel and high-level waste. This report is written in the hopes of providing these individuals and groups with a broad range of factual information and analyses to enable them to reach their own conclusions about spent fuel and high-level waste transportation, and also to inform future planning and decision making by federal agencies and the private sector.

---

[5]*Safety* refers to measures taken to protect spent fuel and high-level waste during handling and transport from failure, damage, human error, and other inadvertent acts. *Security* refers to measures taken to protect spent fuel and high-level waste during handling and transport from sabotage, attacks, and theft.

Although this is a technical report, it was written with the intention that it be accessible to non-experts. The committee has endeavored to keep the use of technical terms and acronyms to a minimum and to provide clear definitions when such terms must be used. A glossary of terms and a list of acronyms are provided in Appendixes D and E, respectively.

1.2 STRATEGY FOR ADDRESSING THE STUDY CHARGES

The original study charge (Sidebar 1.1) was broadly scoped to give the committee flexibility in carrying out this study. The committee found it necessary to make several explicit choices to narrow the scope of its charge to meet the study schedule. In this section, the committee explains these choices to set readers' expectations for the remainder of this report.

The original study charge directed the committee to examine the risks of transporting spent fuel and high-level waste in the United States. Risk is a multidimensional concept: It includes the health and safety risks that potentially arise from exposures of workers and members of the public to radiation from spent fuel and high-level waste transport. Such exposures can have both short-term and long-term health and safety consequences. There is another broad class of risks, referred to as *social risks*, that is described in this report. Social risks arise from social processes and human perceptions and can have economic, institutional, and psychological consequences. The health and safety risks and social risks are collectively referred to as *societal risks* in the statement of task given in Sidebar 1.1.

The committee examined the health and safety and social risks associated with spent fuel and high-level waste transportation activities (Chapter 3) in isolation from the larger systems in which they are embedded. Programs for transporting spent fuel and high-level waste for interim storage or permanent disposal represent the "back ends" of much larger technological systems: Commercial spent fuel transport represents the back end of the nuclear electric power generation system of the United States, whose needs, benefits, and risks are the subject of controversy for some members of the public; research reactor spent fuel transport represents the back end of systems that generate scientific and medical benefits for U.S. society; and defense spent fuel and high-level waste transport represents the back end of systems that generated plutonium for national defense.

A risk-benefit analysis of spent fuel and high-level waste transportation within the context of these larger technological systems, although certainly desirable, would involve the consideration of issues that are well beyond the scope of this study. Such issues would include, for example, national energy policy, global climate change, nuclear nonproliferation, and homeland security. Such an analysis would also have to address the risks and benefits of transporting spent fuel and high-level waste to interim storage or permanent disposal sites versus leaving it at reactor sites for an indeterminate period of time.

While comparisons of the transport versus leave-in-place options are conceptually simple, the analysis is not. Each of these options involves different kinds of risks that have different spatial and time-scale dependencies. The transport option involves the transfer of risk across populations and geographic regions, possibly leading to significant disproportionate impacts: for example, the transfer of nuclear materials such as spent fuel from existing storage sites to new storage or disposal sites that did not previously contain any nuclear materials or activities. Analysis of these options would also have to consider important secondary effects: The transport of spent fuel and high-level waste, for instance, might result in improvements to transportation infrastructure or first-responder training. These improvements could result in a reduction of other types of hazardous materials transport risks.

The committee concluded that while such an analysis would be a useful contribution to the policy process, it could not be carried out in the abstract but would have to examine real scenarios. Instead, the committee operated within the constraints of federal policy decisions, beginning with the Nuclear Waste Policy Act as described in Section 1.3.2, which set the nation on a clear path to transport spent fuel and high-level waste for permanent disposal. The intent of this study is to enhance the technical and societal bases for the transportation of spent fuel and high-level waste, regardless of their ultimate destination.

The committee also decided to focus its examinations on the transport of spent fuel and to give less attention to the transport of defense high-level radioactive waste. The committee judged that the transport of spent fuel posed more important technical and societal challenges because most spent fuel is generated commercially (see Table 1.1); it is being stored at a large number of sites across the United States; and it has been transported on the nation's road and rail systems for several decades.

Defense high-level waste, on the other hand, is government generated and owned and is being stored at only four government sites (Figure 1.1), all of which have direct rail access. High-level waste has radiological properties similar to spent fuel, especially for fission product inventories, which are the greatest contributors to potential external radiation doses during normal transport.[6] However, high-level waste will be transported in an inert solid form, and the process used to solidify this waste (see Sidebar 1.3) eliminates the gaseous fission products that are present in spent fuel. Under

[6]Some long-lived radionuclides, primarily uranium and plutonium, are removed from high-level waste during reprocessing.

TABLE 1.1 Inventories of Spent Nuclear Fuel and High-Level Waste in the United States

Material	Approximate Quantity at End of 2005
Commercial spent fuel	54,000 MTHM[a]
DOE-managed spent fuel[b]	2433 MTHM
Colorado (Fort St. Vrain)	15 MTHM
Hanford Site	2129 MTHM
Idaho National Lab	277 MTHM
Savannah River	27 MTHM
High-level waste	386,000 cubic meters of unprocessed waste,[c] which when processed will consist of[d]
	21,000 cubic meters
	58,000 metric tons[e]
	22,000 canisters[f]

NOTE: MTHM = metric tons of heavy metal.

[a]This quantity is an estimate and was obtained by adjusting the 2002 DOE Energy Information Administration (DOE-EIA) estimate of 47,023 MTHM to account for spent fuel discharges during 2003–2005. Those discharges were estimated using the average of the annual discharges reported by DOE-EIA for the period 1999–2002. The result was rounded to the nearest 1000 MTHM.

[b]Data from DOE, written communication. Includes production reactor fuel, naval spent fuel, foreign and domestic research reactor fuel, and DOE-managed commercial spent fuel.

[c]Data from DOE (2005a) for Savannah River and written communications to the National Academies for Hanford and Idaho.

[d]Data on the processed waste from DOE (2002a, Table A-27).

[e]This is the processed mass of the waste, not MTHM.

[f]Canisters contain high-level waste that has been processed (vitrified) in glass matrices. The canisters are 24 inches (60 centimeters) in diameter by 10 or 15 feet (3 to 4.5 meters) in length. This number is an estimate because only a small fraction of the high-level waste at DOE sites has been immobilized (see Sidebar 1.3).

current plans, high-level waste will be transported to the federal repository in the same types of packages used to transport commercial spent fuel, but high-level waste shipments will comprise fewer than 20 percent of the total planned number of shipments to the federal repository at Yucca Mountain under the "mostly rail" option now favored by DOE (see Table 3.8).

The committee describes the high-level waste locations and inventories and the plans for disposing of it elsewhere in this chapter. It also reviews the risk estimates for transport of high-level waste to the federal repository as part of its risk examinations in Chapter 3. The committee judged that its focused examination of spent fuel transport would likely identify and bound

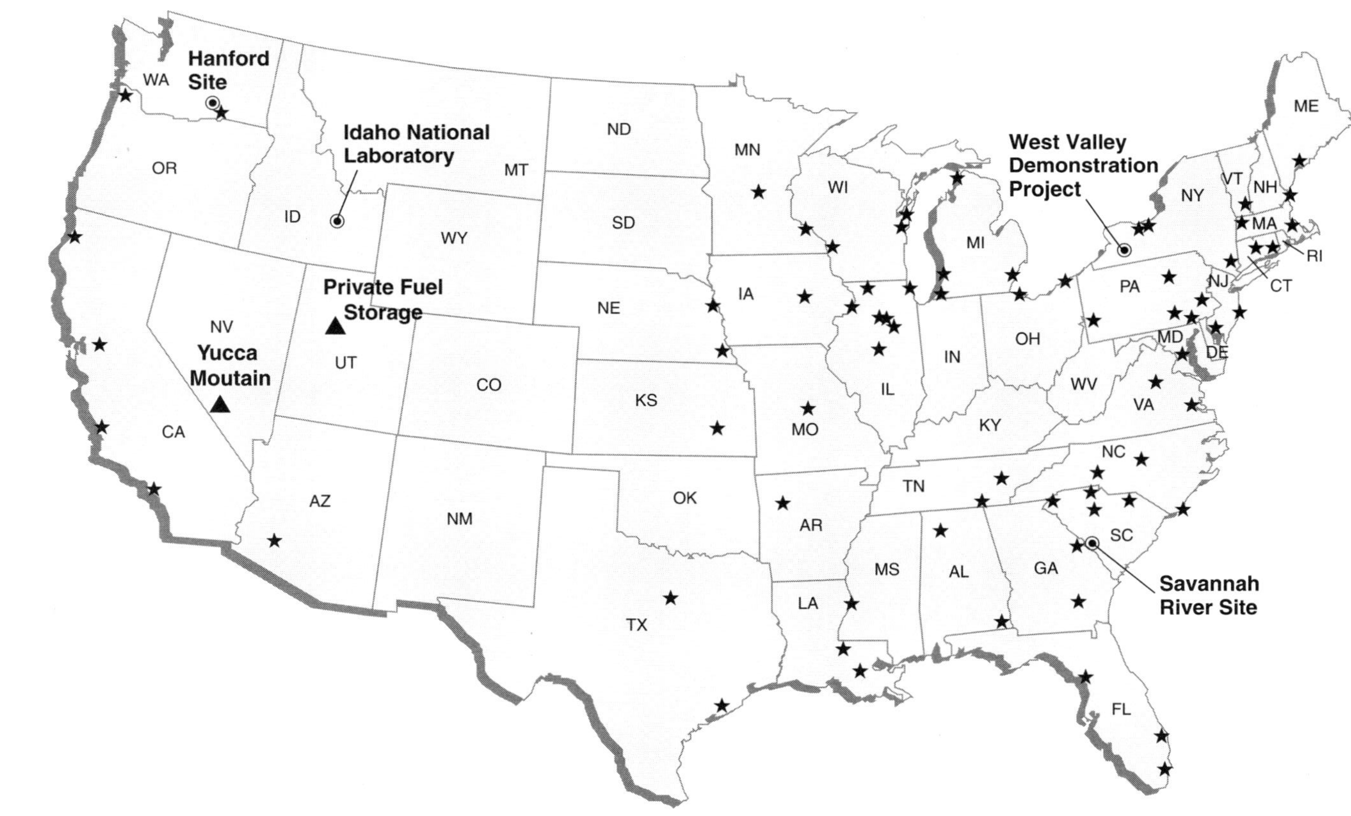

FIGURE 1.1 *Continues*

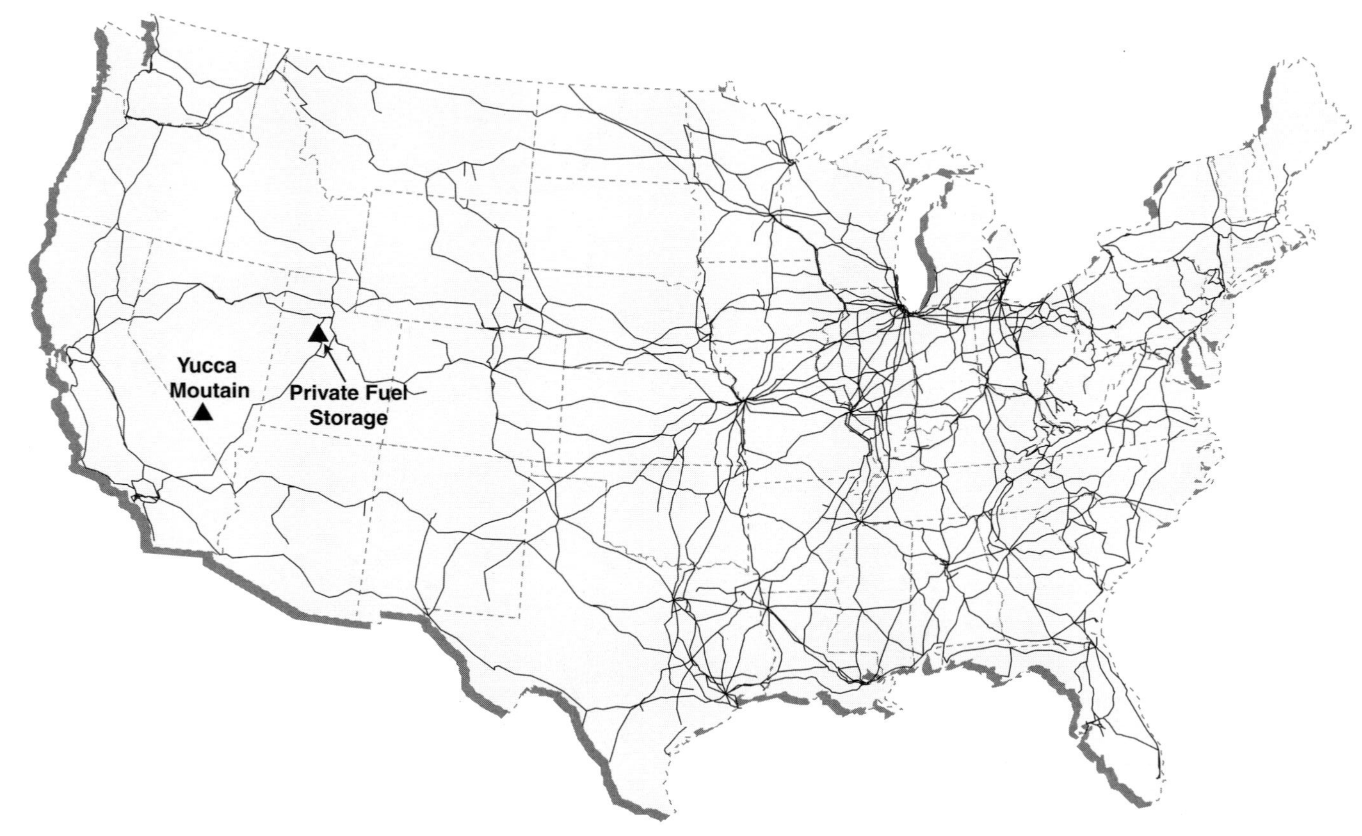
Yucca
Moutain
Private Fuel
Storage

FIGURE 1.1 *Top:* Locations of current spent fuel and high-level waste storage sites, Yucca Mountain, and Private Fuel Storage. *Middle:* National railroad transportation grid. *Bottom:* National interstate highway system. SOURCE: modified from DOE (2002a).

SIDEBAR 1.3 Spent Nuclear Fuel and High-Level Radioactive Waste

The nuclear fuel for most commercial reactors consists of pellets of uranium dioxide surrounded by a zirconium oxide alloy (zircaloy) fuel cladding that is formed into long rods. These *fuel rods* are between 3.5 and 4.5 meters in length and are bundled together into *fuel assemblies* (see figure), each weighing between about 600 and 1500 pounds (275 to 685 kilograms). The uranium dioxide pellets contain two isotopes of uranium: About 3 to 5 percent by weight is uranium-235, which sustains the fission chain reaction in a nuclear reactor, and about 95 to 97 percent is uranium-238, which can capture a neutron to produce plutonium and other heavy elements (known as *actinides*). The fission of uranium-235 and plutonium in an operating reactor generates heat, which is used to produce steam. This steam drives turbines that produce electricity. The fission and neutron-capture reactions also convert uranium, plutonium, and other actinides into nearly 300 other radionuclides in the fuel. These include *fission products* such as strontium-90 and cesium-137 and actinides such as neptunium-237.

As uranium-235 is consumed by fission reactions in an operating reactor, the fuel gradually loses its ability to sustain a chain reaction at full power. After a period of residence in the reactor (typically four to six years for most currently operating reactors), the fuel is considered to be *spent* and is removed from the reactor. At the time of discharge from a reactor, a spent fuel assembly typically generates on the order of tens of kilowatts of heat and radiation doses on the order of thousands of rads per hour (see Sidebar 3.2) at its surface. It must be cooled and heavily shielded to protect workers and the public. A person standing next to an unshielded spent fuel assembly could receive a lethal dose of radiation in a very short time: on the order of minutes for freshly discharged fuel with a high fission product content. Heat production and radioactivity diminish with time as shorter-lived radionuclides decay away. However, some longer-lived radionuclides persist in the spent fuel for hundreds of thousands of years.

High-level radioactive waste is the liquid by-product of the first stage of chemical reprocessing of spent fuel. Civilian reprocessing of commercial spent fuel is used to recover the uranium and plutonium in the spent fuel for recycling into fresh fuel.[a] Reprocessing of fuel from defense production reactors was used to recover plutonium for use in nuclear weapons. High-level waste contains most of the radioactive constituents of the spent fuel except for long-lived radionuclides such as uranium and plutonium. This liquid waste also contains a variety of organic and inorganic chemicals from processing operations.

High-level waste is solidified before it can be transported for interim storage or disposal. The current U.S. technology for solidifying high-level waste is vitrification in borosilicate glass. High-level waste is processed to reduce its volume and remove chemicals that would interfere with the glass-forming process. This processed waste is mixed with molten glass, and the mixture is poured into stainless steel canisters and allowed to solidify (see figure). Twenty-five percent (by mass) or higher waste loadings in the glass can typically be achieved with current technologies. Like spent fuel, the high-level waste canisters generate heat and intense radiation fields. They must be heavily shielded to protect workers and the public.

---

[a]As noted previously, the United States does not currently reprocess commercial spent fuel, but power-reactor fuel is reprocessed by some other countries.

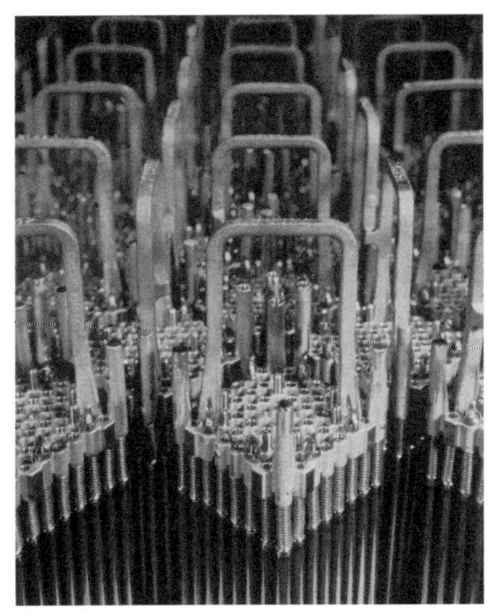

Top: Fuel rods are bundled together into fuel assemblies as shown here. These fuel assemblies are for a boiling water reactor. *Bottom:* Canister of high-level waste from the West Valley Demonstration Project. The canister is 2 feet (0.6 meter) in diameter, 10 feet (3 meters) long, and weighs about 2 metric tons. SOURCES: Top photo courtesy of the Nuclear Energy Institute. Bottom photo courtesy West Valley Nuclear Services Company.

the risks and the technical and societal challenges for transport of high-level waste.

The committee also did not address the *security risks* of spent fuel and high-level waste transportation in this report. Transportation security has received a great deal of media and public attention since the September 11, 2001, terrorist attacks on the United States. In fact, several presenters at the committee's information-gathering meetings highlighted this issue as an important current concern for transportation of spent fuel and high-level waste in the United States.

The committee explored the feasibility of including a substantive examination of transportation security risks in this report. The committee determined that there is a rich literature in existence that could inform such an examination. However, much of this literature is classified or otherwise restricted from public access, and most committee members do not have the necessary security clearances to access it. At the committee's request, staff from the USNRC's Spent Fuel Project Office provided a classified briefing to four committee members and one staff member with security clearances. This briefing provided an overview of current USNRC-sponsored studies to assess the vulnerability of transportation packages to certain types of terrorist attacks. This briefing confirmed the committee's initial view that adequate information exists to undertake a substantive examination.

The committee also requested written guidelines from the USNRC on the public disclosure of information from these and related vulnerability studies. Commission staff were supportive of this request but were unable to provide the necessary guidelines in time for use in this study.[7] The committee concluded that, given these information-access difficulties and the lack of written guidelines for using such information, it could not provide a substantive examination of transportation security in this report.

As will be discussed in Chapter 3, the committee judged that transportation security remains a critical technical issue with important societal implications for spent fuel and high-level waste transportation in the United States. While the committee could not examine transportation security in this report, it judged that this issue could be addressed in a substantive fashion by a future committee if it is given unrestricted access to the classified literature on this topic.

1.3 BACKGROUND ON SPENT FUEL AND HIGH-LEVEL WASTE

Spent nuclear fuel and high-level radioactive waste (Sidebar 1.3) are the by-products of commercial nuclear energy generation, defense plutonium

[7]The guidelines were still under internal review within the Commission when the committee held its last meeting in July 2005.

production, and research and medical activities that utilize nuclear reactors or fission product nuclides. In the United States, these waste by-products are now being stored at more than 70 sites in 31 states (Figure 1.1). Current national policy, which is embodied in the Nuclear Waste Policy Act (NWPA; see Section 1.3.2), calls for these materials to be transported and permanently disposed of in an underground repository that is licensed, constructed, and operated by the federal government.[8] The federal government is now attempting to site and construct a repository for this purpose at Yucca Mountain, Nevada (see Figure 1.1). The federal government is also required by the NWPA to accept ownership of commercial spent fuel for transport to and disposal in this repository.

While this study is silent on policy decisions related to the storage and disposal of spent fuel and high-level waste, the decisions themselves are important to understand because they influence many of the issues that are addressed in this report. The objective of this section is to provide more detailed background information on spent fuel and high-level waste, plans for their long-term disposition, and an overview of the regulations that govern their transport across the nation's highways and railways. This background information will support the more detailed discussions of transportation concerns in subsequent chapters. Knowledgeable readers may wish to skip this section and turn directly to Chapter 2.

1.3.1 Origin of Spent Fuel and High-Level Waste

The Atomic Energy Act (AEA) of 1954[9] opened the way for privately owned companies to build and operate commercial nuclear power plants in the United States. The nation's first commercial nuclear power reactor began operations at Shippingport, Pennsylvania, in late 1957, slightly more than three years after the AEA became law. Over the ensuing three decades, more than 100 power reactors were licensed to operate within the United States. Many more reactors were planned, but their construction was never realized because of an electrical supply overcapacity and the rapid escalation in nuclear plant construction costs due to increased construction times and high interest rates.

The development of commercial nuclear reactor designs, siting requirements, and regulation was based on a number of expectations and requirements. Two of these—situating reactors near populated areas and reprocessing of nuclear fuel—have had a significant influence on shaping programs transporting commercial spent fuel and also gave rise to many

[8]Referred to in this report as the "federal repository."

[9]P.L. 83-703, August 30, 1954.

of the current and future transportation concerns that are addressed in this report.

Commercial power reactors were designed to meet stringent siting criteria that permitted their construction near power consumers, an important requirement in the 1960s and 1970s because of technology limitations on electric power transmission distances.[10] Metropolitan sprawl over the past four decades has greatly increased population densities around many commercial nuclear power plants, especially in the eastern and midwestern United States. It has also increased population densities along highways and rail routes, many of which are linked through major cities. Much of the U.S. commercial spent fuel inventory is now being stored at sites near major populations, many spent fuel shipments originate from highly populated areas, and most shipments will pass through population centers on their way to temporary storage or permanent disposal.

It was also envisioned that the commercial nuclear power industry would operate under a *closed fuel cycle* in which spent fuel would be reprocessed to recover its reusable contents. This could include the recovery of uranium and plutonium[11] for recycling into fresh reactor fuel and other radionuclides for use in industry, medicine, and research. The liquid waste product from this operation, *high-level radioactive waste*, was to be immobilized in solid matrices and eventually disposed of in a permanent repository (see Sidebar 1.3). Power plant operators would be required to store spent fuel on-site in *spent fuel pools* (see Sidebar 1.4) only for about six months after its discharge from a reactor. The spent fuel would then be transported off-site to be reprocessed.

Commercial facilities to reprocess spent fuel were constructed at West Valley, New York, in the 1960s and Morris, Illinois, in the 1970s (Figure 1.1). The construction of a third facility at Barnwell, South Carolina, was eventually withdrawn from the licensing process. The West Valley facility operated from 1966 to 1972 and reprocessed both commercial and defense spent fuel. The reprocessing facility at Morris never opened because of design problems. The facility did, however, accept about 700 metric tons

---

[10]Proximity of electrical production to consumption is dictated by electrical generating and transmission technologies. The U.S. electrical transmission system utilizes mostly alternating-current circuits so that transformers can be used to control voltages. However, capacitance build-up in alternating-current transmission lines limits transmission distances to about 300 miles. High-voltage direct-current transmission lines, which allow electrical power to be moved over longer distances, did not come into common use until the 1970s and are now used primarily to transmit electricity among different geographic regions of the United States.

[11]As described in Sidebar 1.3, plutonium, like uranium-235, can be used as a fuel for commercial power reactors because it is *fissile*. That is, it can undergo fission after capturing a thermal neutron.

of spent fuel for reprocessing from commercial power plants. That spent fuel remains in pool storage today.

In early 1977, President Jimmy Carter made a policy decision not to provide further federal support for reprocessing of commercial spent fuel because of the perceived risk of diversion of plutonium.[12] This decision was made to set an example for other nations in the hope that they would abandon reprocessing. The decision was later reversed in 1981 during the administration of President Ronald Reagan, but no U.S. commercial reprocessing facilities have been constructed as a result of this decision.[13]

The decision by President Carter would have far-reaching implications for the management, transport, and disposal of spent fuel and high-level waste. Most significantly, it changed the operating basis for the U.S. commercial nuclear power industry to an *open fuel cycle* in which spent fuel is discarded rather than recycled.

The Atomic Energy Commission (AEC), the predecessor agency to the USNRC and DOE, had established a program in the mid-1950s to investigate options for disposing of high-level waste from commercial reprocessing. However, decisions about locating and constructing a disposal facility were still decades away, and little thought had been given up to that point to disposing of spent fuel directly. Lacking a near-term disposal option, power plant operators were forced to make provisions for interim storage of spent fuel at plant sites.

By the late 1970s, spent fuel pools at the oldest operating commercial nuclear power plants were approaching their design storage capacities. Operators were able to increase storage capacities at some plants by replacing the original storage racks, which were designed with open spaces between fuel storage cells for water circulation, with dense racks that reduced this open space. This *re-racking* process allowed plant operators to increase storage densities in their pools up to about a factor of five (Emit et al., 2003). This step postponed but did not eliminate the need for additional spent fuel storage at most commercial nuclear power plants.

To relieve the growing shortage of spent fuel storage space, the commercial nuclear industry also developed "dry" systems for storing spent fuel that had been out of the reactor for at least five years (Sidebar 1.4). These

---

[12]Fresh commercial nuclear fuel typically contains between 3 and 5 percent uranium-235 and 95 to 97 percent uranium-238. In an operating reactor, plutonium-239 is formed when uranium-238 captures a neutron.

[13]France, Russia, and the United Kingdom reprocess spent fuel from power reactors, and Japan is constructing a reprocessing facility. The United Kingdom shut down its facility for reprocessing oxide fuel (Thorp) in 2005 when a radioactive leak was discovered in one of its operating cells. It is not clear whether this facility will be ever be restarted. Reprocessing of metallic (Magnox) fuel continues.

SIDEBAR 1.4 Spent Fuel and High-Level Waste Storage

Immediately after removal from a nuclear reactor, spent fuel assemblies are stored in deep water-filled pools called *spent fuel pools* (see figure). The water provides radiation shielding and cooling and also captures non-fixed radioactive material on the external surfaces of the fuel rods. The fuel is stored in racks that contain neutron absorbers (e.g., boron, hafnium, cadmium) to help prevent criticality events (i.e., self-sustaining nuclear reactions such as those that occur when the fuel is in the reactor). The pool water is circulated through heat exchangers for cooling and ion-exchange filters to capture radioactive contaminants.

When a spent fuel pool reaches its storage capacity, the older fuel in the pool may be moved to other pools or placed into dry casks, as shown in the figure. These casks are typically constructed of steel and concrete and are designed to be placed outdoors on reinforced concrete storage pads at reactor sites. The casks are loaded by placing them directly into the spent fuel pool. Once loaded, the cask is closed, water is drained out, the fuel is dried, and the cask is filled with an inert gas. The exterior surfaces of the cask are also decontaminated to remove radioactive materials picked up from the pool water. The cask is moved by cranes and shielded transport vehicles to the storage pad.

Dry casks provide passive cooling of the spent fuel through a combination of heat conduction, natural convection, and thermal radiation. Shielding against radiation is provided by the cask materials: Concrete, lead, depleted uranium, or steel are used to shield penetrating beta and gamma radiation, and polyethylene, concrete, and boron-impregnated metals or resins are used to shield neutrons (neutrons are created in spent fuel by spontaneous fission and alpha particle interactions). Criticality control is provided by the basket that holds the spent fuel assemblies within individual compartments in the cask. The basket may contain boron-doped metals to absorb neutrons.

Standard industry practice is to place in dry storage only spent fuel that has cooled for five years or more after removal from the reactor. Dry casks are designed for specific types of spent fuel and for specific fuel burn-ups. The latter is a measure of the degree to which uranium-235 has been utilized in the reactor, which determines the amount of heat generation and radiation in the spent fuel. Dry casks are designed either for storage only or for both storage and transport. The former are referred to as single-purpose storage systems, and the latter as dual-purpose systems.

Dry cask storage first began in the United States in the mid 1980s and utilized single-purpose systems. Today, dry storage facilities utilize dual-purpose systems that are also suitable for rail transport. A large number of designs are available commercially.

Top: View into a spent fuel pool showing racks for storage of spent fuel. Workers are manipulating a spent fuel assembly. SOURCE: Nuclear Energy Institute. *Bottom:* Spent fuel storage (gray) and transportation (white) packages on a storage pad at a power plant. The smaller photo shows one of the storage packages being moved to the storage pad. The workers in the figure provide scale. SOURCE: Southern Company

systems store spent fuel in large steel and concrete casks designed to be placed outdoors on reinforced concrete pads at power plant sites. Each cask can typically store between about 10 and 18 metric tons (11 and 20 short tons[14]) of spent fuel. The first dry cask storage facility was established at the Surry Nuclear Power Plant (Virginia) in 1986. By 2004, dry storage facilities had been established at 29 U.S. power plants. This number is expected to increase in the future, especially if there are further delays in the construction of a federal repository or if away-from-reactor interim storage is not licensed or constructed.

Nuclear power plant operators in some states have encountered opposition from state regulatory agencies and the public to the establishment of dry storage facilities at their plant sites. Partly as a result, a consortium of operators (Private Fuel Storage, LLC) was formed to construct and operate a centralized dry cask storage facility on Goshute tribal lands in the desert southwest of Salt Lake City (see Figure 1.1) to relieve the growing storage pressures at plant sites. If constructed, this facility could store up to 40,000 metric tons of commercial spent fuel from multiple power plants. A license application for this facility was submitted to the USNRC in 1997. On September 9, 2005, following extensive hearings by the Atomic Safety Licensing Board, the Commission found that there were no further adjudicatory issues to be resolved. It authorized staff to issue a license to construct and operate this facility under the conditions in 10 CFR 72.40.[15]

The consortium expects to begin shipping spent fuel to this facility primarily by dedicated train[16] before the end of this decade. To ship by train, however, the consortium must first build a rail line across lands managed by the federal Bureau of Land Management (BLM). BLM cannot approve the use of its land for this purpose until the Secretary of Defense submits a report to Congress that evaluates the impacts of such construction on military training, testing, or operational readiness.[17] Additionally, the State of Utah strongly opposes this facility and is considering an appeal or court action to block the licensing decision.

Approximately 54,000 metric tons (60,000 short tons) of spent fuel were in storage at commercial power plants nationwide at the end of 2005

---

[14]A short ton is 2000 pounds (about 900 kilograms); a metric ton is 1000 kilograms.

[15]Title 10 Part 72 of the Code of Federal Regulations: Licensing Requirements for the Independent Storage of Spent Nuclear Fuel, High-Level Radioactive Waste, and Reactor-Related Greater than Class C Waste. Part 72.40 describes the conditions under which a license can be issued.

[16]Dedicated trains are trains that transport only spent fuel and high-level waste and no other cargo. These are described in more detail in Chapter 5.

[17]This provision was added to the fiscal year 2000 defense bill by a member of the Utah congressional delegation.

(Table 1.1). The U.S. commercial nuclear industry generates about 2000 metric tons (2200 short tons) of spent fuel each year, approximately 20 metric tons (22 short tons) from each of the 103 currently operating reactors. Under current U.S. policy, all of this spent fuel will continue to be stored at power plant sites or at a centralized storage facility such as Private Fuel Storage, LLC, until it can be transported to a federal repository for permanent disposal.

The federal government holds substantial inventories of spent fuel and high-level waste. The government operated defense reactors and reprocessing plants at sites near Hanford, Washington, and Savannah River, South Carolina, to produce plutonium for nuclear weapons. These facilities operated into the 1980s and produced more than 100 metric tons of plutonium. The government operated a facility at the Idaho National Laboratory[18] to reprocess naval and test reactor spent fuel. As noted previously, the commercial facility at West Valley, New York, also reprocessed a small amount of defense spent fuel. The volumes of currently stored high-level waste from these reprocessing operations are shown in Table 1.1. About 386,000 cubic meters of high-level radioactive waste[19] is stored at Hanford, Savannah River, and Idaho. Current U.S. policy calls for all of this high-level waste to be processed by immobilizing it in glass matrices (a process known as vitrification) and stored on-site until it can be shipped to a federal repository.

There are also about 2129 MTHM (metric tons of heavy metal) of defense reactor spent fuel in storage at the Hanford Site. This fuel was irradiated in one of the plutonium production reactors (the N-reactor) at that site but was never reprocessed. The fuel, much of which is highly corroded from decades of storage underwater, is being dried and placed in storage canisters. It too will eventually be shipped to a federal repository.

The federal government (through DOE) also supplies reactor fuel to university, government, and foreign research reactors, the latter under the Atoms for Peace Program.[20] Some of this fuel will eventually be returned to DOE. Foreign spent fuel is being transported from overseas by ship to the

---

[18]Formerly named the Idaho National Engineering and Environmental Laboratory.

[19]This waste was generated from the reprocessing of about 170,000 metric tons of spent fuel from plutonium production reactors. The original volumes of high-level waste were much greater. Volume reductions were obtained through various processing methods.

[20]The Atoms for Peace Program began under the Eisenhower Administration in 1954. The U.S. government supplied research reactor technology and nuclear fuel to foreign nations that agreed to forgo the development of nuclear weapons. Research reactors were built in 41 countries. In 1996, DOE issued an Environmental Impact Statement (DOE, 1996a) on the management of this fuel and issued a Record of Decision (DOE, 1996b) to return this fuel to the United States for eventual disposal. See Chapter 4.

Naval Weapons Station in Charleston, South Carolina.[21] From there it is offloaded and transported by either train or truck to the Savannah River Site, and some is transported onward to the Idaho National Laboratory by truck. Domestic research reactor spent fuel is being shipped to Savannah River and Idaho by reactor operators. Additional details on this program are provided in Chapter 4. Under current U.S. policy, all DOE's spent fuel and high-level waste will eventually be disposed of in a federal repository.

The DOE Naval Nuclear Propulsion Program is responsible for managing spent fuel from the U.S. Navy. This spent fuel is from nuclear submarines, ships, and training reactors belonging to the U.S. Navy. The spent fuel is offloaded at Navy facilities and shipped by commercial train to a naval spent fuel storage facility at the Idaho National Laboratory using U.S. government-owned transport packages and rail cars. The Naval Propulsion Program also will be responsible for shipping its naval spent fuel to the federal repository. At the end of 2005, about 19.5 MTHM of naval spent fuel was in storage at the Idaho site.

Under current U.S. policies, all of the spent fuel and high-level waste in the United States will be permanently disposed of in a federal repository. These policies are described in the next section of this chapter.

1.3.2 Disposal of Spent Fuel and High-Level Waste

The AEC began to consider options for disposal of high-level waste in the early 1950s. Conferences were held in 1954 at Woods Hole, Massachusetts, and Washington, D.C., to explore options for disposing of this waste in the oceans and on land. In 1955, the AEC signed a contract with the National Academy of Sciences to establish a committee of leading scientists to conduct additional conferences on methods for disposing of radioactive waste and to recommend a program of research.

The first AEC-sponsored Academy conference was held at Princeton University in September 1955 to discuss the land disposal of radioactive waste. The proceedings from this conference were published in the National Research Council report entitled *The Disposal of Radioactive Waste on Land* (NRC, 1957). Although the main topic of consideration at this conference was disposal options, issues related to transportation cost, feasibility, and safety figured prominently in one of the discussions.

Government research on land disposal options continued through the 1970s and culminated in an unsuccessful effort to establish a disposal facility in a salt cavern near Lyons, Kansas. Attention then shifted to salt

[21]In the past, one shipment of foreign research reactor fuel was received at Concord Weapons Station in northern California for transport to Idaho National Laboratory. However, DOE has no plans to receive future fuel shipments at this site.

disposal in the Delaware Basin in Texas and New Mexico. These investigations eventually led to the establishment of what was to become the Waste Isolation Pilot Plant (WIPP) near Carlsbad, New Mexico. This disposal facility, which is used to dispose of transuranic waste[22] from defense programs, was opened in 1999.[23]

The slow pace for addressing the waste disposal problem prompted action from the U.S. Congress in the 1980s. In 1982, Congress passed the Nuclear Waste Policy Act,[24] which established a federal responsibility and a federal policy for the disposal of spent fuel and high-level waste and a schedule for siting, constructing, and operating a federal repository. The NWPA established that the federal government is responsible for disposal of spent fuel and high-level waste and that the generators and owners of spent fuel and waste are responsible for paying the costs of such disposal.

The NWPA vests authority with the Secretary of Energy for carrying out the federal government's spent fuel and high-level waste disposal program, including transportation. It established a Nuclear Waste Fund[25] to cover the costs of transportation and disposal. It also authorized the Secretary of Energy to enter into contracts with owners of spent fuel and high-level waste of domestic origin. These contracts allow the Secretary of Energy to take title to spent fuel and waste for transportation and disposal.

The Nuclear Waste Policy Act established a process to identify two repository sites in different regions of the United States (one in the eastern or midwestern United States and the other in the western United States) for disposal of the nation's spent fuel and high-level waste. After the NWPA was passed, DOE initiated a screening program that eventually resulted in the selection of three potential repository sites in the western United States: Hanford, Washington; Deaf Smith County, Texas; and Yucca Mountain, Nevada.

In 1987, Congress amended the Nuclear Waste Policy Act. This amended act directed DOE to terminate its characterization activities at all sites except Yucca Mountain. It capped the disposal capacity of Yucca Mountain at 70,000 MTHM (77,000 short tons) until after a second repository is oper-

---

[22]Transuranic waste is a by-product of nuclear weapons production activities. It contains long-lived radioactive transuranic elements such as plutonium in concentrations greater than 100 nanocuries per gram.

[23]The land withdrawal act for the Waste Isolation Pilot Plant (P.L. 102-579) specifically prohibits the transport of spent fuel and high-level waste to WIPP or disposal in the WIPP repository.

[24]The Nuclear Waste Policy Act of 1982, P.L. 97-425 (January 7, 1983) and amendments (P.L. 100-203, Subtitle A [December 22, 1987]; P.L. 100-507 [October 18, 1988]; and P.L. 102-486 [October 24, 1992]).

[25]The Nuclear Waste Fund was established by the U.S. Treasury and is funded through a 1.0 mil (0.1 cent) per kilowatt-hour fee on nuclear-generated electricity (see DOE, 2001a).

ating. This capacity has been allocated by DOE for disposal of both commercial spent fuel and DOE-owned spent fuel and high-level waste (Table 1.2). It also directed DOE to postpone the identification of a site for a second repository.

After the passage of the amended act, DOE initiated what would ultimately be a 15-year program to characterize the Yucca Mountain site to determine its suitability to host a repository. In 1998, DOE issued a viability assessment for this site in which it concluded (DOE, 1998a, summary, p. 2) that "Yucca Mountain remains a promising site for a geologic repository" and "work should proceed to support a decision in 2001 on whether to recommend the site to the President for development as a repository."

In late 2001, then Secretary of Energy Spencer Abraham recommended to President G. W. Bush that a federal repository be developed at Yucca Mountain. The President forwarded the recommendation to Congress, which in July 2002—and over the objections of Nevada—authorized DOE to submit an application to the USNRC for a license to construct and

TABLE 1.2 Spent Fuel and High-Level Waste Disposal at Yucca Mountain

Material	Quantity[a]	Number of Shipping Sites
Commercial spent fuel	29,000 cubic meters 63,000 MTHM[b]	73 (72 commercial power plant sites and one commercial storage facility)
Defense spent fuel and high-level waste	7000 MTHM	From sites shown below
Naval spent fuel	900 cubic meters 65 MTHM	1 (INL)
Other DOE spent fuel[c]	1000 cubic meters 2435 MTHM	4 (Hanford, INL, SRS, Fort St. Vrain)
High-level waste	When processed: 21,000 cubic meters 58,000 metric tons 22,000 canisters	4 (Hanford, INL, SRS, West Valley)

NOTE: INL = Idaho National Laboratory; SRS = Savannah River Site.

[a]The quantities of commercial and defense waste listed in this table represent the current legislated capacity of Yucca Mountain. Quantities are given in terms of volumes and masses.

[b]Metric tons of heavy metal.

[c]Includes defense and research reactor spent fuel.

SOURCE: DOE (2002a, Appendix A).

operate a repository. DOE planned to submit this application to the USNRC before the end of 2004 and to begin operating the repository by 2010.

DOE was unable to submit the application in 2004, however, and has encountered setbacks that could delay its plans to establish a repository at Yucca Mountain. In 2004, the Federal Court of Appeals for the District of Columbia struck down part of the Environmental Protection Agency's (EPA's) health and safety standards for Yucca Mountain (Title 40, Part 197 of the Code of Federal Regulations). This remanded standard must be reissued[26] by the EPA before the USNRC can issue a license for Yucca Mountain. DOE also encountered problems in establishing the database of information that will be used to support its license application for Yucca Mountain. The Nuclear Waste Policy Act requires that this database, referred to as the Licensing Support Network, be established at least six months before the license application is docketed by the USNRC. DOE brought this database on-line in 2004, but the USNRC has so far refused to certify that it is complete.

The Nuclear Waste Policy Act committed the federal government to begin disposing of commercial spent fuel in the federal repository by January 31, 1998. However, siting such a repository turned out to be a more arduous process than envisaged by Congress when the NWPA was passed. After the 1998 deadline passed, commercial power plant operators began filing lawsuits against DOE to recover the additional costs incurred for extended on-site storage of spent fuel. The U.S. government settled with one of the plaintiffs (Exelon Nuclear Power Corp.) in 2004. Sixty cases are still pending before the courts, and discussions are under way to settle some of these cases. The Exelon settlement commits the federal government to pay the utility $80 million immediately, with further annual payments as costs are incurred for continued storage of spent fuel at its sites. The federal government will pay Exelon a total of about $300 million if Yucca Mountain opens by 2010, and possibly more if the opening is delayed beyond that date. Settlement costs for the entire nuclear industry could cost taxpayers[27] billions of dollars. DOE is under pressure from the nuclear industry and Congress to move forward with Yucca Mountain or establish one or more centralized interim storage sites to reduce the growing spent fuel inventories at commercial power plant sites as well as the federal government's future monetary liabilities.

The Private Fuel Storage facility could be used as an interim step to-

---

[26]The EPA issued a new draft standard for public comment on August 8, 2005. The final standard had not yet been issued when this report was being finalized for publication in December 2005.

[27]Settlements are being paid out of the federal government's judgment fund, not out of the Nuclear Waste Fund.

ward permanent disposal at Yucca Mountain. The facility's location in Utah was selected in part because of its proximity to Nevada. Even though Private Fuel Storage, LLC, is now poised to receive a license from the USNRC, it may still face several obstacles to opening. These include opposition from the State of Utah as well as other states and communities along likely transportation routes.

1.3.3 Spent Fuel and High-Level Waste Transportation

Small-quantity spent fuel shipping programs have been carried out routinely by both the federal government and the private sector for several decades. The primary objective of these programs has been to move spent fuel to interim storage. The federal agency responsible for government transport of spent fuel is DOE. This agency has transported foreign research reactor spent fuel, DOE research reactor spent fuel, naval spent fuel, and commercial spent fuel from some shut-down power reactors (e.g., Three Mile Island Unit 2) to centralized interim storage sites in South Carolina and Idaho. University and non-DOE government research reactor operators have also transported spent fuel to these interim storage sites. Commercial nuclear power plant operators have transported spent fuel between reactor sites to consolidate storage.

Transportation of spent fuel and high-level waste takes place under a number of federal, state, and local statutes and regulations.[28] The complexity of the regulatory environment is illustrated schematically in Figure 1.2 and in tabular form in Table 1.3. The principal federal regulators are DOT and the USNRC.[29] DOT is responsible for regulating the safety of hazardous material shipments, including radioactive material shipments, under several statutes, including the Department of Transportation Act (49 USC 1655) and the Hazardous Materials Transportation Act (49 USC 1801–1812).[30] The USNRC is responsible for licensing and regulating the receipt, possession, transfer, and use of source materials, byproduct materials, and special nuclear materials (see glossary in Appendix D) under the Atomic Energy Act (42 USC Chapters 6–8) and the Energy Reorganization Act (42 USC 5841).

---

[28]See Clark County Nuclear Waste Division (2004) for a summary of state requirements concerning the transportation of radioactive waste.

[29]Under 10 CFR Part 150, the USNRC retains authority for licensing and regulating spent fuel storage and transport in USNRC Agreement States. All states retain their authorities for carrier safety and emergency response as shown in Table 1.3.

[30]Shipments made for national security purposes by the Department of Defense or DOE may be exempted from DOT regulations if they comply with the security escort requirements in 49 CFR 173.7(b).

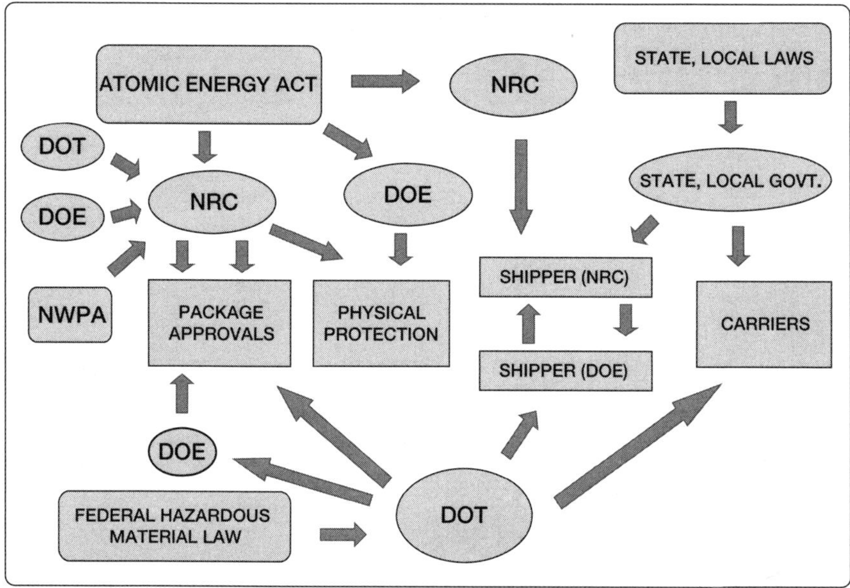

FIGURE 1.2 Schematic illustration of the regulatory bases for spent fuel and high-level waste transportation in the United States. As illustrated by the figure, spent fuel and high-level waste transport is regulated by the federal government and states under a number of statutes. SOURCE: Earl Easton, USNRC.

DOT and the USNRC have signed a memorandum of understanding (MOU) (44 FR 38690, July 2, 1979) that delineates responsibilities for regulating the transport of radioactive materials. This MOU gives USNRC the primary responsibility, in consultation with DOT, for the development of standards and regulations for the design, performance, and inspection of transportation packages for fissile materials, which include spent fuel and high-level waste. USNRC also has the primary responsibility for approval of domestic and foreign package designs used to transport spent fuel solely within the United States.[31] The MOU recognizes the USNRC's responsibility for imposing DOT regulations and conducting inspection activities for shipments of spent fuel by its licensees. In addition, DOT routing regulations (49 CFR 397.101) recognize the USNRC's responsibility for providing physical protection requirements for spent fuel shipments.

[31]While DOT has the responsibility for approving packages for import and export shipments, it relies on USNRC's review under the MOU as the basis for approving or revalidating the use of foreign-designed spent fuel packages.

TABLE 1.3 Principal Responsibilities for Spent Fuel and High-Level Waste Transportation in the United States

	USNRC Licensee Spent Fuel (AEA)		DOE Spent Fuel (AEA, NWPA)		
	Commercial Reactors	Research Reactors	Commercial Reactors	Foreign Research Reactors	Naval Reactors[a]
Package approvals and inspections	USNRC	USNRC	USNRC	DOE and USNRC[b]	DOE and USNRC[b]
Highway route selection criteria[c]	DOT	DOT	DOT	DOT	DOT
Carrier safety	DOT, states	DOT, states	DOT, states	DOT, states	DOT[d]
Emergency response	Federal, state, tribal, and local governments	Federal, state, tribal, and local governments	Federal, state, tribal, and local governments	Federal, state, tribal, and local governments	Federal, state, tribal, and local governments
Route security approval	USNRC	USNRC	DOE	DOE and USNRC[e]	DOE
Physical protection	USNRC	USNRC	DOE and USNRC[f]	DOE	DOE

NOTE: AEA = Atomic Energy Act; NWPA = Nuclear Waste Policy Act.

[a]Naval Reactors has voluntarily chosen not to designate shipments of defense-related spent fuel as national security shipments under 49 CFR 173.7 and instead chooses to comply voluntarily with DOT regulations.

[b]DOE seeks technical reviews from the USNRC on package designs. The USNRC review provides the basis for DOT approval of foreign package designs under 49 CFR 171.12 or for USNRC certification of U.S. package designs.

[c]DOT does not have route selection criteria for rail shipments.

[d]Shipments are subject to the Federal Rail Safety Act, which governs railcars and track. DOE also voluntarily complies with DOT inspection requirements.

[e]DOE has made a practice to seek approval from USNRC under a reimbursable agreement for the foreign research reactor spent fuel shipments that it handles. USNRC approval is not required when DOE has title and possession of spent fuel.

[f]DOE is required under the Nuclear Waste Policy Act to follow USNRC prenotification requirements. These are described in Appendix C.

The MOU recognizes DOT as having the primary responsibility, in consultation with USNRC, for issuing safety requirements for the transport of radioactive materials, including labeling of packages; placarding of vehicles; equipment maintenance requirements; carrier personnel qualifications; procedures for loading, unloading, handling, and storage during transport; and special transport controls (excluding safeguards controls) for radiation safety during transport. DOT also has the primary responsibility for inspecting transportation activities by carriers for both USNRC licensee and non-USNRC licensee activities (e.g., shipments by DOE and USNRC Agreement State licensees).

As shown in Table 1.3, responsibilities for regulating the transport of spent fuel and high-level waste are somewhat different for USNRC licensees[32] and DOE. USNRC licensees are required to use USNRC-certified packages for domestic shipments or a DOT-certified package for import-export shipments. DOE has authority under DOT regulations (49 CFR 173.7), unless otherwise specified in law, to certify packages for the domestic transport of its own spent fuel and high-level waste—for example, its shipments of spent fuel from West Valley to Idaho.

DOE's import shipments of foreign reactor fuel, unless designated as national security shipments, must be made in DOT-approved packages. This could include either a USNRC-certified package, a DOE-certified package, or a foreign package design that is revalidated by DOT. However, under a cost-reimbursable agreement, DOE has sought USNRC review of foreign package designs that can be used as the basis for DOT revalidation. DOE has also made a policy decision to seek USNRC approval of the physical protection measures used for its shipments of foreign research reactor spent fuel. Research reactor shipments are discussed in Chapter 4. A description of some of the USNRC regulations for package certification and associated package tests is provided in Section 2.1.

States and local governments also play important roles in spent fuel and high-level waste transportation. States have an important responsibility for enforcing DOT highway safety regulations concerning federal motor carrier safety and hazardous materials transportation. Highway shipments of spent fuel and high-level waste are subject to state inspection, and state enforcement officials can stop and inspect vehicles and inspect the premises of motor carriers to check for compliance with federal and state requirements regarding equipment, documentation, and driver fitness. States can also require carriers to obtain special permits to operate these vehicles and

[32]The Atomic Energy Act gives the USNRC the authority to issue licenses to private and government (except DOE) organizations to possess radioactive materials, conduct operations involving the emission of radiation, and dispose of radioactive waste. Operators of nuclear reactors for research and power generation are USNRC licensees since these facilities involve the emission of radiation and the generation of radioactive materials.

charge fees for such permits.[33] Rail shipments of spent fuel and high-level waste are not subject to state regulation, but they are subject to inspection by DOT's Federal Railroad Administration.

Federal, state, tribal, and local governments and shippers share the responsibility for emergency response and preparedness. The Federal Emergency Management Agency within the Department of Homeland Security is responsible for providing a national incident response plan. State, tribal, and local governments are responsible for providing the first line of government response to accidents and incidents within their jurisdictions and can enlist the assistance of other agencies and organizations as circumstances require.

More detailed information on responsibilities and regulations is provided in other chapters. Chapter 4 provides a detailed description of highway and rail routing regulations. Chapter 2 describes package performance standards and regulations. Chapter 5 provides descriptions of other regulations governing the transport of spent fuel and high-level radioactive waste to a federal repository.

[33]However, the federal hazardous materials transportation law is explicit that federal rules preempt state rules in cases of conflict (49 USC 5125), consistent with the goal of nationally uniform regulation. DOT has administrative authority to determine when state rules are to be preempted. DOT has determined that state requirements for special permits for highway shipments of radioactive materials are preempted if they require documentation or prenotification in excess of federal requirements. DOT also has determined that state fees imposed on hazardous materials transport are preempted if they are excessive or if the revenue is not used for purposes related to hazardous materials transport.

2
Transportation Package Safety

Package performance—the ability of a transportation package[1] to maintain a high level of containment effectiveness in long-term normal use and under extreme mechanical forces and thermal loading conditions (i.e., thermomechanical conditions) generated during severe accidents—is a crucial issue for transportation safety and key to understanding and quantifying transportation risks. The regulatory requirements for the design, fabrication, certification, and maintenance of these packages are substantially more rigorous than those for transporting most other types of hazardous materials (e.g., hazardous chemicals). Accordingly, the packages used to transport spent fuel and high-level waste are designed to withstand extreme thermomechanical conditions without a significant loss of containment. The slogan "safety is built into the package" is commonly used by package manufacturers and vendors to describe the ability of these packages to maintain their containment effectiveness under most conceivable accident conditions.

This chapter provides a summary of investigations carried out in the United States and several other countries to examine the performance of transportation packages. The chapter provides a brief overview of the re-

[1]Transportation containers loaded with spent fuel or high-level waste are referred to as *packages* in international standards, whereas the containers themselves without their contents are referred to as *packagings*. For simplicity, the term *packages* is used throughout this report. The terms *casks* and *flasks* (the latter term is commonly used in the United Kingdom) are sometimes used synonymously to refer to such packages.

quirements for spent fuel and high-level waste transportation packages; describes some key investigations of package loading conditions that have furthered technical understanding of package performance; and offers findings and recommendations about current standards and regulations and about improvements in package performance.

There is extensive literature on package performance in response to extreme thermomechanical conditions. Some of this work appears in the "gray" literature,[2] which may or may not be peer reviewed. There is also extensive classified literature on package performance in response to potential terrorist acts.[3] As discussed in Chapter 1, the committee did not review this classified literature or perform an in-depth examination of package performance in response to terrorist acts. The committee does, however, comment on this issue in Chapter 5.

2.1 TRANSPORTATION PACKAGE DESIGNS AND REGULATIONS

Packages for the transport of spent fuel and high-level waste are designed to meet three basic requirements both during normal conditions of transport[4] and during a range of hypothetical accident conditions established in 10 CFR Part 71:[5]

1. Prevent an unsafe configuration (i.e., accidental criticality[6]) of spent fuel.
2. Prevent or limit the release of radioactive contents.
3. Limit dose rates on external package surfaces to acceptable levels.

A wide range of package designs have been developed to meet these general performance requirements.

---

[2]The U.S. Interagency Gray Literature Working Group defines gray literature as "foreign or domestic open source material that usually is available through specialized channels and may not enter normal channels or systems of publication, distribution, bibliographic control, or acquisition by booksellers or subscription agents." (Gray Information Functional Plan, January 18, 1995, accessed at *http://www.osti.gov/graylit/whatsnew.html*). This literature includes technical reports, conference proceedings, and business documents.

[3]In addition, some unclassified literature on this topic was removed from public circulation after the September 11, 2001, terrorist attacks.

[4]Normal conditions of transport subject packages to minor mishaps due to rough handling or exposure to weather. Such conditions would not be expected to compromise the vital containment functions of the package.

[5]Title 10, Part 71 of the Code of Federal Regulations, Packaging and Transport of Radioactive Material. The hypothetical accident conditions are described in subpart 73 (i.e., 10 CFR 71.73).

[6]That is, a configuration that allows the establishment of a self-sustaining nuclear chain reaction as occurs in an operating nuclear reactor.

Most packages consist of a hollow cylindrical body that is open at one end (Figure 2.1). The body is typically constructed of multiple layers of the following materials: Steel is used to provide structural strength and durability; steel, lead, depleted uranium, or concrete is used to provide shielding against gamma radiation; and water, borated polymers, or concrete is used to provide shielding against neutrons.

The package closure system consists of one or two steel lids that are attached to the open end of the package body with steel bolts. Elastomer or

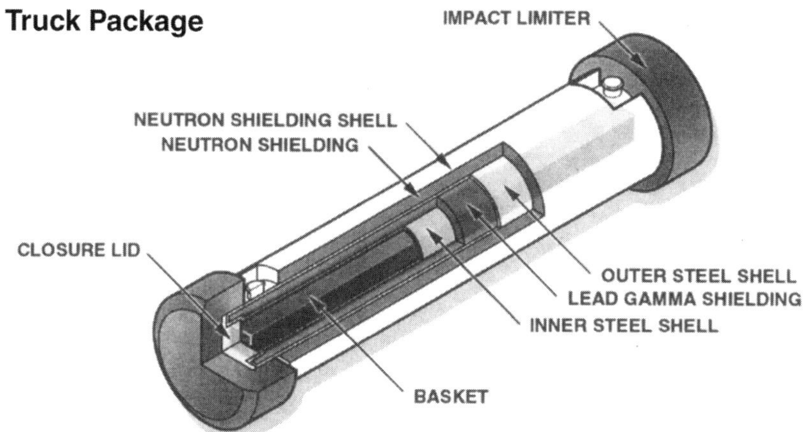

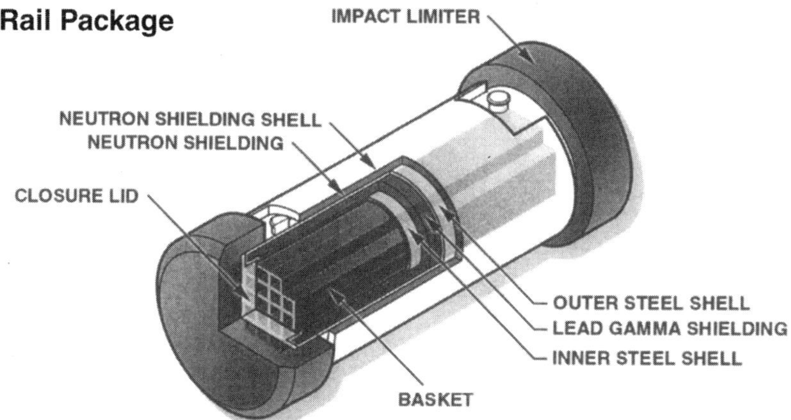

FIGURE 2.1 Generic truck and rail spent fuel packages. In these particular designs, lead is used as the shielding material. Other materials such as steel and concrete are used in other package designs. SOURCE: Sprung et al. (2000).

metal seals are used between the body and lids to obtain airtight seals. The package body and lids may contain sealable openings, such as pipes and tubes, to allow for the removal of water, the addition of inert gases, leak testing of lid seals, and monitoring of internal pressures after the package is closed. The lid, lid bolts, pipes, and valves are usually recessed to protect them from damage.

The package is also designed with impact limiters to absorb mechanical forces generated in the event of transport accidents and to provide thermal protection for the lid seals in case of fires. Impact limiters, which are usually attached to the ends of packages, are typically constructed of wood, rigid foam, or honeycombed metal. Metal fins may also be machined into (or welded onto) the exterior surfaces of the package body and closures for additional impact protection and heat dissipation.

The interior of the package contains a basket to hold the spent fuel assemblies in a fixed configuration to ensure criticality control and minimize damage to the fuel during transport. In packages designed to hold multiple spent fuel assemblies, the baskets are typically constructed of materials containing neutron absorbers (e.g., borated metals) to provide a further margin of criticality safety. In some package designs, these baskets are placed into a stainless steel canister with a welded lid, which in turn is placed into a transportation overpack. Packages containing such canisters are sometimes referred to as *canister-based packages*. Packages without such canisters are sometimes referred to as *bare-fuel packages* because the fuel basket is placed directly into the package body.

Packages designed for transport by legal-weight trucks[7] typically carry between about 0.5 and 2 metric tons (0.55 and 2.2 short tons) of spent fuel. These packages are about 3 feet (0.9 meter) in diameter and weigh up to about 25 metric tons (28 short tons) when loaded. Packages designed to be transported by train can hold about 10 to 18 metric tons (11 to 20 short tons) of spent fuel. The packages typically have diameters of about 8 feet (2.4 meters) and can weigh 150 metric tons (165 short tons) or more when loaded.

The number and types of spent fuel assemblies that can be carried in a package are, of course, determined by its size. Package size, in turn, depends on transportation mode (i.e., rail versus truck). Legal-weight truck package size is limited primarily by highway weight regulations, whereas rail package size is usually not weight limited but instead limited by railcar clearance requirements. Regulatory limits on radiation levels on the exterior surfaces of packages (10 CFR 71.47) also restrict the age and burn-up of spent fuel that can be carried in the package.

---

[7]A truck having a total gross weight (i.e., including cargo) of 80,000 pounds (36,300 kilograms) or less.

The International Atomic Energy Agency (IAEA) has established standards for the safe transportation of radioactive materials. These standards were first issued in 1961[8] and have undergone several revisions and amendments; the latest edition was issued in 2005 (IAEA, 2005a). The IAEA standards establish recommended requirements for a number of different package types, each designed to transport specific quantities and types of radioactive materials:

- *Excepted packages* are designed for transport of very small quantities of radioactive material such as radiopharmaceuticals.
- *Industrial packages* are designed for transport of low-specific-activity materials such as uranium ore and low-level radioactive wastes.
- *Type A packages* are designed for transport of materials of limited radioactivity—for example, uranium hexafluoride and fresh nuclear fuel.
- *Type B packages* are designed for transport of larger quantities of radioactive material including spent fuel, high-level waste, and mixed oxide fuel.[9]
- *Type C packages* are designed for air transport of quantities of radioactive material exceeding a defined (large) threshold including, for example, plutonium and mixed oxide fuel.

This report is concerned with Type B packages.

IAEA safety standards are recommendations and are not legally binding on member states. However, the United States and many other member states adopt these standards, either in whole or in part, in their own national regulations. U.S. regulations for the packaging and transport of radioactive materials are provided in 10 CFR Part 71 (see footnote 5). The Nuclear Regulatory Commission (USNRC) and Department of Transportation (DOT) recently modified these regulations (USNRC, 2004a) to reflect the 1996 amended version of the IAEA standards (IAEA, 2000).[10,11]

---

[8]The first comprehensive set of regulations for transporting radioactive materials by common carrier were established by the Interstate Commerce Commission in 1947–1948. These regulations were drafted by the National Research Council's Subcommittee on Shipment of Radioactive Substances (see NRC, 1951). These and subsequent U.S. regulations served as an important basis for the establishment of IAEA standards (see Pope, 2004).

[9]Mixed oxide fuel, or MOX, contains uranium and plutonium.

[10]The updated regulations did not incorporate the IAEA standards for Type C packages, because U.S. federal law mandates more stringent requirements for air shipments of plutonium.

[11]The IAEA issued updated safety standards in 2003 and 2005 (see *http://www-ns.iaea.org/standards/documentpages/transport-of-radioactive-material.htm*). USNRC and DOT staff were discussing whether to update U.S. regulations to bring them into conformance with these updated standards when this report was being finalized for publication in December 2005.

Type B packages that are to be used for transporting spent fuel and high-level waste in the United States are required to be certified by the USNRC. To receive a certification, the applicant (i.e., the package manufacturer or vendor) must demonstrate to the Commission's satisfaction that the package design will meet the testing requirements in 10 CFR Part 71, the key elements of which are described in Sidebar 2.1. Under these regulations, packages are required to survive a free-drop test onto an essentially unyielding surface (Sidebar 2.2) as well as a puncture test, an immersion test,[12] and a thermal test with less than the specified loss of containment effectiveness (see Sidebar 2.1).

The USNRC permits quantitative analysis (e.g., computer simulations using finite element models), scale-model (typically one-quarter or one-half scale; see Sidebar 2.3), and full-scale testing of packages or package components, and comparisons with existing approved package designs to be used to demonstrate compliance with the regulations. Testing of full-scale packages[13] is not a requirement of the regulations.

Testing of full-scale spent fuel packages is not carried out routinely because of the cost. A full-scale package can cost more than a million dollars, and it is generally not reusable after undergoing full-scale testing in accordance with USNRC regulations. The paucity and costs of suitable testing facilities are also impediments. There are no package testing facilities in the United States capable of handling large truck or rail packages. A new facility was recently opened in the Horstwalde region in Germany, which is located near Berlin.[14] Two full-scale 9-meter (30-foot) free-drop tests on rail packages were carried out at this facility in late September 2004 in conjunction with the 14th International Symposium on Packaging and Transportation of Radioactive Materials (PATRAM 2004). A subgroup of the committee's members visited this facility and witnessed a full-scale test of a 180 metric ton (198 short ton) rail package (Figure 2.2).

USNRC regulations also require that transportation packages be de-

[12]Package immersion is not discussed at much length in this chapter because the committee judges it to be of a lower concern than the thermomechanical conditions generated during truck and train accidents.

[13]There is sometimes confusion about what is meant by "full scale" in regard to package testing. The regulatory tests described in Sidebar 2.1 are full-scale tests because they are carried out using actual Type B packages. These tests are referred to as *certification tests*. The term full-scale can also apply to tests made on actual Type B packages under simulated accident conditions such as those described in Section 2.3. These tests are referred to as *demonstration tests*.

[14]This facility is operated by the German Federal Institute for Materials Research and Testing (BAM).

signed and manufactured using an approved quality assurance program.[15] Packages must be designed to standards that include conservative assumptions and design margins for material properties such as yield stress and ductility. This requirement provides for a built-in "safety margin" (see Sidebar 2.4) and offers increased confidence that packages will survive thermomechanical conditions somewhat more severe than regulatory requirements.

The USNRC has certified several spent fuel storage and transportation package designs for use in the United States. These include storage-only packages as well as packages that are designed for both transportation and storage (see JAI, 2005).

IAEA standards and USNRC regulations for Type B packages (see Sidebar 2.1) were not derived from a comprehensive bounding analysis of all possible extreme thermomechanical conditions resulting from package mishandling and accidents. Rather, their development, which dates from the early 1960s as noted previously, was based on then-available data on typical impacts (e.g., drops from cranes, other mechanical accidents, vehicular collisions) and thermal environments (e.g., fires from spilled fuel) to which a package might be exposed in the course of transport (see Pope, 2004). As such, they do not necessarily reflect the most extreme conditions that might be encountered during spent fuel or high-level waste shipments.

Nevertheless, during the committee's information-gathering meetings, several industry presenters asserted that it is very unlikely that a certified Type B package for spent fuel or high-level waste would fail under any credible loading conditions that might be encountered during transport. This assertion is based on the confidence that these presenters place in the combination of rigorous regulatory requirements for package certification, the built-in margins of safety in current package designs (see Sidebar 2.4), and the worldwide decades-long record of spent fuel transport without a significant package failure (see Chapter 3). However, other presenters pointed out that the spent fuel transport experience in the United States is limited, and that the planned large future shipping campaigns to a federal repository could expose transportation packages to a wider range of loading conditions and longer-term use than have been experienced to date.

---

[15] 10 CFR 71.101 defines the term *quality assurance* as those "planned and systematic actions necessary to provide adequate confidence that a system or component will perform satisfactorily in service. Quality assurance includes quality control, which comprises those quality assurance actions related to control of the physical characteristics and quality of the material or component to predetermined requirements." Quality assurance requirements apply to the design, fabrication, assembly, testing, maintenance, repair, modification, and use of the proposed package.

SIDEBAR 2.1 U.S. Regulations for Type B Transport Packages

Type B packages for transporting spent fuel and high-level waste are designed to withstand severe accident conditions without a loss of containment or an increase in external radiation *to levels that would endanger emergency responders or the general public.* USNRC regulations for Type B packages contain requirements for both "normal conditions of transport" and "hypothetical accident conditions." Under normal transport conditions, the regulations require that Type B packages maintain their containment effectiveness as given by the following three conditions:

1. No loss or dispersal of radioactive contents to a sensitivity of 10^{-6} A_2 per hour.[a]
2. No substantial reduction in packaging effectiveness.
3. No substantial increase in external surface radiation levels.

For hypothetical accident conditions, the regulations allow for some degradation of the packages' containment effectiveness as given by the following four conditions:

1. No escape of krypton-85 exceeding 10 A_2 in 1 week.
2. No escape of other radioactive material exceeding a total amount A_2 in 1 week.
3. No external radiation dose rate exceeding 10 millisieverts per hour (1 rem per hour) at 1 meter (about 40 inches) from the external surface of the package.[b]
4. Compliance with these requirements may not depend on filters or mechanical cooling systems.

These release and radiation limits are designed primarily to protect emergency responders and members of the public in case of an accident. The values for A_2 are tabulated in 10 CFR Part 71 and are radionuclide specific. For krypton-85, the A_2 value from Table A-1 in 10 CFR Part 71 is 10 terabecquerels (TBq; approximately 270 curies).[c] The regulations do not place any limits on the physical form of the radioactive materials (e.g., particulate size) that could be released in a hypothetical accident.

The USNRC is responsible for certifying the design of Type B packages. The requirements for certification (10 CFR 71.73) are derived directly from IAEA standards:

> Evaluation for hypothetical accident conditions is to be based on the sequential application of the tests specified in this section, in the order indicated, to determine their cumulative effect on a package or array of packages . . . . [E]xcept for the water immersion tests, the ambient air temperatures before and after the tests must remain constant at that value between –29°C and +38°C which is the most unfavorable for the feature under consideration. The initial internal pressure within the containment system must be the maximum normal operating pressure, unless a lower internal pressure, consistent with the ambient temperature presumed to precede and follow the tests, is more unfavorable.

---

[a]The regulations also provide guidance on determining A_2 activity values for mixtures of radionuclides. See the glossary (Appendix D).

[b]These increased doses would result from structural damage to the package shielding.

[c]The fission product krypton-85 was selected for use in this regulation because it is the only noble gas that exists in significant quantities in irradiated fuel that has been cooled for several years. While it is not a significant contributor to dose, it is likely to be among the most mobile of radionuclides in a spent fuel package.

The following tests are specified in the regulations (10 CFR 71.73; see figure below):

- A *free-drop* test in which the specimen is dropped through a distance of 9 meters (about 30 feet) onto a flat, essentially unyielding horizontal surface (see Sidebar 2.2), with the package striking the surface in the position expected to produce maximum damage. A package dropped from this height strikes the ground at a speed of about 13 meters per second (48 kilometers per hour, or about 30 miles per hour).
- A *puncture* test in which the specimen used in the free-drop test is dropped through a distance of 1 meter (about 40 inches) onto the upper end of a 6-inch (15.2-centimeter) diameter solid, vertical, cylindrical mild steel bar mounted on an essentially unyielding horizontal surface. The package is dropped onto the bar in a position that is expected to produce maximum damage.
- A *thermal* test in which the same specimen is fully engulfed in a hydrocarbon-fuel fire with an average flame temperature of at least 800°C (about 1475°F) for a period of 30 minutes.
- An *immersion* test in which a separate, undamaged specimen is subjected to a pressure head equivalent to immersion in 15 meters (about 50 feet) of water. Additionally, 10 CFR 71.61 also specifies that for packages designed for transport of spent fuel with activity exceeding 37 petabecquerels (PBq; 10^6 curies), the undamaged containment system must be able to withstand an external water pressure of 2 megapascals (290 pounds per square inch) for one hour without collapse, buckling, or in-leakage of water. This pressure corresponds to a water depth of about 200 meters (650 feet), typical of maximum water depths on the U.S. continental shelf.

The velocity of package impact in the free-drop test described above is lower than many real world crashes (e.g., the Central Electricity Generating Board's [CEGB's] full-scale crash described in Section 2.3.3). Nevertheless, the impact forces on the package are much higher in the free-drop test because the impact surface is "essentially unyielding."

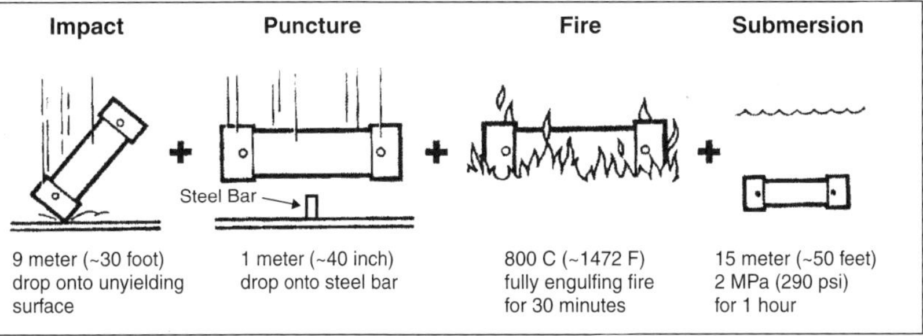

Illustration of the hypothetical accident conditions in 10 CFR Part 71. SOURCE: Modified from a USNRC circular.

SIDEBAR 2.2 What Is an "Essentially Unyielding" Surface?

An *essentially unyielding surface* is used for impact tests on Type B packages under both International Atomic Energy Agency standards (IAEA, 2000) and U.S. regulations (10 CFR Part 71). The standards and regulations require that the impacting surface be essentially unyielding, that is, sufficiently massive and stiff to produce maximum damage to the specimen being tested. Such surfaces are usually constructed of a thick reinforced concrete slab with a steel plate floated onto its surface (i.e., slid onto the concrete surface while still wet). The concrete provides a large reaction mass and the steel plate provides stiffness. For tests of Type B packages, the slab and plate should have a combined mass at least 10 times that of the specimen being tested (IAEA, 2002).

An article such as a spent fuel transportation package that is dropped onto such an unyielding surface will be subjected to higher impact forces and will consequently experience greater deformation than if the same article is dropped onto a surface that is itself deformed by the impact (i.e., a yielding surface). An article dropped onto an essentially unyielding surface in the 9-meter (30-foot) regulatory drop test (Sidebar 2.1), for example, will be traveling at about 13 meters per second (30 miles per hour, or 48 kilometers per hour) when it impacts the surface. The deformation of the article caused by this impact will be essentially identical to the deformation resulting from the head-on collision between two identical copies of that article if each is traveling at 13 meters per second. This is comparable to the difference between a collision of a moving vehicle and a parked car versus a head-on collision between two moving vehicles. The head-on collision between moving vehicles results in much greater damage.

Gonzales et al. (1986) compared the hardness of an essentially unyielding target to several other target types in a series of impact tests. These tests involved impacts of a 2500 kilogram (5500 pound) cylindrical steel test unit resembling a

They also questioned the adequacy of regulations that do not require full-scale testing to demonstrate package performance.

There is, in fact, a good deal of quantitative information available on the performance of transportation packages under extreme loading conditions. This information comes from modeling and full-scale testing studies carried out in the United States and Europe over the past three decades. A summary of selected studies is provided in the following sections.

2.2 PACKAGE PERFORMANCE MODELING STUDIES

Computer simulation models are routinely used to estimate the thermo-mechanical behaviors of truck and rail packages under a range of extreme loading conditions that would not be practicable to obtain from actual spent fuel transportation package onto four targets: desert soil, a concrete runway, a concrete highway, and an unyielding target. The unyielding target was constructed of a 56 metric ton (62 short ton), 3.6-meter (11.8-feet) thick reinforced concrete slab with a steel face plate. The table below shows the results of these tests for an impact speed of 13 meters per second (30 miles per hour), which corresponds to the impact speed for the 9-meter (30-foot) regulatory free-drop test.

In the table, the second column shows how far the test unit penetrated into each of the targets, and the third column shows the maximum strain in the test unit. Penetrations ranged from 48 centimeters (19 inches) for the soil target to 0 centimeters for the unyielding surface. The maximum strains experienced during the impact of the test unit on the unyielding surface were several times greater than for the other target types. The reason for this is simple: The other targets absorbed some of the impact forces, whereas the test unit absorbed essentially all of the impact forces when impacted against an unyielding surface. In fact, in only one case—impact onto the unyielding target—was the strain great enough to produce permanent deformation of the test unit.

Experimental Results Obtained by Gonzalez et al. (1986)

Target	Test Unit Penetration, centimeters (inches)	Maximum Strain, microstrains[a]
Soil	48 (19)	90
Concrete highway	10 (4)	400
Concrete runway	0.6 (0.25)	500
Unyielding target	0	3500

[a]Microstrains are measured in parts per million, for example, centimeters per 10^6 centimeters.

testing of full-scale articles. "Generic" truck and rail packages, which incorporate the salient design features of certified packages, are typically investigated with such models. The loading conditions used in these models are usually derived from historical accident reconstructions, but hypothetical accident scenarios may also be used to investigate conditions that exceed those of any known historical accident. The committee selected a number of modeling studies for discussion in the following subsections. These are summarized in Table 2.1.

2.2.1 Modal Study

The modal study (Fischer et al., 1987) examined the expected responses of spent fuel transportation packages to thermomechanical conditions de-

SIDEBAR 2.3 Scale Modeling

The term *scale modeling* is conventionally taken to mean the testing of either an article that is a miniaturized version of a large structure (see figure), or a large component of a complex structure. The major benefit of scale modeling is that it reduces material, fabrication, and testing costs, thus allowing more variables to be explored, and a multiplicity of data to be obtained, for the same cost as a single test on the full-scale structure. This is of particular value when the structure is very expensive, the article needs to be tested to a point at which it has no further utility, and/or the loads that must be applied exceed the capacities of readily available testing apparatuses. Typically models of at least one-quarter scale are used in engineering tests of spent fuel transport packages, although models as small as one-eighth scale have been used for initial scoping tests.

The failure condition for the full-scale structure generally can be estimated from the results of scale-model testing through the application of well-known scaling laws. However, there are a number of pitfalls in this procedure. The most important is that the weak points in a complex structure are generally associated with joining processes—bolts, welds, and adhesives—that may not scale precisely. For example, in a weldment, the heat-affected zone sizes and mechanical properties, and the residual stresses that are induced, will depend on the number of weld passes that are made and the heat input rate that is used, and these parameters depend on the thicknesses of the materials that are joined. Similarly, the strength of an adhesive joint will depend on the sizes and surface quality of the materials involved, as well as the thickness of the adhesive layer.

Structures that utilize many component materials that are off-the-shelf items (e.g., bolts, seals, gaskets) may be hard to scale because these items may not be available in the exact size or quality that is needed. Also, bolts and other fasteners that may be required for the actual structure are not always available in the same metallurgical form in scaled-down sizes. Additionally, there are scaling issues in monolithic and composite structures. Metallic parts may have different thermomechanical properties because of differences in heat treatments, which are affected by article sizes. Scaling of composite laminates can be done in a practical manner only by reducing the number of plies, essentially making it into a different material.

Finally, cracks and other potential initiators of brittle failure also do not scale because it is their absolute size that is of first-order importance, with their relative sizes generally being of second-order importance. In its simplest form, crack instability and growth is generally governed by a parameter of the form $K = \sigma[\pi c]^{1/2} \beta(c/h)$, where K is the stress intensity factor, c denotes the crack size, σ is the applied stress, and β is a function of the ratio of the crack size to one of the component dimensions (h). For small- to moderate-sized cracks, β is on the order of unity. Although the mathematical relationships for corrosion pits, dents, and gouges are not as well established as in fracture mechanics, their qualitative behavior will be similar to cracks. This can make it very difficult to duplicate the failure mechanism in a small-scale test.

This does not mean that scale modeling is without value. Rather, it means only that direct application of the results of small-scale testing to the full-scale article should be done with great caution. In cases where some features of the test article cannot be modeled accurately at reduced scale (e.g., the valve assemblies used on spent fuel transport packages), it may be possible to combine a simplified reduced-scale model to determine decelerations and then separately test the full-scale component when subjected to the appropriately scaled decelerations. Another approach is to use scale modeling as a test bed for the calibration and validation of a computational analysis simulation of the structure. Then, with the further assurance gained from viable predictions of the results of a small but representative set of independent "proof-of-concept" tests made on a full-scale structure, the computer simulation can be used with confidence, and in a highly cost-effective manner, for further evaluations of the performance of the structure in a broad range of anticipated and accident service conditions.

Scale modeling is routinely used in spent fuel transport package testing and certification. A good technical discussion of scaling laws and properties for materials in pristine condition is provided in Donelan and Dowling (1985). Recent work on scaling laws for materials with flaws is provided by Bazant (2004).

Full-scale (background) and ½-, ¼-, and ⅛-scale models (foreground) of the Magnox flask. SOURCE: Magnox Electric Ltd.

FIGURE 2.2 The German Federal Institute for Materials Research and Testing (BAM) carried out two 9-meter regulatory drop tests in September 2004 at its recently completed testing facility near Berlin. One of the tests was conducted on the180 metric ton (198 short ton) Mitsubishi rail package shown in these photos. The top photo shows the package orientation for this test. The package was hoisted and dropped onto the steel plate embedded in the floor. The bottom photo shows the package after the test. SOURCE: Photo by K.D. Crowley.

SIDEBAR 2.4 Margin of Safety

To reduce the possibility of catastrophic failure, engineering structures in which the weight of the structure is not a critical concern (e.g., buildings, dams, bridges, power plants, storage tanks) are generally designed with large "margins of safety." These safety margins are generally achieved in two ways. First, conservative values of the mechanical properties of the materials used in the structure (i.e., values that underestimate the potential strength of the materials) are selected for use in the design. Second, "worst-case" assumptions are made on the applied loads that the structure must resist, and/or an arbitrary "factor of safety" is introduced to artificially inflate the expected loads.

Factors of safety range from as low as 1.4 for natural gas transmission pipelines located in unpopulated areas to 10 or more in especially sensitive applications, with a value of 3 being a typical choice for noncritical structures. The cumulative effect of conservative design choices and the imposition of a factor of safety on each structural component means that the actual margin of safety, while not specifically known, could be well above the safety margin for each individual component. In a typical design, loads sufficient to cause structural failure could be more than four to five times greater than the load anticipated in actual, normal service.

Large safety margins are necessary because of the multitude of uncertainties that could affect the integrity and durability of a structural design. These uncertainties include the potential for fabrication defects from inadequate workmanship and less than satisfactory material properties and joining techniques; in-service mechanical damage, fatigue and environmental degradation (e.g., corrosion); and unexpectedly severe operating conditions due to accidents or sabotage. Structures that are designed with large safety margins are very likely to be able to resist catastrophic failure over their intended service lifetime, even when the expected operating conditions are substantially exceeded once or many times during service.

In contrast, in engineering structures for which the weight of the structure is a critical concern (e.g., aircraft), the safety margins are generally much lower. Airplanes are typically designed and tested to achieve a margin of safety of only about 50 percent over the maximum anticipated operating loads. In these applications, considerable effort is placed on analytical projections of times to failure based on anticipated defects or other forms of damage, and rigorously scheduled inspections to determine if or when damage has come to exist at critical locations within the structure. This approach has come to be known as damage tolerance methodology.

There are built-in margins of safety for the design of spent fuel transportation packages. USNRC guidance for the design of spent fuel transportation packages (USNRC, 1978) adopts portions of the American Society for Mechanical Engineers (ASME) code for boilers and pressure vessels. The ASME code specifies the use of highly ductile materials that accommodate unusually high stresses through deformation rather than fracture. The code also specifies maximum stress limits for these materials that are well below their yield strengths. Vessels that are properly designed, manufactured, and maintained to these codes should perform as intended even under conditions that exceed design specifications because of this built-in margin of safety.

TABLE 2.1 Summary of Selected Studies on Package Performance Modeling

Test Description	Results in Brief	Reference
Modeling the performance of a generic truck and rail package under mechanical and thermal loading conditions derived from historical accident records	A very small number of rail accident scenarios could result in package releases in excess of regulatory limits	Modal study: Fischer et al. (1987)
Modeling the performance of two generic truck packages and two generic rail packages for impacts at various orientations against unyielding and yielding surfaces at 30, 60, 90, and 120 miles per hour (48, 96, 144, and 192 kilometers per hour)	No package penetration at any orientation or impact speed against yielding or unyielding surfaces; package seals may leak during high-speed impacts against unyielding surfaces in excess of regulatory test limits	Spent fuel shipment risk reexamination study: Sprung et al. (2000)
Comparison of the thermomechanical conditions from severe historical accidents to regulatory test limits and other modeling studies	No accident produces mechanical loads in excess of the 9-meter regulatory free-drop test. Two accidents could have produced thermal loads in excess of the IAEA 30-minute fully engulfing fire test	Fischer et al. (1987); Ammerman et al. (2002, 2003); Ammerman and Ginn (2004)
Modeling of thermal performance of a rail package in the July 2001 Howard Street tunnel fire near Baltimore, Maryland	No package releases would have occurred	USNRC (2003a)
Analysis of the thermal conditions in the December 1984 Summit Tunnel fire near Manchester, England, and the likely effect on a fuel package	Thermal conditions would have exceeded those of the regulatory thermal test such that package seals might have failed should a package have been exposed in this fire	UK Department of Transport (1986)

rived from the historical record of truck and train accidents in the United States. These data were compiled from government and private databases. The data included accident speeds and impact angles, the hardness of impacted objects (i.e., other vehicles, wayside terrain), and the frequency and duration of accident-associated fires. The data were used to develop a suite of historical accident scenarios and associated accident probabilities that

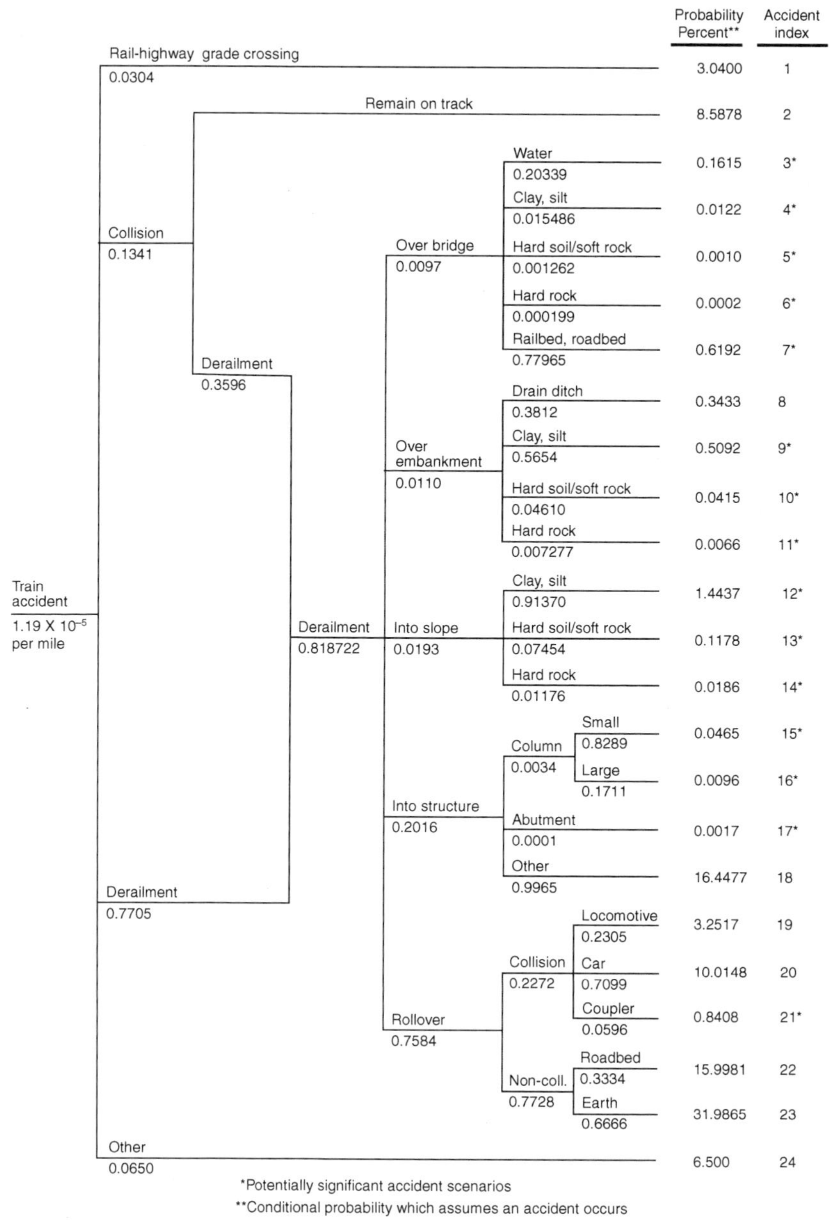

Train accident 1.19 X 10^{-5} per mile

Event	Branch	Derailment	Location	Impact	Sub-branch	Probability Percent**	Accident index
Rail-highway grade crossing 0.0304						3.0400	1
Collision 0.1341	Remain on track					8.5878	2
	Derailment 0.3596		Over bridge 0.0097	Water 0.20339		0.1615	3*
				Clay, silt 0.015486		0.0122	4*
				Hard soil/soft rock 0.001262		0.0010	5*
				Hard rock 0.000199		0.0002	6*
				Railbed, roadbed 0.77965		0.6192	7*
			Over embankment 0.0110	Drain ditch 0.3812		0.3433	8
				Clay, silt 0.5654		0.5092	9*
				Hard soil/soft rock 0.04610		0.0415	10*
				Hard rock 0.007277		0.0066	11*
		Derailment 0.818722	Into slope 0.0193	Clay, silt 0.91370		1.4437	12*
				Hard soil/soft rock 0.07454		0.1178	13*
				Hard rock 0.01176		0.0186	14*
			Into structure 0.2016	Column 0.0034	Small 0.8289	0.0465	15*
					Large 0.1711	0.0096	16*
				Abutment 0.0001		0.0017	17*
				Other 0.9965		16.4477	18
Derailment 0.7705			Rollover 0.7584	Collision 0.2272	Locomotive 0.2305	3.2517	19
					Car 0.7099	10.0148	20
					Coupler 0.0596	0.8408	21*
				Non-coll. 0.7728	Roadbed 0.3334	15.9981	22
					Earth 0.6666	31.9865	23
Other 0.0650						6.500	24

*Potentially significant accident scenarios

**Conditional probability which assumes an accident occurs

FIGURE 2.3 The event tree for train accidents from the 1987 modal study (Fischer et al., 1987). The numbers shown at each branch are probabilities for the accident branch based on an analysis of historical data. The accident scenarios marked with an asterisk were determined to produce consequences that would approach or exceed regulatory limits. SOURCE: Fischer et al. (1987, Figure 2-5).

SIDEBAR 2.5 Radioactive Material Releases in Severe Transportation Accidents

There are two barriers to the release of radioactive materials from spent fuel packages into the environment during a severe accident. The first is the package itself. As described in this chapter, Type B packages used to transport spent fuel are designed to withstand severe accidents without a significant loss of containment or an increase in external radiation to levels that would endanger emergency responders or the general public.

The second barrier is the fuel rod cladding. The cladding for most commercial nuclear fuel is made from a zirconium metal alloy, referred to as *zircaloy*, which is fabricated into long (3.5 to 4.5 meter [11.5 to 14.75 feet]) tubes (Sidebar 1.3). These tubes, which contain the uranium dioxide fuel pellets, are pressurized and sealed to resist collapse or leaks when placed into the high-pressure operating environment of the reactor core. Fuel rods are bundled together into fuel assemblies using metal structural supports. These supports also help to prevent the fuel rods from collapsing and buckling in a severe transportation accident.

The release of significant quantities of radioactive materials from a loaded spent fuel transport package into the environment during a severe accident would occur only if the package *and* one or more fuel rods were breached (small amounts of radioactive contamination from the external surfaces of the fuel rods [crud] could be released from the package if the package seals were compromised, even if the fuel rods maintained their integrity). Fuel rod breaching could potentially occur by two processes: mechanical rupture or thermal creep. The former could occur if impact forces exceed the mechanical strength of the cladding, causing it to buckle. The latter could occur at elevated temperatures due to time-dependent elongation of the cladding along fracture planes. High burn-up fuel rods may be more susceptible to breaching because of cladding embrittlement resulting from their longer residence in the reactor.

If breached, the fuel rods would depressurize, and radioactive material could

could be displayed as "event trees" (Figure 2.3). Quantitative analyses were carried out to assess the effects of the loading conditions represented by these event trees on generic train and truck transportation packages that were similar in design to the packages in service in the mid-1980s. A more complete discussion of this study is provided in Chapter 3.

The analysis involved a two-stage screening process. Phase 1 screening used dynamic linear stress analysis and standard transient heat-transfer models to screen out those accident scenarios in which the thermomechanical conditions did not exceed the regulatory testing requirements in 10 CFR Part 71 (Sidebar 2.1). For these scenarios, any radioactive material releases from the packages were assumed to fall below regulatory limits. Approximately 99.4 percent of all truck accidents and 99.7 percent of the rail accident scenarios analyzed fell into this category (Fischer et al., 1987, be released into the interior of the transportation package by depressurization flow. Two types of radioactive materials could be released: (1) gaseous materials (e.g., radioactive noble gases such as krypton-85 and volatile materials such as cesium-137) produced by fission reactions while the fuel is in the nuclear reactor; and (2) fine particles of the fuel itself, referred to as *fuel fines*, which are created by mechanical fracturing. The particle-size distributions of the fuel fines will depend on the burn-up of the fuel and the magnitude of the mechanical forces on the fuel pellets during the accident.

Only fuel fines smaller than the size of the cladding breach can be released from the fuel rod into the package; larger particles can also clog the cladding breach and reduce the quantity of fine-particle releases into the package. Most of the released fines would be deposited onto the interior surfaces of the package. Some of the remaining airborne fines (which typically comprise only a few percent of the fines released into the package; see Sprung et al., 2000, p. 7-30) and radioactive gases could be released into the environment, but only if *all* of the package barriers (i.e., the package lids and, if present, the inner canister) were breached. Volatile components such as cesium-137, if present, would condense upon cooling. The quantity of materials released from the package would depend on the size of the breaches and the presence of a driving force (e.g., depressurization) to propel material out of the package. Once air pressure between the package interior and outside environment was equalized, further material releases would occur by much slower diffusion processes.

The process for the release of radioactive materials from transportation packages containing high-level waste is similar to that for spent fuel with three notable exceptions. First, high-level waste does not contain fission-produced noble gases; those gases were removed from the waste during processing. Second, high-level waste to be transported in a vitrified (glass) form is contained in stainless steel canisters (see Sidebar 1.3), rather than zircaloy cladding. Third, the canisters are not pressurized, so there are no large depressurization forces to drive radioactive material releases from the canister into the package or the environment.

p. 9-2). Phase 2 screening involved more sophisticated analyses of package responses and radiological releases for the relatively small number of accident scenarios that exceeded the 10 CFR Part 71 testing limits. This screening employed nonlinear dynamic stress analysis models to estimate package deformation and transient thermal models that took into account the phase change accompanying the melting of lead shielding in the package at high temperatures. These analyses assumed that the packages contained five-year-cooled pressurized water reactor fuel having a burn-up of 33,000 megawatt-days per metric ton, which was typical of spent fuel at that time.[16] The analyses also considered breaches of the spent fuel cladding due

[16]Present-day fuel burn-ups are typically between 50,000 and 60,000 megawatt-days per metric ton.

to both impact and thermally induced creep (see Sidebar 2.5). The radiological effects considered included releases of radioactive materials from the package as well as increased radiation doses resulting from damage to the package shielding.

Based on the phase 2 screening, Fischer et al. (1987) concluded that roughly 0.3 to 0.6 percent of extreme accidents would result in radioactive releases that approach or slightly exceed the regulatory limits in 10 CFR Part 71 (Sidebar 2.1), with less than 0.001 percent of the truck and 0.012 percent of the rail accident scenarios actually having releases that would exceed regulatory limits. Because these conditions pushed the capabilities of the computer codes, there was no attempt made to model the releases for the most extreme accidents. Instead, estimates of the releases for these very extreme accidents were extrapolated from the release behavior during less extreme accidents.

2.2.2 Reexamination Study

In a "reexamination study," Sprung et al. (2000) updated the modal study analyses using different package designs and modeling approaches. This study examined the performance of four generic transportation packages: steel-lead and steel-depleted uranium truck packages and steel-lead and monolithic steel train packages. These packages were similar in design to the USNRC-certified packages that were in use in the late 1990s. A detailed description of this study is provided in Chapter 3.

Mechanical performance was estimated using a three-dimensional finite element code (PRONTO 3D[17]), which was developed by Sandia National Laboratories for modeling large deformations in nonlinear mechanical behavior for materials subjected to very high strain rates. The code was used to estimate the mechanical response of the four packages to end, center-of-gravity over corner, and side impacts onto unyielding and yielding surfaces at speeds of 30 (the impact speed for the regulatory free-drop test; see Sidebar 2.1), 60, 90, and 120 miles per hour (about 48, 96, 144, and 192 kilometers per hour). The package impact limiters were assumed to be in place but fully crushed before impact occurred. This is a very conservative assumption; in an actual accident the impact limiters would be expected to absorb most of the impact forces by crushing, as they are designed to do.

Impacts onto unyielding surfaces provide the most rigorous test of package performance. Based on their modeling analysis, Sprung et al. (2000)

---

[17]PRONTO 3D is a Lagrangian finite element code developed by Sandia National Laboratories that is roughly comparable to LS DYNA 3D, a code that is used worldwide for transport package design and verification analyses.

concluded that package impacts onto an unyielding surface would have produced strains lower than those required for package penetration at all of the modeled impact speeds and orientations. The models indicate that the seals on the truck packages maintained their integrity in all but possibly the 120 mile per hour impact; however, for the latter impact, the seal leak areas would have been small. The models also suggest that for rail packages, some seal leakage could occur for some impact orientations at impact speeds onto unyielding surfaces as low as 60 miles per hour and possibly at all orientations at speeds of 120 miles per hour.

A one-dimensional heat transport code was used to estimate the time required to cause the failure of the elastomeric package seals and rupture fuel rods when the package was subjected to a fully engulfing optically dense fire at 800°C (1472°F) (the regulatory thermal test; see Sidebar 2.1) and 1000°C (1832°F). Failure temperatures for elastomeric seals were estimated from data available in the literature. These data suggested that the seals would experience rapid degradation at temperatures exceeding 350°C (662°F). For the 800°C fire (1472°F), it was found that the minimum time to the 350°C (662°F) seal degradation temperature was just over one hour for one of the truck packages, with the maximum time being almost 2.5 hours for a rail package. For a 1000°C (1832°F) fire, the minimum and maximum times to degradation were about 0.6 and 1.4 hours, respectively, for a truck package and a rail package. As noted previously, packages are required by USNRC regulations to withstand a 30-minute, fully engulfing fire (see Sidebar 2.1).

2.2.3 Historical Accident Reconstructions

Additional investigations have been undertaken to reconstruct the thermomechanical conditions from a number of historical accidents that, had they involved spent fuel or high-level waste transportation packages, could have provided a severe test of package performance. It should be emphasized that none of these accidents actually involved shipments of spent fuel or high-level waste. The modal study (Fischer et al., 1987) developed estimates of the thermomechanical conditions for four severe accidents selected from a database of 400 train and truck accidents in the United States that were known to have produced extreme loading conditions. These are summarized in Table 2.2. The authors concluded that only one of the four accidents—the September 1982 Livingstone, Louisiana, train derailment and fire—could have resulted in any releases of radioactive materials. Whether releases would have occurred depended on where the package was placed on the train relative to the location of the fire, which was allowed to burn for several days.

Ammerman and colleagues (Ammerman et al., 2002, 2003; Ammerman

TABLE 2.2 Severe Accident Scenarios Examined in the 1987 Modal Study

Date	Location	Description	Conclusions About Accident Severity by Fischer et al. (1987)
January 1979	Hunter, Alabama	Five railcars plunged off a rail bridge into the muddy bottom of a river about 23 meters (75 feet) below	No radioactive material releases or increases in external radiation expected
March 1981	San Francisco, California	A tractor trailer traveled through a bridge railing and fell onto a soil surface about 19.5 meters (64 feet) below the bridge	No radioactive material releases or increases in external radiation expected
April 1982	Oakland, California	A truck fire in a highway tunnel involving about 33,300 liters (8800 gallons) of gasoline	No radioactive material releases or increases in external radiation expected
September 1982	Livingston, Louisiana	A train derailment and fire fed by plastics and petroleum products; fires burned for several days	Package releases could have exceeded regulatory limits depending on where the package was located in the fire

SOURCE: Fischer et al. (1987).

and Ginn, 2004) from the Sandia National Laboratories examined the thermomechanical conditions for 12 historical accidents, some of which had been identified by the State of Nevada as potentially being severe enough to compromise the containment effectiveness of spent fuel transportation packages (Table 2.3). Accident loading conditions were reconstructed from National Transportation Safety Board reports and newspaper accounts. These conditions were compared to the loads experienced by transportation packages during regulatory testing (e.g., the 9-meter drop test; 30-minute thermal test); to the accident scenarios estimated in the modal study (Fischer et al., 1987); and to the 2000 reexamination study estimates (Sprung et al., 2000).

The authors concluded that the thermomechanical conditions described in Table 2.3 were encompassed by the event trees used in the Sprung et al. (2000) study (see Chapter 3). They also concluded that none of these acci-

TABLE 2.3 Severe Accident Scenarios Examined by Sandia National Laboratories

Date	Location	Description	Conclusions About Accident Severity by Ammerman and Colleagues
June 1983	Greenwich, Connecticut	Two trucks and two cars plunged off an interstate highway bridge into a river about 21 meters (70 feet) below	Impact would have been less severe than the 9-meter drop test onto an unyielding surface
August 1985	Checotah, Oklahoma	Transported military ordnance (2000-pound Mk-84 bombs) exploded after a truck accident	Detonation of military ordnance would not have caused package failure
July 1986	Miamisburg, Ohio	A train carrying yellow phosphorus and molten sulfur derailed and caught fire	Fire would not have exceeded the 30-minute regulatory thermal test
April 1987	Amsterdam, New York	Several cars and a truck plunged off an interstate highway bridge, falling about 24 meters (80 feet) into a rain-swollen creek	Impact would have been less severe than the 9-meter drop test onto an unyielding surface
December 1988	Memphis, Tennessee	A tanker truck carrying about 9500 gallons of propane caught fire and exploded	Package would have experienced only superficial damage
February 1989	Helena, Montana	A runaway train collided with a locomotive at 15 to 25 miles per hour, causing two large explosions from hazardous cargo	Fire would not have exceeded the 30-minute regulatory thermal test
February 1989	Akron, Ohio	One railcar carrying butane ruptured, releasing its contents in the form of a fireball	Fire would not have exceeded the 30-minute regulatory thermal test
May 1989	San Bernardino, California	A train derailed at high speed (100 miles per hour); a gas pipeline failed catastrophically following cleanup of the derailment	Impact would not have caused package breach or release of contents

continues

TABLE 2.3 Continued

Date	Location	Description	Conclusions About Accident Severity by Ammerman and Colleagues
July 1989	Freeland, Michigan	A derailed freight train carrying flammable materials burned for several days	Fire could have exceeded the 30-minute regulatory thermal test, but conditions would not have exceeded those shown by Sprung et al. (2000) to be necessary to cause package seal failure
October 1989	Oakland, California	The upper level of a viaduct collapsed onto the lower deck	Collapse of viaduct onto a truck package would not have been severe enough to cause seal failure, but package shield could be somewhat compromised
December 1994	Cajon, California	A runaway freight train struck the rear of another train at a speed of about 45 miles per hour	No significant damage to package would have occurred
February 1996	Cajon Junction, California	The derailment of a freight train caused a fire that burned for several days	Fire conditions would not have exceeded those shown by Sprung et al. (2000) to be necessary to cause package seal failure

SOURCE: Ammerman et al. (2002, 2003); Ammerman and Ginn (2004).

dents would have produced thermomechanical conditions that exceeded the regulatory test conditions in 10 CFR Part 71.

The USNRC undertook a detailed thermal analysis of the July 2001 fire in the Howard Street tunnel in Baltimore, Maryland, that resulted from the derailment of a train carrying hazardous materials. The fire was fed by a tanker railcar carrying about 28,600 gallons (106,300 liters) of liquid tripropylene. A National Institute for Standards and Technology (NIST) study of the fire (McGrattan and Hamins, 2003) used detailed numerical simulations to develop estimates of temperatures in the tunnel for the most severe portion of the tripropylene fire, which occurred between the time of its ignition and the rupture of a water main within the tunnel about three hours later.[18] The study estimated that peak temperatures in the narrow flaming region in the tunnel reached about 1000°C (1832°F) and that the tunnel walls reached peak temperatures of about 800°C (1472°F). The hot gas layer near the fire had average temperatures of about 500°C (932°F).

Tunnel temperatures were also estimated by analyzing oxide layer thickness, metal loss, and metal melting on railcar components recovered after the fire (Garabedian et al., 2002). This analysis suggested that gas temperatures in excess of 800°C (1472°F) existed for more than 30 minutes near the fire source. At 20 meters (66 feet) from the fire source, the analysis suggested that maximum surface temperatures of 600°C (1112°F) could have been reached for much less than 30 minutes.

USNRC staff modeled the thermal behavior of a specific USNRC-approved spent fuel package[19] subjected to these estimated "extraregulatory" thermal conditions. The package and its cradle were modeled using a two-dimensional finite element code for two scenarios: first assuming a one-railcar (20-meter, or 66-foot) separation between the package and fire source, as would be required by DOT regulations had spent fuel and hazardous materials been transported together; and second assuming a 5-meter (16.4-foot) separation. Both scenarios were analyzed for 150 hours of fire exposure at the maximum temperature conditions estimated by the NIST model.

The committee received a briefing from USNRC staff on the results of this analysis, which can be summarized as follows (see also Bajwa, 2002; USNRC, 2003a): For the first scenario, the temperature of the fuel element cladding exceeded regulatory limits of 570°C (1058°F)[20] after about 166 hours of fire exposure. For the second scenario, the fuel cladding would have reached 570°C (1058°F) after 37 hours of exposure. Calculations were also carried out to estimate the stresses on the welded canister resulting from fire exposure. Those calculations indicated that the welded canis-

---

[18]The NIST study noted that the distribution of tripropylene fuel within the tunnel, and thus the duration of the tripropylene fire, are difficult to estimate. The study suggests that the tripropylene fire was extinguished sometime between 3 and 12 hours after ignition either from a lack of fuel or from water suppression. Smoldering of combustible materials contained in closed boxcars on the train continued for several days after the tripropylene fire was extinguished.

[19]The Holtec Hi-Star MPC package was modeled. This rail package is designed to hold five-year-old pressurized water reactor spent fuel assemblies with maximum burn-ups of 45,000 megawatt-days per metric ton. It has a bolted external closure and an internal welded canister. For the purposes of the USNRC analysis, the spent fuel assemblies were assumed to generate the maximum internal heat (20 kilowatts) allowed by the package design.

[20]The 570°C (1058°F) regulatory limit was established to prevent fuel cladding failure from thermal creep during storage. The actual burst temperature for zircaloy fuel rods is about is about 750°C (1382°F) (see USNRC, 2003b).

ter would have maintained its integrity. Consequently, USNRC staff concluded that no radioactive material would have been released from this package in this fire.

It is noteworthy that this analysis contains several significant "conservatisms" (i.e., assumptions that resulted in more dire predictions of package performance than might have occurred in an actual fire): The maximum fire temperatures in the tunnel were assumed to have been maintained for 150 hours—more than 6 days. The actual duration of the tripropylene fire in the Howard Street tunnel was estimated by NIST (McGrattan and Hammins, 2003) to last from 3 to 12 hours. A two-dimensional thermal model was used in the analysis. This model ignored axial direction heat transfer, which could have reduced the peak temperatures.[21] Also, the package was assumed to have the maximum allowed internal heat load from spent fuel decay heat.

The USNRC is extending its Howard Street tunnel fire analyses to examine the performance of two additional spent fuel packages: a TN68 rail package mounted on a railcar and NAC-LWT truck package in an ISO (International Organization for Standardization) container mounted on a railcar. Both are bare-fuel packages (see Section 2.1). These packages are currently certified by the USNRC for use in the United States, and the transport of truck packages by rail, which is one of the scenarios being examined, is allowed under current regulations.

A draft report containing the Howard Street tunnel fire analyses (Adkins et al., 2005) was made available to the committee in early September 2005, after the committee held its last meeting for this study.[22] Just prior to the committee's final meeting in July 2005, the State of Nevada also provided a preprint of a paper describing a thermal analysis of a generic steel-lead-steel

[21]Marvin Resnikoff of Waste Management Associates criticized the USNRC's analysis on the basis that it did not use a three-dimensional thermal model and did not explicitly model the bolts and seals on the external closure. While the committee agrees that additional details in the models would have been informative, it also judges them unlikely to have changed the results, given that the modeling predicted that there were no failures of the internal welded canister of the package.

[22]This analysis assumed that the packages were located 20 meters (66 feet) from the fire source and that the fire burned for seven hours, a shorter time than the original analysis. According to the draft paper, the analysis shows that the maximum temperatures on the seals of the TN68 and NAC-LWT packages would have exceeded their rated service temperatures, making it possible for the release of radioactive materials to occur. An analysis was also carried out to estimate the radioactive releases from these packages. They were characterized in the paper as "very small—less than an A_2 quantity" (see Sidebar 2.1) and consisting of non-fixed radioactive material (crud) from the external surfaces of the fuel rods. The draft paper indicates that the fuel cladding would have maintained its integrity.

truck package exposed to a fully engulfing hydrocarbon fire (Greiner et al., 2005).[23] Because these papers were provided so late in the study, the committee was unable to analyze, discuss, and integrate them into this report.

The United Kingdom Department for Transport (1996) analyzed the thermal loading conditions in the Summit rail tunnel fire near Manchester, England, on December 20, 1984. The fire resulted from the derailment of 10 tank cars carrying gasoline. The fire burned for about four days (Figure 2.4) and completely destroyed several tanker cars. The analysis showed that fire conditions in the tunnel exceeded those required in the regulatory thermal test, which suggested that there could have been releases of radioactive materials had a spent fuel transportation package been involved in the derailment and fire. As a result of this analysis, an operational rule was established that prohibited English trains carrying spent fuel packages and trains hauling flammable materials from crossing in rail tunnels.

2.3 FULL-SCALE PACKAGE TESTING UNDER EXTREME CONDITIONS

Full-scale testing on transportation packages under severe extraregulatory conditions has been carried out in both the United States and the United Kingdom. In the United States, these tests have been carried out under the sponsorship of the Atomic Energy Commission and its successor agencies, the Energy Research and Development Administration (ERDA) and the USNRC. In the United Kingdom, one test has been carried out by the British Central Electricity Generating Board (CEGB). In addition, the USNRC plans to carry out an additional test on a rail package when funds are made available by Congress. These studies are described in the following sections, and the results are summarized in Table 2.4.

2.3.1 Sandia National Laboratories Air-drop Tests

Two air-drop tests were conducted by Sandia National Laboratories in 1975 to provide a demonstration of the ruggedness and survivability of shielded containers in a manner that was thought to be better appreciated by the general public than a regulatory test (Waddoups, 1975). The test

[23]According to this analysis, the elastomeric seal for this generic package would reach its melting temperature (referred to as the "temperature of concern" in the paper) in about two hours if the impact limiter is attached to the lid end of the package and about 0.7 hour without the impact limiter. The paper did not provide an analysis of the consequences of exceeding the seal melting temperature in terms of possible releases of radioactive materials from the package.

FIGURE 2.4 Photos from the December 1984 Summit Tunnel fire near Manchester, England. The top photo shows two fire plumes emerging from tunnel ventilation shafts. The bottom photo is an interior view of the tunnel. Part of the tunnel ceiling has collapsed onto one of the tank cars. SOURCE: Photos taken by a member of the West Yorkshire Fire Brigade or Manchester Fire Brigade (used with permission of *www.todchat.com*).

TABLE 2.4 Full-scale Package Tests Described in this Chapter

Test Description	Results in Brief	Reference
600-meter (2000-foot) air drop of two small packages onto hard soil	Less severe damage observed than for a 9-meter free-drop test onto an unyielding surface	Waddoups (1975)
Crash of a truck carrying a 20 metric ton (22 short ton) package mounted on a trailer into a massive reinforced concrete barrier at 98 kilometers per hour (61 miles per hour) and 135 kilometers per hour (84 miles per hour)	Superficial package damage for 98 kilometers per hour (61 miles per hour) test; deformation of package with small amount of water leakage observed for 135 kilometers per hour (84 miles per hour) test	Huerta (1977)
Crash of locomotive into 25 metric ton (28 short ton) package mounted on a trailer at 130 kilometers per hour (81 miles per hour)	Package was deformed, and a small leak was detected when the package was pressurized	Huerta and Yoshimura (1983)
Crash of a 68 metric ton (75 short ton) package mounted on a railcar into a massive reinforced concrete barrier at 131 kilometers per hour (82 miles per hour)	Superficial package damage	Huerta (1981)
9 meter free-drop tests of a package onto its side and corner	Water spray from lid-body joint at impact releasing up to a few liters of water	IME (1985)
Crash of a locomotive into a package mounted on a railcar at 160 kilometers per hour (100 miles per hour)	Superficial damage with an internal pressure drop corresponding to the loss of about 0.5 liter (0.1 gallon) of water through the package seal	IME (1985)
Full-scale testing of a rail package mounted on a rail carrier car placed at 90 degrees to a simulated rail crossing, subjected to a collision with a locomotive and several freight cars traveling at 60 miles per hour, followed by a fully engulfing, optically dense, hydrocarbon fire for a duration of one-half hour post-collision	Test has not yet been carried out	USNRC (2003c, 2004b,c,d, 2005a,b)

involved 600-meter (almost 2000-foot) air drops of two "obsolete"[24] packages: a Pratt and Whitney 1 package[25] and an OD-1 Oak Ridge Research Reactor Spent Fuel Carrier.[26] These drops were made onto a hard prairie (a hard, dry, sandy silt soil) at the Sandia Edgewood Test Range in New Mexico.

The packages impacted the prairie at speeds exceeding 100 meters (350 feet) per second and created deep impact craters (in one case exceeding 2 meters [7 feet] in depth). One of the packages experienced superficial damage, while the other experienced some bulging and shifting of its internal lead shielding. Waddoups (1975) noted that the impact velocities for these drop tests reached speeds of about 230 and 246 miles per hour (103 and 110 meters per second), many times greater than the 30 mile per hour (about 13 meter per second) speeds in the 9-meter regulatory drop test. However, the hard prairie surface at the test site was not "essentially unyielding," as evidenced by the deep craters created by the impacts. Based on a comparison of damage to one of the packages, Waddoups (1975, p. 15) concluded that "the 30-foot drop test onto an unyielding surface is a more severe environment than the 2000-foot drop onto hard soil."

2.3.2 Sandia National Laboratories Crash Testing

ERDA (predecessor agency to the Department of Energy) sponsored a full-scale testing program at Sandia National Laboratories to obtain a better understanding of the behavior of transport packages in severe accident environments (Jefferson and Yoshimura, 1977). This program had two primary objectives: (1) assess and demonstrate the validity of analytical modeling and scale modeling for predicting the damage to transport packages in accidents; and (2) develop quantitative information on the conditions in extreme accident environments.

This full-scale test program was carried out in three separate phases: (1) use of computational methods to predict the conditions in accident environments and the potential damage to shipping containers in such

---

[24]Both packages were considered obsolete because they were not designed to meet fire standards with an acceptable loss of shielding. They also were not as rugged as then-licensed packages.

[25]This package had a 0.622-meter (25-inch) outside diameter, 0.9065-meter (36-inch) outside height, and a weight of 3054 kilograms (3.4 short tons). The package is smaller than many of the packages currently in use to transport commercial spent fuel.

[26]This package had a 0.8-meter (32-inch) outside diameter, 1.2-meter (48-inch) outside height, and a weight of 7410 kilograms (8.2 short tons). The package also is smaller than many of the packages currently in use to transport commercial spent fuel.

environments; (2) determination of physical damage mechanisms through scale-model testing; and (3) full-scale testing of representative hardware to validate the computational analysis methodology.

Several criteria were considered in selecting the test scenarios. These included the desire to expose transport packages to realistic and severe accident environments, tractability of the scenarios to mathematical analysis and scale-model testing, cost-effectiveness, and the likelihood of successful execution. The last criterion eliminated scenarios that were difficult to replicate such as skids into barriers. The cost criterion prompted the use of out-of-service transport packages, used tractors, and a military surplus locomotive in the tests. Three full-scale test scenarios were eventually selected:

1. Impacts of tractor-trailer rigs carrying spent fuel transport packages into a concrete barrier at nominal speeds[27] of 100 kilometers (62 miles) per hour and 130 kilometers (about 80 miles) per hour.
2. Impact of a locomotive into a spent fuel transport package mounted on a truck trailer at a simulated grade crossing at a nominal speed of 130 kilometers (about 80 miles) per hour.
3. Impact of a spent fuel transport package mounted on a railcar into a concrete barrier at a nominal speed of 130 kilometers (about 80 miles) per hour, followed by exposure to a fire.

Prior to performing these tests, Sandia carried out both analytical and one-eighth scale-model tests to predict the response of the vehicles and transport packages under each of these impact conditions. Analyses were conducted using "lumped parameter models" of the transport systems in which the vehicle system and package are represented as a series of loads and couplings. A limited amount of finite element modeling was also carried out to elucidate the details of package deformation.

Scale-model testing was carried out in two phases. First, scale-model packages were impacted directly against rigid barriers to identify and quantify potential damage mechanisms. Then, scale models of the entire transport system were tested to understand total system response. The latter tests helped researchers determine the appropriate vehicle-package configurations for the full-scale testing described in the following sections.[28]

---

[27]Actual test speeds varied slightly from these nominal speeds in some cases.

[28]Videos of these tests are available on the Department of Energy's web site at *http://www.ocrwm.doe.gov/newsroom/videos.shtml.*

Tractor-trailer Impact Tests

The tractor-trailer impact tests (Huerta, 1977) were carried out using an obsolete spent fuel transport package weighing 20,500 kilograms (45,000 pounds) that was mounted on a trailer in a head-on position. The trailer was attached to a standard tandem-axle tractor. The package contained an unirradiated fuel assembly and was filled with water. Conventional balsa-wood impact limiters were mounted on each end of the package. The tractor-trailer was crashed into a massive (626 metric tons [690 short tons]) reinforced concrete barrier backed by more than 1500 metric tons (1650 short tons) of soil. The target was described by Jefferson and Yoshimura (1977, p. 13) as "essentially unyielding" and of a weight greatly exceeding what would be encountered along normal truck routes.

The tests were carried out at Sandia's sled test-track facility on January 18 and March 16, 1977 (Figure 2.5). For each of the two tests, a tractor-

FIGURE 2.5 High-speed (135 kilometers per hour [84 miles per hour]) crash of a spent fuel package mounted on a truck trailer into a massive barrier carried out at Sandia National Laboratories in 1977. The truck cab was destroyed in the crash, but the package remained attached to the trailer. SOURCE: Sandia National Laboratories.

trailer was accelerated to the target by a rocket sled mounted on guide rails behind the trailer. The sled was disengaged from the trailer prior to impact to allow the vehicle to coast into the barrier at the predetermined speed. The same transport package was used in both full-scale tests. It was instrumented with accelerometers, triaxial strain gauges near the front (impact) end, and passive water pressure sensors inside the package to measure peak pressures. The tests were recorded by high-speed photography.

On the first test, the trailer impacted the concrete barrier at a speed of 98 kilometers per hour (61 miles per hour). The tractor and the front end of the trailer were completely destroyed. The package remained attached to the trailer throughout the test, although the front package tie-down failed and the rear tie-down was damaged. The package suffered only superficial damage. There was no water leakage from the package, and the fuel assembly was intact and undamaged. The package experienced a peak deceleration of about 18 times the acceleration of gravity (i.e., 18 *g*'s) based on a velocity-time analysis of the crash photos (see Sidebar 2.6).

On the second test, the vehicle hit the concrete barrier at 135 kilometers per hour (84 miles per hour). At this speed the vehicle had approximately double the kinetic energy[29] of the first test. Nonetheless, the response of the tractor-trailer was similar to the first test, and the package remained attached to the trailer. The front impact limiter was partially crushed and displaced, allowing the package to impact the rigid barrier. The front end of the package was slightly deformed and the package length was reduced by about 6 centimeters (2.4 inches). The impact created a 0.95-centimeter (0.4-inch) gap between the lead shielding and the outer shell at the back of the package. A small amount of water seepage (two drops per minute) was observed at the package head (Jefferson and Yoshimura, 1977, p. 29). Mechanical means had to be employed to remove the package head and a large force applied to remove the fuel assembly because the package had deformed. Some of the fuel rods were buckled by the impact.

Rail Grade-crossing Impact Test

The grade-crossing test (Huerta and Yoshimura, 1983) involved a crash of a locomotive traveling at 130 kilometers per hour (81 miles per hour) into a tractor-trailer holding a spent fuel transport package at a simulated grade crossing (Figure 2.6). The test was carried out on April 24, 1977,

---

[29]The kinetic energy of a body in motion is equal to $\frac{1}{2} mv^2$, where m is the body mass and v is the body velocity. Because the package masses are the same in both tests, the ratio of the kinetic energies is equal to the ratios of the test velocities squared: that is, $(135/98)^2 \approx 1.90$.

SIDEBAR 2.6 Impact Severity

The *accelerations* and *strains* that are routinely measured during package impact tests provide quantitative information for the verification of computation simulation analyses. Analyses of this kind are particularly important for comparing the relative severity of full-scale crash tests and regulatory free-drop tests.

The tests described in this chapter were designed to generate large forces on transport packages by accelerating them to a known speed and then impacting them against rigid barriers. The forces generated by the impact can be calculated using Newton's equation:

$$\text{Force } (F) = \text{mass } (m) \times \text{acceleration } (A).$$

During an impact, the package undergoes a rapid and negative change in *A* (i.e., it decelerates) as its velocity goes to zero. It is possible to determine a nominal (average) value for *A* directly by measuring the change in position of the package as a function of time. This measurement is typically made using photographs taken during the crash by high-speed cameras, which record at up to 3000 frames per second. Instruments, called accelerometers, also can be mounted on the package to provide this information. Unfortunately, the highest decelerations and forces occur locally at the point of impact where measurements are very difficult.

Estimates of *A* as a function of time are usually given in the form of a smooth curve (see figure), where acceleration is expressed relative to the acceleration imparted by Earth's gravity field. The units of measurement are expressed in *g*'s, (1 g = 9.8 meters per second squared, or 32 feet per second squared). Because the mass of the package is known and does not change during the test, the average forces (*F*) acting on the package can be determined directly using the above equation. The peak of the curve provides a good estimate of the peak force of the test.

The forces imparted during the test will cause both the package and the barrier to deform. If the deformations are below the *elastic limit* they will be temporary, and the deformed materials will return to their original shapes after the forces are removed. If the deformation is above that limit, the deformations will be permanent. This *elastic limit* is material specific but tends to be less than 1 percent strain for steel objects.

using a 2545-kilogram (56,000-pound) stainless steel and lead package[30] containing an unirradiated fuel assembly. The package was mounted to the trailer with heavy steel bands. The trailer in turn was attached to a used gasoline tractor. A military surplus locomotive weighing 109,000 kilo-

[30]This package was constructed of a 2.54-centimeter (1-inch) thick outer stainless steel shell and a 1.9-centimeter (0.75-inch) thick inner stainless steel shell with a 21.3-centimeter (8.37-inch) thick sandwich of lead shielding. The package head was attached by eight 2.54-centimeter (1-inch) diameter stainless steel bolts.

The amount of deformation can be measured directly by installing strain gauges on the package. These devices provide an estimate of the maximum local (i.e., where the strain meter is installed) changes in dimension (strain) of the object. Strain is usually expressed in units such as microstrains (see table in Sidebar 2.2).

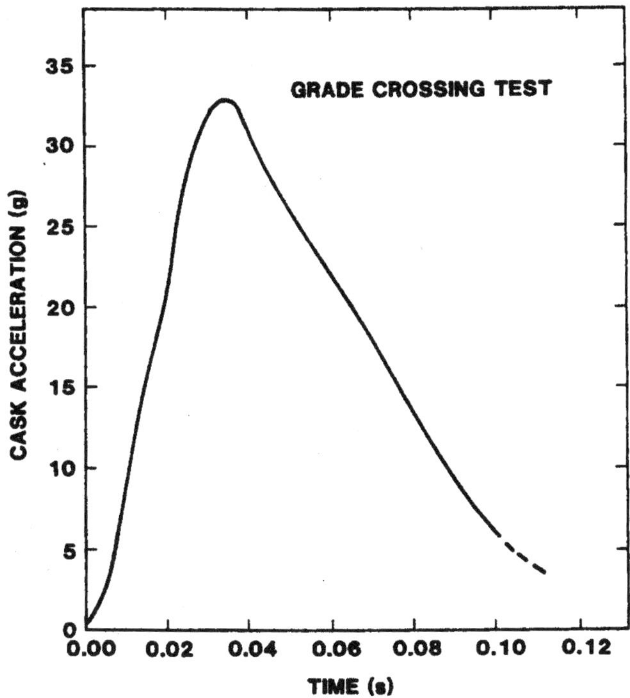

Curve showing the variation in package acceleration as a function of time after impact in the Sandia National Laboratories grade-crossing test. SOURCE: Huerta and Yoshimura (1983, Figure 24, p. 25).

grams (240,000 pounds) was used in the tests. The package was instrumented with strain gauges and accelerometers, and the crash was recorded by high-speed photography.

The test geometry was such that the heavy locomotive frame, which was constructed of I-beams and welded steel plates, impacted the package below its centerline. Upon impact, the package plowed through about 3 meters (10 feet) of the lighter locomotive superstructure above the heavy frame and then became detached from the trailer, which became wrapped around the front of the locomotive. The impact forces launched the pack-

FIGURE 2.6 High-speed (130 kilometers per hour [81 miles per hour]) crash of a locomotive into a package mounted on a truck trailer carried out at Sandia National Laboratories in 1977. The top photo shows the impact of the locomotive and the trailer. The end of the package can be seen near the center of the photo. The bottom photo shows the package after the test. SOURCE: Sandia National Laboratories.

age into the air. It hit the ground about 46 meters (150 feet) from the point of impact and tumbled for another 15 meters (49 feet). The package attained a maximum horizontal velocity in flight of about 60 meters per second (about 50 miles per hour).

The impact produced two indentations in the package where it was struck by the frame of the locomotive. The cooling fins on the package were crushed and the outer package shell was bulged inward by about 2.5 centimeters (1 inch). A small leak in the package head was detected when the package was pressurized following the test (Jefferson and Yoshimura, 1977, p. 37). The inside cavity of the package was undeformed, however, and although some of the fuel rods had bowed slightly, the assembly was otherwise undamaged. The maximum deceleration on the package was about 33 *g*'s based on a velocity-time analysis of the high-speed photos. Two accelerometers mounted on the package gave peak readings of about 90 *g*'s and 200 *g*'s, but these readings may have been affected by package rotation, which reached a peak of about 1500 rotations per minute. The peak strain readings were below the yield strain for the package material.

Railcar Impact Tests

The third full-scale Sandia test involved the high-speed crash of a railcar-mounted spent nuclear fuel package into the same concrete barrier used in the truck crash tests (Huerta, 1981). The railcar system used in this test was constructed around 1960 but was no longer in use at the time of the tests. The package weighed about 68,000 kilograms (150,000 pounds). It was mounted in a steel-frame railcar with a package encasement system of about the same weight.

The transport package was larger than that used in the rail-crossing test, but it had a similar construction.[31] It was designed to carry 10 spent fuel assemblies in water. For the purposes of this test, the package was loaded with nine mock assemblies and one unirradiated assembly, and it was filled with water. The package was mounted in the railcar with its closure end facing forward (i.e., toward the impact end of the railcar). The railcar and package were extensively instrumented with strain gauges and accelerometers. The crash was recorded by high-speed photography using both stationary and railcar-mounted cameras.

The test was conducted on September 27, 1977 (Figure 2.7). The railcar impacted the barrier at a speed of 131 kilometers per hour (82 miles per

[31]It had two stainless steel shells, 3.5 centimeters (1.375 inch) thick on the outside and 0.95 centimeter (0.375 inch) thick on the inside. The lead shielding was sandwiched between these shells. The package head was attached to the body with 24 high-strength bolts.

FIGURE 2.7 High-speed (131 kilometers per hour [82 miles per hour]) crash of a spent fuel package mounted in a railcar into a massive barrier carried out at Sandia National Laboratories in 1977. The railcar was destroyed in the crash, but the package sustained only superficial damage. SOURCE: Sandia National Laboratories.

hour). The impact crushed both the front end of the railcar and a spacer device that was placed at the forward end of the package. The package suffered some external damage to its cooling fins, but was otherwise undeformed. The package remained leaktight after the test. The fuel rods themselves were undamaged, although one of the support brackets was slightly distorted.

High-speed photography was used to estimate the deceleration-time curves for the package. Maximum decelerations were calculated to be 32 *g*'s. The strain gauge data from the package and fuel showed that maximum strains were below the elastic limits for these materials, which is consistent with the observation of no permanent deformations of those objects.

2.3.3 British Central Electricity Generating Board Tests

In the early 1980s, the British CEGB undertook a testing program aimed at improving understanding of the performance of spent fuel transport packages. The program had three objectives:

1. Understand how to assess package impact resistance and demonstrate compliance with regulatory requirements.
2. Estimate the probabilities of transport accident scenarios.
3. Demonstrate to the public that the Generating Board's packages that meet regulatory requirements will withstand severe accident conditions.

The impact performance of transportation packages was investigated over a period of four years through a carefully planned progression of analytical studies, scale-model testing, drop testing, and a full-scale crash test. A discussion of these tests is provided in Blythe et al. (1984) and an Institution of Mechanical Engineers (IME, 1985) report.

CEGB selected the Magnox package for this testing program. This package has been used since the 1950s for transporting Magnox fuel[32] to the Sellafield site for reprocessing. By the early 1980s this package was being used for most of CEGB's fuel movements to Sellafield as well as shipments to Sellafield by the Scotland Electricity Board and British Nuclear Fuels Limited (BNFL). The package has undergone several design improvements since being introduced into service. It has a monolithic cuboid body with a bolted lid with welded steel fins for cooling. A photograph of this package is shown in Sidebar 2.3. The package can hold up to 400 Magnox fuel elements and 1 metric ton (1.1 short tons) of water for heat transfer and radiation shielding. The loaded weight of the package is about 48 metric tons (53 short tons). The packages are transported mostly by train on specially designed railcars called flatrols.

Analytical studies carried out under the CEGB project indicated that because of their massive construction, the package body and package lid would be unlikely to sustain major damage in an accident. Any damage would likely be minor (e.g., bent cooling fins). These studies also suggested that any package releases would most likely be caused through impacts that result in bolt extension and decompression of the elastomer seals at the lid closure (Dallard, 1985, p. 49).

Analytical studies and scale-model testing undertaken by CEGB had shown that the maximum deformation would be sustained in a drop test in

---

[32]Magnox fuel contains uranium metal encased in a magnesium alloy "can." The United Kingdom now operates 6 Magnox reactors, down from a peak of 26. The first reactor (Calder Hall) began operating in 1956 and is now shut down.

which there is no package rotation at impact. To achieve this condition, the drop tests were designed so that the center of gravity of the package was located directly above the impact point. The drop tests were designed to produce maximum deformation of the lid-body joint, because this would have a direct effect on containment integrity. Tests were conducted using one-quarter- and one-half-scale models and a full-scale package (shown in Sidebar 2.3). The packages were instrumented by transducers to measure forces on the bodies, lid displacements, and internal water pressures. Strain gauges were fitted to the closure bolts to measure elongations. High-speed photography was used to capture a detailed visual record of each test.

Two full-scale tests were carried out: In the first, the package was dropped onto the corner of the lid, and the second public demonstration test, the package was dropped onto the lid edge (Figure 2.8). The same package was used for both tests. The "public demonstration" drop test was conducted in March 1984. The package was filled with steel bars to simulate the Magnox fuel, and it also contained the other internal components and water in a manner that would be typical in an actual package shipment. It was pressurized to 100 pounds per square inch (6.9 bars) for leak testing.

A detailed discussion of the results of these tests is given in IME (1985). There was generally good agreement between the measured accelerations and displacements in the scale-model and in the full-scale drop tests. It was concluded that package behavior can be characterized accurately in scale-model tests "provided that all the important features are accurately represented in the models" (Barnfield and Donelan, 1985, p. 82; see Sidebar 2.3).

A small decrease in internal package pressures (ranging from about 1.5 to 6 pounds per square inch [0.1 to 0.4 bar]) was measured in both the scale-model and the full-scale drop tests. High-speed photography of the full-scale and half-scale tests showed a water spray from the lid-body joint lasting about 20 milliseconds. The water loss from one of the full-scale tests was estimated to be on the order of a few liters. Calculations based on this observation suggested that the associated radiological releases would have been less than about 5 percent of the amount permissible under IAEA regulations for accident conditions.[33]

The next phase of the CEGB study involved the identification of package transport impact hazards along transport routes. The objective of this hazard analysis was to estimate the probability of occurrence of various accident scenarios for use in designing the full-scale impact tests. Most spent fuel transport in the United Kingdom is carried out using rail, so the analysis focused on the identification of potential hazards along current and future rail routes used to transport packages to Sellafield. CEGB recog-

---

[33]These are given by the A_2 values described in Sidebar 2.1.

FIGURE 2.8 Preparation of a Magnox spent fuel package (behind the workers on the scaffold) for the March 1984 CEGB free-drop test. The package was subjected to a 9-meter (30-foot) drop onto an unyielding surface. SOURCE: Magnox Electric Ltd.

nized that most people would have difficulty comparing the severity of a 9-meter drop test required by the regulations to a rail accident that might occur when a package was being transported at high speeds on the British rail system. This test was designed to provide a graphic demonstration of such a rail accident.

Information was collected on topography, geology, tunnel and bridge abutments, mobile hazards such as road vehicles and aircraft, and previous railway collisions. This information was used to construct event trees (e.g., see Figure 2.3) that could be used to assess the probability of future accident occurrences. The hazard analysis identified a range of potential events and estimated their probabilities of occurrence. The most likely accident scenarios range from derailments (5 chances in a hundred per year [5×10^{-2} per year]) to the impact of a package onto rock from a height exceeding 20 meters (70 feet) (1 chance in 100 million per year [1×10^{-8} per year]).

Two scenarios were considered initially for the full-scale crash test (Hart et al., 1985a, p. 116) based on the hazard analysis: (1) package-flatrol impact following a fall of 20 meters (65 feet); and (2) a package-flatrol striking a bridge or tunnel abutment at a speed exceeding 20 meters per second (about 45 miles per hour), which was the speed limit for trains carrying spent fuel packages at the time these tests were carried out.[34] A third scenario was added because it was frequently mentioned as a cause of public concern: a derailed package-flatrol struck by a train traveling at a closing speed of greater than 20 meters per second (45 miles per hour). Other possible scenarios were eliminated from consideration because they were thought to have a very low probability of occurrence. These included aircraft impacts, explosions, and blasts.

The third scenario (train crash into a derailed package-flatrol) was finally selected for demonstration because model tests showed that of the three scenarios, this one would inflict the most damage on the package closure. The testing was limited to one full-scale crash because of cost and logistical considerations.

The crash test was carried out using the heaviest locomotive in service, which had an operating weight of about 140,000 kilograms (310,000 pounds) and a maximum operating speed approaching 45 meters per second (100 miles per hour). Three passenger coaches were hooked to the locomotive to "add realism to the test" (Collins et al., 1985, p. 206), even though calculations suggested that they would not increase its severity.

The test configuration chosen was similar to that used for the drop test: an impact on the package-closure joint with the center of mass of the locomotive aligned with the center of mass of the package to minimize rotational forces. The same package body used in the drop tests was used for this crash test, but it was fitted with a new lid. The package was mounted on a flatrol railcar, which was turned on its side and laid diagonally across the track so that the front coupler on the locomotive was aligned with the package-closure joint (Figure 2.9). The locomotive and package were instrumented with accelerometers and strain gauges, and the test was recorded using high-speed photography.

The test was carried out at a test track at Old Dalby in Leicester,

[34]This speed limit has since been raised to 65 miles per hour (104 kilometers per hour).

FIGURE 2.9 Configuration for the July 1984 full-scale crash test of a Magnox spent fuel package carried out in Leicester, England by the CEGB. The Magnox package was mounted on a flatrol railcar, which was turned on its side and laid diagonally across the track as shown in this photograph. The wires leading away from the package are connected to instruments for monitoring conditions during the early phases of the crash. SOURCE: Magnox Electric Ltd.

England on July 17, 1984, and was shown live on national television.[35] In the crash, which was described as "first and foremost a visual spectacle" (Hart et al., 1985b, p. 234; Figure 2.10), the package suffered only superficial damage. The peak recorded acceleration, about 49 *g*'s, occurred about 10 milliseconds after impact. The pressure within the package was checked after the crash and was found to have decreased by 0.012 megapascal, corresponding to a fluid loss of about 0.5 liter (Hart et al., 1985b, p. 234).

A significant observation from this test is that the peak force on the package during the test (29 meganewtons [MN]) was considerably less than the peak force in the 9-meter drop test (75 MN) on the same package. In other words, this visually spectacular crash was actually a much less severe mechanical test of package containment than the 9-meter free-drop test used in the IAEA standards and USNRC regulations (see Sidebar 2.1).

[35]A video of this test is available on the Department of Energy's Web site at *http://www.ocrwm.doe.gov/newsroom/videos.shtml.*

FIGURE 2.10 Full-scale crash test of a Magnox spent fuel package carried out in Leicester, England by the CEGB in July 1984. This photograph was taken just after the locomotive, which was traveling at 100 miles per hour (45 meters per second), collided with the package and flatrol. The collision has lifted the package and flatrol off the ground. SOURCE: Magnox Electric Ltd.

2.3.4 Package Performance Study

In 1999, the USNRC initiated a five-year project, referred to as the Package Performance Study, which had the following three objectives (USNRC, 2003c, p.4):

1. Assess whether finite element analysis is a valuable tool for characterizing package and fuel response in extreme thermomechanical environments.
2. Demonstrate the inherent safety of spent fuel package design using public outreach as a significant element.
3. Refine dose and risk estimates to the public and workers through the collection of additional empirical data and improved transportation statistics.

USNRC staff embarked on an effort to obtain public input to inform this study. Initially, the staff held public meetings and solicited comments through its Web site to identify concerns that could be addressed in this study. An analysis of these comments by Sandia National Laboratories (Sprung et al., 2001) identified several common concerns: Some commenters asserted that the regulatory testing requirements in 10 CFR Part 71 were unconvincing as a demonstration of a transport package's performance under severe accident conditions. They wanted more realistic full-scale testing to demonstrate package performance. Some members of the public also thought that the accident statistics used in the analytical studies of package performance in severe accidents (e.g., the 1987 modal study) needed to be reanalyzed in light of increased truck traffic and vehicular speed limits on interstate highways.

USNRC staff then held additional public meetings to help them assess the accuracy of the Sprung et al. (2001) analyses of public comments and to obtain additional comments on a set of proposed package tests. These comments were used to develop a set of draft protocols for extraregulatory tests of rail and truck packages (USNRC, 2003c). Two draft protocols were proposed:

1. High-speed (75 miles per hour [120 kilometers per hour]) impact tests involving drops of rail and truck packages from a tall tower onto an essentially unyielding surface. The rail package drop would be a package-corner-over-center-of-gravity test with the impact limiter in place. The truck package drop would be onto the package body in a so-called back-breaker orientation that bypassed the impact limiters.
2. A fire test for both the rail and the truck packages that would involve a fully engulfing, optically dense hydrocarbon fire of a greater-than-30-minute duration.

Both of these tests are extraregulatory in the sense that they would exceed the test requirements in 10 CFR Part 71 (Sidebar 2.1). The 9-meter free-drop test, for example, produces, neglecting windage, a terminal speed of 13.3 meters per second, which is approximately 30 miles per hour (or approximately 48 kilometers per hour). This impact speed is much lower than the 75 mile per hour (120 kilometers per hour) impact speed envisioned in the staff's draft test protocol.

USNRC staff proposed to perform full-scale testing on two packages that are currently certified in the United States. These packages are similar in construction to the generic packages that were used in Sandia's spent fuel shipping risk reexamination (Sprung et al., 2000).

These draft testing protocols were put out for additional public comment. USNRC staff identified several groups of suggestions from the public comments on the draft testing protocols (USNRC, 2004c): full-scale testing should be conducted to regulatory limits; tests should be based on realistic accident scenarios; they should be designed to test packages to failure; and they should address terrorist acts. USNRC staff commented in its recommendations to the Commissioners (USNRC, 2004c, p. 3) that having realistic accident scenarios and testing to failure are incompatible goals. Staff also noted that terrorism was being addressed in other studies currently under way within the USNRC that "are not suitable for the public participatory approach."

USNRC staff rejected the suggestion to test packages to failure as unrealistic and unworkable for the following reasons: There is no readily agreed-upon definition of package failure among various stakeholders; package failure is a design-specific issue that would have little generic application to risk insights; and there are no realistic accident scenarios that are sufficiently severe to lead to package failure.

USNRC staff also decided to eliminate extraregulatory testing. In its place, staff recommended to the Commissioners that one or more of the following tests be carried out on full-scale truck and train packages:

- A test of a rail package to the regulatory limit. This would subject a rail package with impact limiters to the regulatory tests in 10 CFR Part 71.
- A test of a truck package with its impact limiters to the same regulatory limits, except that the 2-megapascal pressure test might be eliminated if the truck and rail packages are tested in combination.
- A full-scale crash demonstration of a rail package with impact limiters and its railcar with a simulated bridge abutment at about 75 miles per hour (120 kilometers per hour), followed by a fire from a ruptured tank car.
- A full-scale crash test of a truck package with impact limiters and its trailer by a locomotive traveling at about 75 miles per hour (120 kilometers per hour) on a grade crossing followed by a fire.

The staff proposed to the Commissioners a testing protocol involving various combinations of these tests, as well as an option to perform the extraregulatory tests originally outlined in USNRC (2003c). In December 2004, the Commissioners approved a modified full-scale test that will be carried out when funds are made available by Congress (USNRC, 2004d). USNRC staff was directed to plan for a demonstration test involving a single rail spent fuel package involved in a "viable" transportation accident. The Commissioners directed that

> the test should consist of a simulated rail crossing with a train traveling at an appropriate speed colliding at a ninety degree angle with a transportation cask on its rail carrier car in a normal transportation configuration. . . . The test will consist only of the collision and the natural results of that collision. No separate fire testing or immersion testing will be conducted on the cask.

The Commissioners directed USNRC staff to prepare an information paper that provided details and estimated costs of the test.

The requested information paper was submitted to Commissioners in March 2005 (USNRC, 2005a). The staff proposed a test involving a full-scale package of the kind that is likely to be used by the Department of Energy (DOE) in its Yucca Mountain transportation program. The package would contain surrogate fuel elements and would be mounted on a rail carrier car placed at 90 degrees to a simulated rail crossing. The rail package would be subjected to a collision with a locomotive and several freight cars traveling at 60 miles per hour (96 kilometers per hour). In accordance with the Commissioners' December 2004 directive, no fire or immersion testing was included in the proposed testing plan.

The Commissioners approved this proposed test in June 2005 (USNRC, 2005b). In addition, the Commissioners reversed their earlier instructions concerning the thermal test, directing the staff to "add a fire test scenario . . . involving [a] fully engulfing, optically dense, hydrocarbon fire for a duration of one-half hour post-collision." The Commissioners directed staff to inform it of the details and estimated costs for the proposed fire test. They also noted that the proposed test plan provided by staff (USNRC, 2005a) "is not the final word on this issue, as the project is subject to additional modifications and Commission direction once additional information becomes available." The Commissioners directed staff to engage with higher-level management in DOE to request financial support for the demonstration test and to request increased funding from Congress if DOE is unable to provide support.

2.4 CRITICAL ASSESSMENT OF PACKAGE PERFORMANCE

It is clear to the committee that package performance under severe accident conditions is a major concern for transportation safety among many members of the public, especially those who live and work along shipping routes. Finding a way to resolve this issue continues to be a challenge to regulators in the United States and may eventually become a challenge for DOE and the private sector in their commercial spent fuel transportation programs.

The packages used to transport spent fuel and high-level waste are designed to contain their radioactive contents under normal transport conditions and to withstand accident conditions without an increase in external radiation to levels that would endanger emergency responders or the general public. The package performance standards developed by the IAEA and embodied in USNRC regulations (Sidebar 2.1) were developed to help ensure that properly manufactured, certified packages provide such containment effectiveness. The 9-meter regulatory free-drop test onto an essentially unyielding surface, for example, is a more severe mechanical test of package performance than has been produced in any of the severe full-scale crash tests described in this chapter (see Section 2.3).

The safety of transporting spent fuel and high-level waste depends to a great extent on the "inherent" safety of transportation packages to contain their contents even under severe accident conditions. The committee observes that while the full-scale crash tests described in this chapter have imposed mechanical conditions on packages that are less severe than the regulatory free-drop test, some of these tests have resulted in small releases from package containments.[36] However, these releases would not have exceeded regulatory limits, which have as their goal the protection of emergency responders and the general public. The committee judges that the regulatory free-drop test imposes mechanical forces that are severe enough to bound conditions likely to be encountered in foreseeable real-life accidents. At the same time, the committee understands that any releases of radioactivity could have substantial social risk implications (see Chapter 3).

The accident reconstruction and modeling studies described in Section 2.2.3 suggest that there may be a very small number of credible accident conditions involving very long duration, fully engulfing fires[37] that are potentially capable of damaging the seals on transportation packages if such fires are allowed to burn in an uncontrolled manner for long periods of time. Such damage could potentially lead to the release of radioactive material from the package through the processes described in Sidebar 2.5.

The potential for radioactive material releases from packages involved in such fires and the consequences of such releases are incompletely understood at present. USNRC staff has completed a solid technical analysis of the response of one package design to a realistically extreme thermal event resulting from the Howard Street tunnel fire. Additionally, staff are completing analyses of the performance of two additional package types for this fire exposure (see footnote 22 in this chapter). The committee judges that additional analyses of this type are needed to better understand package performance for realistic accident conditions involving very long duration fires.

---

[36]The releases described in this chapter consisted of water that leaked from the package interiors. This water may have been slightly contaminated with radioactive material from the fuel rod cladding (crud), but the fuel itself remained intact. It is important to note that in the United States, packages now in use to transport commercial spent fuel are filled with inert gas, not water. Inert gas is a much less efficient medium than water for transporting radioactive contamination out of leaking packages.

[37]The committee uses this term to describe fires that burn for periods of hours (or longer). Such fires could produce thermal loading conditions that exceed those for the regulatory thermal test specified in 10 CFR 71.73. Some of the real-world accidents described in this chapter involved fires that burned for hours to days.

The number of additional analyses might be small in number, especially if they focused on fire scenarios that would essentially bound expected real-world accident conditions,[38] and if they examined the performance of a representative set of package designs that are likely to be used in future large-quantity shipping programs. The analyses have to address both the containment effectiveness of packages in response to very long duration fires and the consequences of any radioactive releases that result from package seal failures.

The purpose of these analyses would be to inform changes, if needed, to regulatory and operational practices to reduce the likelihood of occurrence and potential consequences of accidents involving very long duration fires. At least two options are available to regulators and implementers if such changes are needed:

1. Operational steps to reduce the likelihood of occurrence of long-duration fires during spent fuel transport, and/or
2. Design or testing requirements to improve the thermal resistance of transportation packages.

The first option would clearly be preferable from a cost and implementation standpoint. The objective of taking such operational steps would be to reduce the potential for accidents between trucks or trains carrying spent fuel and other vehicles carrying large quantities of flammable materials, especially flammable liquids, in places where it might be difficult to mount an effective firefighting response.[39] Operational controls are used routinely by railroads to control transportation hazards. Such controls would probably be more easily implemented and enforced when shipments of spent fuel and high-level waste are made by dedicated train.

The second option would be more difficult and expensive to implement, because it very likely would require expensive changes to regulatory testing requirements and some existing package designs. However, there would be no reason to change the testing requirements or package designs if effective operational controls could be implemented.

The committee received several comments at its meetings on the testing of transportation packages under extreme loading conditions, especially

[38]The historical accident record provides perhaps the best available information to identify bounding accident scenarios. For example, the Livingston, Louisiana, fire (see Table 2.2) and Summit Tunnel fire (see Section 2.2.3 and Figure 2.4) are possible examples of bounding scenarios.

[39]Tunnel fires are of special concern because of access restrictions that can delay or prevent an effective firefighting response. Such delays could allow the fires to burn in an uncontrolled manner for long periods of time.

with respect to the USNRC's Package Performance Study. Many commenters asserted that the regulatory testing requirements in 10 CFR Part 71 are unconvincing as a demonstration of a transport package's performance under severe accident conditions. They wanted realistic full-scale testing to be conducted to demonstrate package performance. The State of Nevada suggested that comprehensive full-scale testing would improve the overall safety of the package and vehicle system and enhance confidence in risk analysis techniques. Full-scale testing was proposed as a way to potentially increase the acceptance of spent fuel shipments by state and local officials and members of the general public and to "potentially reduce adverse social and economic impacts caused by public perception of transportation risks" (Hall, 2003, p. 3).

Full-scale testing can be used to determine how packages will perform under both regulatory and credible extraregulatory conditions. To be of value, however, the committee believes that extraregulatory tests must be designed to closely mimic conditions that would reasonably be expected to be encountered in actual service. The test program undertaken by the CEGB in the 1980s (Section 2.3.3) provides an excellent example of the appropriate use of full-scale testing. In that program, the full-scale test was the end point of a much larger deliberative investigation of the conditions that could be encountered during transport of spent fuel. The full-scale test was used both to validate the analytical and scale-modeling work that had been carried out beforehand, and at the same time to provide a demonstration of the containment effectiveness of the packages in a way that could be appreciated by the general public.

The USNRC's Package Performance Study has elements of an appropriate full-scale testing program. The design for a full-scale test was developed through a deliberative process that involved technical analysis and provided many opportunities for public comment. However, none of the tests that emerged from this process were selected by the USNRC for execution. In fact, the scenario that was selected (a collision between a locomotive and a package mounted on a railcar placed at 90 degrees to the rail crossing in an upright position; see Section 2.3.4) duplicates many features of the rail-crossing impact test carried out by Sandia National Laboratories more than 25 years ago. One very important difference between the Sandia test and the currently proposed test is the packages selected for testing. The Sandia test used an obsolete truck transport package. The proposed test would use a rail package that is currently certified and of the type to be used by DOE for transport of spent fuel to the federal repository. If executed properly, this full-scale test will be valuable for validating scale-model and computer simulations of package performance and thereby increasing confidence in current regulatory testing approaches. The test is also likely to be visually powerful and could provide the public with useful information about the performance of transportation packages.

The committee also endorses package testing studies that integrate full-scale testing, scale-model testing, and finite element and associated structural analysis methods to demonstrate compliance with regulatory standards. These analysis methods allow a much wider variety of failure scenarios to be examined than is possible with full-scale testing alone. The analysis codes (e.g., LS DYNA, PRONTO) have been well benchmarked for these applications. Full-scale testing has two primary uses in this integrated test regime. First, it can be used to validate the computer codes for each package design, thus providing additional confidence in package performance under real-world conditions. Second, it provides a direct demonstration of a package's ability to meet specific regulatory test conditions.

Several participants at the committee's information-gathering meetings strongly urged that testing to failure should be carried out on all package designs that will be used to transport spent fuel and high-level waste to Yucca Mountain. However, they were not clear on what would constitute a failure. When pressed for specifics, the committee understood that what was generally meant was that full-scale testing should be carried out to destruction, presumably to establish the ultimate strength of transportation packages. The State of Nevada specifically recommended that the USNRC undertake an evaluation of the costs and benefits of destructive testing of a randomly selected production model cask (Hall, 2003, p. 3).

The principal argument in favor of destructive testing is that it can provide information on the magnitudes of thermomechanical loads required to eliminate containment effectiveness, thus establishing a "safety factor" for a given package design. There are significant drawbacks to this approach, however. Most importantly, each of the tests would involve only one thermal or mechanical loading condition out of the set of many possible conditions, and the selected loading condition would likely far exceed what could reasonably be expected to occur even in the most extreme real-world accidents. Moreover, a separate package might have to be acquired for each test, which would greatly increase the costs of a testing program for even a single package design.

It is important to recognize that *any* transportation package could be destroyed if no limits are placed on the loads that act upon it. Moreover, the failure of a package, in the sense that it can no longer perform its intended containment function, will generally occur under conditions that are much less severe than needed for destruction. Consequently, even if costs were not prohibitive, testing to destruction would provide little or no insight into the conditions that would cause a loss of package containment under real service conditions.

Even full-scale testing to failure could be problematical. Because examining the multitude of possible accident conditions and failure scenarios through full-scale testing is clearly impractical, the committee judges that the bounding approach that the IAEA has established is entirely appropriate. The approach is one that reflects four plausible accident-like conditions: free drop, puncture, thermal exposure, and water immersion (see Sidebar 2.1).

The important question to be answered by testing is not whether a package could be made to fail; as noted previously, it would certainly be possible to design tests that would accomplish this goal. Rather, the question that needs to be answered is whether there are *credible* accident conditions that would result in releases of radioactivity to the environment that would endanger emergency responders or the general public. It is clear from the modeling and full-scale tests described in this chapter that transportation packages are extremely rugged. The committee judges that packages designed, fabricated, used, and maintained under current regulatory standards are very unlikely to encounter loading conditions under real-world conditions, with the possible exception of very long duration fires, that would lead to releases in excess of regulatory limits. The committee recognizes, however, that even minor releases from package containment might have important social implications. These are discussed in Chapter 3.

2.5 PACKAGE PERFORMANCE FINDINGS AND RECOMMENDATIONS

The committee offers the following findings and recommendations based on the analysis of package performance provided in this chapter.

FINDING: Transportation packages play a crucial role in the safety of spent fuel and high-level waste shipments by providing a robust barrier to the release of radiation and radioactive material under both normal transport and accident conditions. International Atomic Energy Agency package performance standards and associated Nuclear Regulatory Commission regulations are adequate to ensure package containment effectiveness over a wide range of transport conditions, including most credible accident conditions. However, recently published work suggests that extreme accident scenarios involving very long duration, fully engulfing fires might produce thermal loading conditions sufficient to compromise containment effectiveness. The consequences of such thermal loading conditions for containment effectiveness are the subject of ongoing investigations by the Nuclear Regulatory Commission and other parties, and this work is improving the understanding of package performance. Nonetheless, additional analyses and experimentation are needed to demonstrate a bounding-level understand-

ing of package performance in response to very long duration, fully engulfing fires for a representative set of package designs.

RECOMMENDATION: The Nuclear Regulatory Commission should build on recent progress in understanding package performance in very long duration fires. To this end, the agency should undertake additional analyses of very long duration fire scenarios that bound expected real-world accident conditions for a representative set of package designs that are likely to be used in future large-quantity shipping programs. The objectives of these analyses should be to

- Understand the performance of package barriers (spent fuel cladding and package seals);
- Estimate the potential quantities and consequences of any releases of radioactive material; and
- Examine the need for regulatory changes (e.g., package testing requirements) or operational changes (e.g., restrictions on trains carrying spent fuel) either to help prevent accidents that could lead to such fire conditions or to mitigate their consequences.

Strong consideration should also be given to performing well-instrumented tests for improving and validating the computer models used for carrying out these analyses, perhaps as part of the full-scale test planned by the Nuclear Regulatory Commission for its package performance study.

Based on the results of these investigations, the Commission should implement operational controls and restrictions on spent fuel and high-level waste shipments as necessary to reduce the chances that such fire conditions might be encountered in service. Such effective steps might include, for example, additional operational restrictions on trains carrying spent fuel and high-level waste to prevent co-location with trains carrying flammable materials in tunnels, in rail yards, and on sidings.

FINDING: The committee strongly endorses the use of full-scale testing to determine how packages will perform under both regulatory and credible extraregulatory conditions. Package testing in the United States and many other countries is carried out using good engineering practices that combine state-of-the-art structural analyses and physical tests to demonstrate containment effectiveness. Full-scale testing is a very effective tool both for guiding and validating analytical engineering models of package performance and for demonstrating the compliance of package designs with performance requirements. However, deliberate full-scale testing of packages to destruction through the application of forces that substantially exceed credible accident conditions would be marginally informative and is not justified given the considerable costs for package acquisitions that such testing would require.

RECOMMENDATION: Full-scale package testing should continue to be used as part of integrated analytical, computer simulation, scale-model, and testing programs to validate package performance. Deliberate full-scale testing of packages to destruction should not be required as part of this integrated analysis or for compliance demonstrations.

3
Transportation Risk

The original statement of task for this study (Sidebar 1.1) directs the committee to examine the "principal risks" for transporting spent fuel and high-level waste; determine how well these risks are understood; and compare them to other risks that confront members of society. Those tasks are addressed in this chapter.

As noted in Chapter 1, risk is a multidimensional concept: It includes *health and safety risks* that arise from exposures of workers and members of the public to radiation from shipments of spent fuel and high-level waste. It also includes *social* risks that arise from social processes[1] and people's perceptions,[2] even in the absence of radiation exposures. The health and safety risks and social risks are collectively referred to as *societal risks* in the statement of task given in Sidebar 1.1.

A great deal of work has been carried out over the past four decades to understand the risks (Sidebar 3.1) arising from the transport of spent fuel, both in the United States and abroad. Although the principal focus of this

---

[1]*Social process* is defined as "a characteristic mode of social interaction" (*Webster's Third New International Dictionary*). Social interactions shape the communities in which people live by, for example, influencing choices about where to purchase or rent a home, where to work, and where to send children to school.

[2]Perception is defined as the "integration of sensory impressions of events in the external world . . . as a function of nonconscious expectations derived from past experience" (*Webster's Third New International Dictionary*). Perceptions can have a strong influence on peoples' behavior, whether or not such perceptions are an accurate picture of reality.

SIDEBAR 3.1 Background on Risk

Risks for spent fuel and high-level waste transportation arise from conventional vehicular accidents and exposures to ionizing radiation under both normal and accident conditions. Radiation risks are primarily a concern for people who live near, or travel on, spent fuel shipment routes.

Risk considers both the likelihood of occurrence of a specific hazard and its consequences. Frequently, one considers several scenarios that involve different kinds of hazards, each with a different likelihood of occurrence and consequence. One way in which risk can be expressed is in terms of a triplet (Kaplan and Garrick, 1981):

$$\text{Risk} = f\,(\text{scenarios, probability, consequences}),$$

Where, for spent fuel shipments,

- *Scenarios* represents transport conditions that can lead to an exposure to ionizing radiation from either routine operations or severe accidents,
- *Probability* expresses quantitatively the likelihood that a scenario will actually occur during one shipment; it is expressed as a dimensionless quantity that ranges in value from 0 (impossible) to 1 (certain)—for example, a probability of 0.5 indicates that a particular scenario has a 50 percent chance of occurring, and
- *Consequences* describe the undesirable results if the scenario does occur: for example, undesirable health effects.

The risks from spent nuclear fuel transport can be characterized by several measures. For example, risk can be expressed in terms of the expected number of deaths per quantity of spent fuel transported, per number of packages shipped, or per number of package shipments. It also could be expressed in terms of the number of deaths expected for a specific subpopulation exposed to ionizing radiation, for example, the subpopulation of transportation workers. Although they are difficult to quantify, consequences may also include socioeconomic outcomes.

The choice of scenarios and consequences selected for a risk calculation can make a difference in how that risk is understood by potentially affected populations. A risk may be understood as low by one measure in comparison to another, even though the same risk is being considered (NRC, 1996). This has implications for informing decision makers, for communicating about risks with non-experts, and for the legitimacy of risk comparisons in the eyes of interested and affected people (NRC, 1989, 1996). Comparing risks arising from fundamentally different activities also requires care in the selection of appropriate scenarios and consequences. This point is discussed in more detail elsewhere in this chapter.

report is on transportation in the United States, much can be learned from international experiences. Spent fuel and high-level waste are being transported in many other countries, in some cases in much greater quantities than in the United States. The committee has drawn upon the experiences of foreign transportation programs to inform its judgments. This information has been acquired through the review of published documents and from first-hand experience.[3]

Transportation risks can arise both during *normal* transport operations and from *accidents* involving loaded spent fuel or high-level waste shipping packages.[4] Table 3.1 describes the transportation impacts for these two operational scenarios. These risks can also arise from *incidents* (e.g., terrorist attacks) involving packages containing spent fuel or high-level waste; such incidents are not addressed in this report (see Section 1.2).

The health and safety risks to be discussed in this chapter arise from exposures of people who travel, work, or live near transportation routes, and transportation workers themselves, to radiation (Sidebar 3.2) from loaded spent fuel and high-level waste transportation packages. During normal operations, such exposures can occur as the result of *radiation shine*[5] from transportation packages loaded with spent fuel or high-level waste. Although the radiation doses to individuals near transport routes are likely to be very low, large numbers of individuals may receive exposures, producing a "collective dose" that can be used to estimate health impacts (see Sidebar 3.3). Degradation and/or loss of package containment in a severe accident has the potential to increase such radiation exposures and possibly result in the release of radioactive material from the package to the environment, although the committee notes in Chapter 2 that the robust design of transportation packages makes such releases unlikely.

Health and safety risks are frequently characterized in terms of human health effects: for example, injuries and loss of human life. Modern society generally considers such consequences to be the most severe harms that can

---

[3]Committee member Clive Young has extensive experience with the transport of spent fuel in the United Kingdom. In addition, a group of committee members visited Germany and the United Kingdom in September 2004 (see Appendix B) to obtain first-hand information about European transportation programs.

[4]International Atomic Energy Agency regulations for the safe transport of radioactive materials define three levels of severity for transportation conditions: routine conditions of transport, normal conditions of transport, and accident conditions of transport. Routine conditions are free of any transport mishaps; normal conditions can involve minor mishaps due to rough handling or exposure to weather. Accidents subject packages to severe conditions that well exceed normal conditions of transport.

[5]Transportation packages contain heavy shielding to protect workers and the public from the radiation emitted by the spent fuel or high-level waste contained within them. The packages are effective in shielding well over 99 percent of this emitted radiation, but a small amount (below regulatory limits) of radiation, primarily gamma rays, can escape from the interior of the packages and provide external doses to workers and the public. This report uses the term *radiation shine* to refer to this external radiation.

TABLE 3.1 Transportation Risks Examined in this Report

	Transportation Impacts	
Transportation Conditions	Health and Safety Risks	Social Risks
Normal	Health impacts arising from the emission of radiation (i.e., radiation shine) from transportation packages	Direct socioeconomic impacts: loss of economic or social well-being as a direct result of transportation program operations
		Perception-based impacts: anxiety and associated illness; loss of property values; and reduced economic activity
Accidents	Health and environmental impacts arising from elevated radiation and/or the physical release of radioactive material as a result of the degradation or loss of package containment	Direct socioeconomic impacts: Temporary loss of transportation route use and associated business disruptions such as a loss of tourism
		Perception-based impacts: social amplification of the normal impacts as a result of accidents; these can result in secondary or tertiary impacts, including stigmatization of people and places; loss of trust in transportation program management; moratorium on transportation program operations; and/or increased program costs

result from any hazard, especially anthropogenic hazards that can be controlled through regulation. Society places a high value on human life and routinely demands that governments strictly regulate life-threatening hazards. Reductions in severe injuries and loss of human life are generally considered to be primary measures of regulatory effectiveness. Because of this emphasis on protecting human life, risk assessment experts have developed methodologies to quantitatively estimate risks to human life. Sidebar 3.1 provides a description of one way in which such risks are estimated.

Social risks can have both direct socioeconomic and perception-based impacts such as those shown in Table 3.1. These risks may reduce the desirability of living and working in communities associated with spent fuel and high-level waste transportation operations. The social risks of interest in this chapter are harder to measure than the corresponding health and safety risks, and even identifying cause-and-effect relationships can be difficult. The impacts of social risk occur within a much larger sphere of social and economic activities that can mask important effects. The measurement of perception-based impacts can be especially difficult, because it frequently requires the use of surveys to measure people's anticipated, rather than actual, behaviors.

The health and safety risks and the social risks associated with spent fuel and high-level waste transportation can have significant interactions. Increases in radiation exposures or in the incidence of health effects from transportation operations (e.g., because of well-publicized mishaps, accidents, or fatalities) may, over time, increase the perception-based impacts. On the other hand, transportation operations that are carried out without demonstrable health impacts may, over time, reduce the perception-based impacts. A shipping incident or accident that leads to a moratorium on transportation operations might well change the entire profile of social risks associated with a transportation program.

There is another class of nonradiological impacts that are not considered in detail in this chapter: *conventional vehicular impacts* associated with the transportation of spent fuel and high-level waste. These include the health impacts of exhaust emissions from transport conveyances and vehicular accidents that result in fatalities, injuries, and property damage. While these impacts are real and predictable, they generally do not garner the same level of awareness or concern among members of the public as the radiation-based impacts described previously. Moreover, it could be argued that given the higher standards for driver training and equipment maintenance, and the conduct of vehicle and package inspections and operations for spent fuel and high-level waste transportation, conventional vehicular impacts associated with accidents would actually be lower than for other types of hazardous materials or heavy-freight transport. In any case, spent fuel and high-level waste transportation programs are a very small component of the overall transport system for hazardous materials, as measured both by load mass and by volume of traffic.

The committee provides an examination of health and safety risks of spent fuel and high-level waste transportation in Section 3.1. The social risks are examined in Section 3.2, and the risk comparisons called for by the study charge (see Sidebar 1.1) are described in Section 3.3. The committee's findings and recommendations on transportation risks are provided in Section 3.4.

SIDEBAR 3.2 Radiation Dose

Materials that are *radioactive* are unstable (i.e., the nuclei in the atoms of the material possess too much energy) and transform spontaneously (*decay*) through the emission of *radiation*. This radiation may be in the form of energetic particles, such as alpha particles, beta particles, or neutrons, or energy may be emitted in the form of electromagnetic radiation (e.g., gamma rays). Collectively these emissions are known as *ionizing radiation* because they are sufficiently energetic to directly or indirectly ionize the matter (i.e., remove electrons from the atoms) they travel through.

Absorption of radiation energy by a cell of a living organism can alter its chemical and physical state. The absorption of large amounts of radiation can produce short-term or "acute" effects in the cell. The most severe effect would be cell death. Small amounts of radiation (not sufficient to cause cell death) potentially can damage the cell's genetic material (i.e., DNA contained in the chromosomes). If the cell's natural repair mechanisms cannot repair this damage correctly, it may lead to the induction of cancer at some future time. Because of the long time periods involved in their development, such cancers are referred to as "latent."

The following quantities are commonly used to characterize radiation exposures in living organisms:

- *Absorbed dose.* The quantity of ionizing radiation deposited into an organ or tissue, expressed in terms of the energy absorbed per unit mass of tissue. The basic unit of absorbed dose is the rad or its SI (international system of units, also known as the metric system) alternative the gray (Gy; 1 Gy = 100 rad).
- *Equivalent dose.* The absorbed dose averaged over the organ or tissue of interest multiplied by a weighting factor that accounts for the differences in biological effects (per unit of absorbed dose) for different types of radiation. The weighting factor ranges from 1 for X-rays and gamma rays to 20 for alpha particles and some neutrons. The equivalent dose is expressed in units of rem or its SI alternative the sievert (Sv; 1 Sv = 100 rem).
- *Effective dose.* A measure of dose that accounts for the differences in biological effects of different types of radiation and for the varying sensitivity of different organs to the biological effects of radiation. Effective doses are also expressed in rem or sieverts.

Radiation and radioactivity can be found in nature in a number of different forms. The *natural radiation environment* consists of cosmic and solar radiation, external radiation from radioactive materials present in rocks and soil, and inhaled

3.1 HEALTH AND SAFETY RISKS

Two approaches are used in this section to estimate the health and safety risks of spent fuel and high-level waste transportation: (1) an examination of the worldwide record of spent fuel transport (Section 3.1.1); and

and ingested radioactive materials from air, food, and water. These sources have been present since the creation of Earth and provide an effective dose to all living organisms. The table below shows that, worldwide, the average annual effective dose to individuals is about 2.4 millisieverts (mSv; 240 millirem). This annual exposure is a good starting point to use in judging the magnitude of equivalent doses received from man-made sources. The National Council on Radiation Protection and Measurements (NCRP, 1987) estimates that man-made sources of radiation and radioactivity contribute an additional effective dose of 0.6 mSv (60 millirem) annually to an average person living in the United States. Most of this dose is due to medical procedures such as diagnostic X-rays and nuclear medicine procedures. Since the NCRP report was published, new diagnostic medical procedures that utilize ionizing radiation (especially computed tomography scanning) have come into wide use in the United States. Consequently, the average annual doses from medical procedures are probably increasing.

It is important to keep in mind that the effective dose statistics presented in this sidebar are averages. Individuals can receive much more or much less than the average depending on where they live (i.e., location as well as height above sea level), the type of house in which they reside, their occupation, the medical procedures they undergo, and many other factors determined by the individual life-style (e.g., air travel, watching television, diet).

TABLE Worldwide Exposure to Natural Sources of Radiation and Radioactive Material

Source of Exposure	Average Annual Effective Dose, mSv (millirem)	Typical Range of Effective Doses, mSv (millirem)
Cosmic radiation	0.39 (39)	0.3–1.0 (30–100)
External terrestrial radiation	0.48 (48)	0.3–0.6 (30–60)
Inhalation (U, Th, ^{222}Rn, ^{220}Rn)	1.26 (126)	0.2–10.0 (20–1,000)
Ingestion (^{40}K)	0.29 (29)	0.2–0.8 (20–80)
Total	2.42 (242)	1.0–10.0 (100–1,000)

SOURCE: Adapted from UNSCEAR (2000, Annex B, Table 31).

(2) an examination of the principal quantitative risk analyses that have been carried out for spent fuel and high-level waste transport, including the analysis for transporting spent fuel and high level-waste to a federal repository at Yucca Mountain, Nevada (Section 3.1.2).

SIDEBAR 3.3 Collective Dose and Latent Cancer Fatalities

Collective dose is defined as the sum of all radiation doses received by all members of a population at risk (NCRP, 1995). The units of collective dose are usually given as person-sieverts or person-rem. This concept is frequently used in radiation protection applications, both for controlling actual exposures and for estimating potential exposure risks. The use of the collective dose for radiation protection purposes assumes the following (NCRP, 1995):

1. There is a direct proportionality between radiation dose and risk over their respective ranges of concern.
2. Risk is independent of dose rate.
3. A radiation dose leads to an identical risk whether it is administered to a single individual or to a population.

NCRP (1995, p. 1) notes that "[w]hile these assumptions may or may not be valid, they are considered to be conservative and have been generally accepted by the scientific community concerned with radiation protection."

The Department of Energy (DOE) used the collective dose concept to estimate *latent cancer fatalities* (LCFs) in the final Yucca Mountain Environmental Impact Statement (EIS). These cancers are expected to be produced many years after a radiation dose is received (i.e., after a latency period) and are never traceable directly to the received dose. DOE calculated collective doses for populations of workers and the public from prospective exposures to radiation from its transportation program. These doses were calculated using computer programs such as RADTRAN (see Sidebar 3.4). DOE then estimated the number of latent cancer fatalities using the International Commission on Radiological Protection (ICRP, 1991) recommended conversion factors: 5×10^{-4} latent cancer fatality per person-rem (5×10^{-2} per person-sievert) of collective dose for the general public and 4×10^{-4} latent cancer fatality per person-rem (4×10^{-2} per person-sievert) of collective dose for workers; these factors were doubled when doses greater than 20 rem are received over short time periods. The conversion factor for the public is higher because it applies over an entire lifetime, whereas the factor for workers applies only for working ages.

Radiation is a weak carcinogen at the low doses involved in the routine transport of spent fuel and high-level waste. Consequently, the incremental increase in latent cancers from low doses of radiation is very small relative to the natural occurrence of this disease in human populations. Moreover, radiation-induced cancers do not have any special characteristics that allow them to be differentiated from cancers developed from other causes.

3.1.1 Historical Record of Spent Fuel Transport

Spent fuel has been transported routinely in more than a dozen countries, including the United States, for many decades. The quality of record keeping on these shipments varies significantly, especially between the mid-

1940s and 1970s. The committee first provides a brief review of the transportation experience in the United States and then examines experiences in some other countries.

U.S. Transportation Experience

The United States has been transporting irradiated nuclear fuel since World War II. The first irradiated fuel shipments were made by the Manhattan Project as part of the national effort to develop atomic weapons. More than 170,000 MTHM[6] of irradiated fuel were transported within the Hanford (Washington) and Savannah River (South Carolina) sites as part of the nuclear weapons production effort between 1944 and the end of the Cold War in the late 1980s. Most of this transport occurred over very short distances (a few kilometers [miles]) on publicly restricted lands, mostly by rail.

By the early 1960s, civilian spent fuel was being transported routinely on the nation's road and rail systems by the Atomic Energy Commission (AEC).[7] In 1974, the AEC was reorganized,[8] and authority for regulating the commercial transport of radioactive materials transportation was given to the newly established U.S. Nuclear Regulatory Commission (USNRC). The most complete records of spent fuel transportation date from the creation of this agency.

Most spent fuel transport across the nation's public highways and private railroads has involved small-quantity shipments of commercial spent fuel. Estimates of quantities of commercial spent fuel shipments are available from several sources. Pope et al. (1991, 2001) provide commercial spent fuel shipping estimates since 1964. These estimates were developed from Department of Energy (DOE) and Department of Transportation (DOT) databases supplemented with data from the USNRC and private sources. Pope et al. (1991) note that their estimates do not include shipments from six commercial reactors because of the difficulty in obtaining data.

---

[6]Metric tons of heavy metal, where the heavy metal is uranium. This is a commonly used measure of fuel quantity. For comparison purposes, a typical reactor core contains about 100 MTHM (110 short tons) of nuclear fuel. A typical truck transport package typically holds between 0.5 and 2 MTHM (0.55 and 2.2 short tons) of spent fuel; a typical rail package holds between 10 and 18 MTHM (11 and 20 short tons).

[7]The AEC was created by the Atomic Energy Act (also known as the McMahon Act) in 1946 to control and promote the use of nuclear power.

[8]The Energy Reorganization Act of 1974 abolished the AEC and created two federal agencies in its place: the Nuclear Regulatory Commission and the Energy Research and Development Administration. The Energy Research and Development Administration became the Department of Energy after the Energy Reorganization Act of 1977.

TABLE 3.2 Commercial Spent Nuclear Fuel Shipments in the United States, 1964–2004

	Mass of Spent Fuel Shipped (MTHM)		Number of Shipments	
Time Period	Highway	Rail	Highway	Rail
1964–1978	473	348	1565	126
1979–1997	356	1097	1181	153
1998–2004	16	766	102	261
Totals	845	2211	2848	540

NOTE: MTHM = metric tons of heavy metal.

SOURCES: Pope et al. (1991, 2001); USNRC, written communication, 2005.

Additional unpublished information was made available to the committee from Energy Resources International (Supko, 2000; Supko, 2005, written communication) and the USNRC (USNRC, 2005, written communication). The latter communication provided shipping data for 1998–2004 based on shipper notifications required under regulation 10 CFR Part 73.[9] This communication is an update of USNRC's public circular on spent fuel shipments (USNRC, 1998). Updated information has not been released since the September 2001 terrorist attacks. The agency was preparing an updated version for public release when the present report was being finalized in December 2005.

Table 3.2 provides an estimate of commercial spent fuel shipments in the United States since 1964. The committee was not able to assess the completeness or accuracy of these data, except to note that the pre-1979 data are likely incomplete. Information on spent fuel shipments prior to 1964 is not available, although accident and incident reports dating back to the mid-1950s are available, as discussed elsewhere in this chapter.

One conclusion that can be drawn from these data is that the previous U.S. spent fuel shipping experience as measured by the total number of shipments or mass of spent fuel shipped is small compared with anticipated future transportation campaigns. The federal repository and Private Fuel Storage, LLC, programs, for example, plan to ship about 20 and 13 times, respectively, the amount of commercial spent fuel that has been shipped in the United States since 1964. Moreover, both programs plan to ship spent fuel primarily by rail. The planned number of rail shipments to a federal

[9]Code of Federal Regulations, Title 10, Part 73: Physical Protection of Plants and Materials.

repository at Yucca Mountain under the mostly rail scenario (9600 shipments; see Table 3.8) is approximately 18 times the number of rail shipments that have occurred in the United States since 1964.

Spent nuclear fuel shipments in the United States are usually made under the USNRC's or DOT's exclusive use regulations (10 CFR 71.47(b)).[10] Such shipments can be transported using public road and rail systems in the United States only if they do not exceed the following dose limits:

- 2 millisieverts (mSv) per hour (200 millirem [mrem] per hour) (see Sidebar 3.2) on the external surface of the transport package[11] and at any point on the outer surface of the vehicle.
- 0.1 mSv per hour (10 mrem per hour) at any point 2 meters (6.5 feet) from the outer lateral surfaces (but not the top or bottom) of the vehicle.
- 0.02 mSv per hour (2 mrem per hour) in any normally occupied space. This provision does not apply to private carriers if exposed personnel under their control wear approved radiation dosimetry devices.[12]

U.S. agencies do not collect records of radiation exposures resulting from the transportation of irradiated nuclear fuel as is done for personnel exposures in nuclear power plants. Private carriers will keep records for those workers who use radiation monitoring devices in accordance with regulations, but these records are not published. Consequently, the doses received by workers and the public associated with spent nuclear fuel shipments in the United States are not precisely known, although the committee judges that they are likely to be relatively small given the external dose limits allowed by regulations combined with the small numbers of shipments that have been made to date. Estimates of doses to populations and to hypothesized maximally exposed individuals in future shipments have been made by the USNRC, DOE, and DOT based on the projected number and characteristics of shipments and the populations that live in the proximity of planned transportation routes. These estimates have generally assumed that the dose rates from the packages are the maximum allowed

[10]*Exclusive use* is defined in 10 CFR 71.4 as "sole use by a single consignor of a conveyance for which all initial, intermediate, and final loading and unloading are carried out in accordance with the direction of the consignor or consignee. The loading and unloading must be carried out by personnel having radiological training and resources appropriate for the safe handling of the consignment."

[11]A limit of 10 mSv per hour (1000 mrem per hour) applies when the shipment is made in a closed transport vehicle in which the package is secured so that its position remains fixed and there are no loading or unloading operations between the beginning and end of transportation.

[12]In this case, the normal occupational dose limits apply (see Table 3.10).

under USNRC or DOT regulations. This is illustrated elsewhere in this chapter for the planned transportation program to a federal repository at Yucca Mountain.

Information on accidents and incidents involving spent fuel shipments in the United States has been reported since the late 1940s. Data for 1949–1971 are provided in AEC reports. In 1971, DOT established the Hazardous Materials Incident Reporting System (HMIS).[13] This computerized database contains information on incidents involving the interstate transportation of hazardous (including radioactive) materials by air, highway, rail, and water. DOT regulations (49 CFR 171.15) require that all accidents and incidents involving radioactive materials transport that meet one or more of the following criteria be reported to it for inclusion in this database:

- Deaths or injuries requiring hospitalization
- Property damage in excess of $50,000
- Evacuations of the public that last for one or more hours
- Closure of major transportation arteries or facilities
- Changes to flight patterns or routing of aircraft
- Fire, breakage, spillage, or suspected contamination involving radioactive materials
- Unintentional release from a package or any discharge during transportation

USNRC regulations (10 CFR 20.2201-2206; 10 CFR 73.71) also require that thefts, exposures, and releases of radioactive materials be reported.

In 1981, the Transportation Technology Center at Sandia National Laboratories established the Radioactive Material Incident Report (RMIR) database (Weiner and Tenn, 1999). This database contains information about radioactive materials transportation incidents that have occurred in the United States since 1971. It incorporates information from the HMIS in addition to information from the USNRC and other organizations such as state radiological authorities and the media.

The RMIR records *accidents* involving vehicles carrying radioactive materials that involve fatalities or injuries or that involve sufficient damage that the vehicle cannot move under its own power. It also records *incidents* that involve actual or suspected releases or surface contamination that exceeds regulatory limits. This database was discontinued in 1998 due to funding cutbacks. DOT's HMIS is now the primary source of data on hazardous material transportation incidents in the United States.

---

[13]Information about this database can be found on the DOT Web site at *http://hazmat.dot.gov/abhmis.htm*.

TABLE 3.3 Summary of Spent Fuel Shipping Accidents and Incidents, 1949–1996

Time Period	Accidents or Incidents Reported	Radioactive Material Contamination[a]	Surface Contamination[b]	No Description	Vehicular Deaths[c]
1949–1970	14	6	None reported[d]	2	1
1971–1996	58	2[e]	49	0	1

[a]Any detectable loss, dispersal, or escape of radioactive material from the package's containment system.

[b]Detectable non-fixed contamination on external surfaces.

[c]Deaths caused by vehicular accidents, not the release of radiation.

[d]It is unclear whether surface contamination was routinely tested for or reported during this period.

[e]One incident (in 1984) involved an empty package; the other (in 1976) involved a pinhole leak of coolant or moderator on the outside jacket of the package, not the release of spent fuel.

SOURCE: Data compiled by DOE Office of Civilian Radioactive Waste Management.

A summary of available data on transportation accidents and incidents is provided in Table 3.3. These data were compiled by DOE's Office of Civilian Radioactive Waste Management. According to DOE, the AEC reported radioactive material contamination incidents between 1949 and 1970. This contamination involved the package and/or the conveyance and in some cases the surrounding environment. Two additional radioactive material contamination incidents were reported between 1971 and 1996. One involved an empty package, and the other involved a pinhole leak of coolant or moderator. These older incidents apparently involved packages that were designed to transport spent fuel in water for cooling and shielding, and the leaks presumably involved the release of small amounts of this water through holes in pipes, valves, and seals. At present, only spent fuel cooled for more than five years is transported in the United States, and all spent fuel is transported in a dry state. However, some package designs still utilize water jackets for shielding neutrons, but these are physically separated from and are not in contact with the interior of the package.

Most of the reported incidents did not involve package leaks, but rather the detection of non-fixed surface contamination[14] on the transport pack-

[14]Non-fixed contamination adheres to the outer surfaces of the package and can be detected by wiping. The limits for such contamination, which are established by international standards and U.S. regulations (49 CFR 173.443 and 10 CFR 71.87 (i) by reference to 49 CFR 173.443), are as low as reasonably achievable and not to exceed 4 becquerels per square

ages, and these incidents were described as minor. Such surface contamination typically results from inadequate decontamination of packages following the loading of spent fuel.[15] However, there is no confirmation of cause of contamination in the database.

Table 3.4 provides a list of transportation accidents involving spent fuel transport packages between 1971 and 2005 in the United States. There were four accidents involving trucks and five accidents involving trains during this time.[16] None of these accidents resulted in the release of radioactivity. It is important to recognize that all but the December 1971 accident were minor in that they did not result in severe impacts or fires that would test the integrity of transport package containment. However, claims about this safety record in the United States have to be interpreted carefully given that spent fuel transport quantities are quite limited, especially for rail transport.

The U.S. government opened a repository for the disposal of defense transuranic waste in New Mexico in 1999. This repository, the Waste Isolation Pilot Plant (WIPP), had, as of April 2005, received shipments of waste from eight DOE sites across the continental United States. About 3500 truck shipments traveling about 3.5 million truck-miles had been made to the repository as of April 2005[17] using Type B packages of a special design mounted on legal-weight trucks. To date, there have been three highway accidents involving WIPP shipments. None resulted in the release of radioactivity from the transportation package to the environment.

Worldwide Transportation Experience

There is no centralized database for the worldwide shipment of radioactive materials, nor is there an international mandate to collect such infor-

---

centimeter averaged over a 300 square centimeter sampling area for beta, gamma, and low-toxicity alpha emitters, and one-tenth that limit for other alpha emitters. The International Atomic Energy Agency (IAEA) recently issued a technical document resulting from a coordinated research project aimed at evaluating the adequacy of these limits under current radiological protection and transportation practices. The document notes (IAEA, 2005b, p. 84) that "the studies carried out under [the research project] indicate that the present limits on non-fixed contamination on the surfaces of packages and conveyances are conservative."

[15]Contamination occurs when the package is placed into the spent fuel pool for loading and is contaminated with small amounts of radioactive material present in the pool water. The external surfaces of these packages are decontaminated to remove non-fixed contamination before shipment.

[16]It is interesting to note that while the number of rail accidents exceeded the number of truck accidents during this period, the number of truck shipments was much higher (Table 3.2). Between 1979 and 2004, for example, there were roughly three times more truck shipments than rail shipments.

[17]*http://www.wipp.ws/shipments.htm.* Accessed on April 18, 2005.

TABLE 3.4 Summary of Transportation Accidents Involving Commercial Spent Fuel Packages, 1971–2005[a]

Mode	Date	Location	Description
Truck	December 8, 1971	Tennessee	Package thrown free of trailer and landed in ditch following head-on collision with car. No package damage or release. Driver killed
Truck	February 2, 1978	Illinois	Trailer collapse while crossing railroad tracks. No package damage or release
Truck	August 13, 1978	New Jersey	Trailer collapse while empty package was being loaded. Package not damaged
Truck	December 9, 1983	Indiana-Illinois border	Trailer's fifth wheel failed. No package damage or release
Train	March 29, 1974	North Carolina	Empty package struck by a derailed tank car on adjacent track. Superficial package damage. No release
Train	March 24, 1987	Missouri	Train-auto collision at grade crossing. No package damage or release
Train	January 9, 1988	Illinois	Train carrying empty packages derailed. No damage to packages
Train	December 14, 1995	North Carolina	Railway car carrying empty packages derailed. No damage to packages
Train	September 22, 2005	New York	Railcar carrying an empty spent fuel package derailed in a rail yard. Railcar tipped over. No release

[a]This table lists only accidents involving loaded and empty spent fuel packages; it does not include all of the incidents listed in Table 3.3.

SOURCES: Weiner and Tenn (1999); data from the RMIR compiled by Energy Resources International (December 11, 1997) and DOE correspondence; media reports (for the 2005 accident).

mation. However, since 1980, the International Atomic Energy Agency (IAEA)[18] has been collecting such data at the recommendation of its transportation advisory committee.[19] The agency is developing databases on

[18]The IAEA was established under the United Nations in 1957 as part of the "Atoms for Peace" program to promote the safe, secure, and peaceful uses of nuclear technologies.

[19]The Standing Advisory Group on the Safe Transport of Radioactive Material was the predecessor body to the Transport Safety Standards Committee (TRANSSC). Committee member Clive Young is a past chairman of TRANSSC.

radioactive materials shipments, accidents, and radiation exposure. Member states provide information to these databases on a voluntary basis. The response of member states to requests for information to populate these databases has been mixed. Consequently, the databases are incomplete and contain only a limited representation of available data.

As part of this database development effort, the IAEA launched a literature search and a series of informal contacts with 25 member countries in 2000 to obtain information on worldwide shipments of spent fuel[20] and high-level radioactive waste. A summary of this information was provided in a paper published by IAEA staff (Pope et al., 2001; see Table 3.5). This information is characterized by Pope et al. (2001) as "informal and incomplete" because not all countries responded, and some respondents provided incomplete data or data that were inconsistent with other published sources.

Although the data are incomplete, they nevertheless allow several useful observations to be made about the worldwide spent fuel transportation experience. Spent fuel is being transported within and across the borders of many countries. Worldwide, a major purpose for shipping spent fuel is to reprocess it. Reprocessing facilities have been constructed in France (La Hague) and the United Kingdom (Sellafield) for reprocessing domestic and foreign spent fuel. Belgium, Germany, Italy, Japan, and Spain have shipped spent fuel to these facilities for reprocessing, and some return shipments of high-level waste have been made to some of these countries. Spent fuel from Finland also has been shipped to the USSR or Russia for reprocessing. Most of the other spent fuel shipments being made within or between countries are for the purpose of interim storage.

The total quantity of spent fuel shipped worldwide was estimated in the 2001 IAEA study (see Table 3.5) to be between about 73,000 and 98,000 MTHM. The amount of spent fuel shipped in the United States is small in comparison. The total quantity of spent fuel transported worldwide also exceeds the legislated capacity of Yucca Mountain (70,000 MTHM) and is about twice the planned capacity of Private Fuel Storage (40,000 MTHM). Moreover, a majority of the worldwide spent fuel shipments have been by rail, the preferred mode for shipping to Yucca Mountain and Private Fuel Storage. Shipments to reprocessing plants in France and the United Kingdom have been made for more than 35 years. In contrast, the shipment of the first 70,000 MTHM of spent fuel and high-level waste to a federal repository or interim storage facility would likely take place over periods of a little more than two decades.

Spent fuel rail shipments to France are made using general and dedicated trains, whereas shipments within the United Kingdom are made using

---

[20]The IAEA refers to this fuel as "irradiated nuclear fuel."

TABLE 3.5 Worldwide Spent Fuel Transportation Estimates

Country of Origin	Mass of Spent Fuel Shipped (MTHM)[a]	Number of Packages Shipped	Shipping Modes	Destination
Canada	100	187		
Czech Republic	242	65	Rail	To and from Slovakia
Finland	233	65	Highway and rail	USSR or the Russian Federation
France				
Domestic	11,700	2,600	Mostly rail	La Hague
Other Europe	10,000	2,500	Mostly rail	La Hague
Japan	2,940	660	Sea and highway	La Hague
Germany	>25	66	Highway and rail	Domestic
Hungary	258	72		
Italy	81	52	Highway	Domestic
Japan				
1995–1999	161	50	Sea and land	Domestic
2000–2004 (proj.)	1,700		Sea and land	
Russian Federation	3,500	500		Domestic
Slovakia	239–380	635–700		
Sweden	3,300	1,100	Sea	Domestic
Ukraine	1,300	300		
United Kingdom				
Domestic	20,900–43,200	11,300–28,900	Mostly rail	Sellafield
Other Europe	2,860	1,100	Rail	Sellafield
Japan	4,720	1,420	Sea and rail	Sellafield
United States	2,270	3,020	Highway and rail	Domestic
Approximate Totals	73,000–98,000	24,000–43,000		

[a]Numbers are rounded to three significant figures.

SOURCE: Modified from Pope et al. (2001).

dedicated trains.[21] DOE recently announced that it planned to use dedicated trains when possible for shipments to a U.S. federal repository (see Chapter 5). The shipments in Europe share the rails with other freight and passenger trains. This is similar to current plans for spent fuel shipments in the United States. Moreover, trains carrying spent fuel in France and the United Kingdom are on routes that pass through large cities. This too is likely to be the case for spent fuel and high-level waste transport in the United States because many mainline rail routes pass through large cities. Some of the spent fuel being transported to La Hague and Sellafield is cooled for less than a year before being shipped.[22] In contrast, current practice in the United States is to cool commercial spent fuel for at least five years before shipping it, and some of the spent fuel to be shipped to Yucca Mountain will have been cooled much longer than five years.[23] Road and rail shipping distances to La Hague and Sellafield are generally 1000 kilometers (about 600 miles) or less, compared with the several thousand kilometers for shipping spent fuel from the eastern United States to interim storage or a federal repository in the United States. In the United Kingdom, the average transport package transport distance is about 300 miles (480 kilometers).

Data on accidents and incidents are kept by individual countries, but there is no standardized reporting format. In the United Kingdom, for example, accidents and incidents have been tracked since 1958 in the Radioactive Material Transport Events Database. Since 1989, annual reports of accidents and incidents have been issued. The latest report reviewed by the committee was issued in 2003 (Watson and Jones, 2004).

Since 1958, there have been 786 incidents involving the transport of radioactive materials in the United Kingdom. Approximately 24 percent of these have involved transport of spent fuel. These range from derailments of the conveyance to incidents involving non-fixed radioactive contamination above regulatory limits on the external surfaces of transport packages (see footnote 14). There have been no reported accidents involving spent fuel transport packages that have resulted in the release of radioactive material from the containment system to the environment.

The most recent significant spent fuel transport incidents in Europe occurred in 1997–1998, when inspections by the French nuclear safety

[21]As noted in Chapter 1, dedicated trains are trains that transport only spent fuel or high-level waste and no other cargo.

[22]As discussed in Chapter 5, decay heat production in spent fuel drops rapidly in the first five years after its discharge from a power reactor; see Figure 5.2.

[23]See, however, Section 5.2.4 for a discussion of the acceptance order for commercial spent fuel to be shipped to Yucca Mountain.

regulator (Nuclear Installations Safety Directorate) showed that a high percentage of transport packages and conveyances at one reactor and at the rail terminal at Valognes (the receiving terminal for La Hague) contained non-fixed external contamination in excess of regulatory limits. The cause was eventually traced to inadequate decontamination of transport packages after they were loaded with spent fuel at reactors in France, Germany, and Switzerland. There was a three-year moratorium on spent fuel shipments in these countries while the incidents were investigated and new procedures were put into place to eliminate the contamination problems. A subsequent investigation by regulatory authorities concluded that no workers or members of the public had received radiation doses exceeding the relevant regulatory limits as a consequence of these incidents (HSK, 1998).

Two major conclusions can be drawn from the historical record of worldwide transport of spent fuel. The first is that there have been no recorded instances of which the committee is aware of any releases of radioactive material exceeding regulatory limits from any transport package in Western Europe, Japan, or the United States. There are, however, well-documented instances of exposures to radioactivity from inadequate decontamination of the external surfaces of transport packages after they are loaded with spent fuel. However, these releases have been small, and the committee is not aware of any documented instances in which exposures to workers or the public exceeded regulatory limits.

3.1.2 Quantitative Analyses

The AEC and its successor agencies have conducted several assessments of radioactive materials transport risk in the United States. These assessments have used numerical models, informed by expert judgment, to estimate the performance of transportation packages under normal and accident conditions. These studies are summarized in Table 3.6 and are described in this section. Two of these assessments have produced quantitative estimates of risks to workers and members of the public from such transport activities.

The committee was not charged to perform an in-depth technical review of these studies to assess the technical quality of the assumptions, models, results, and uncertainties. The committee does, however, provide comments where appropriate on the assumptions used in these studies and their impact on the resulting consequence estimates. All of these studies have undergone some level of technical review by the issuing organizations, but there are differences in assumptions and approaches among the studies that must be considered when comparing results.

TABLE 3.6 Description of the Analytical Assessments of Transportation Risk Described in This Chapter

Study Reference	Objective	Summary of Results	
WASH-1238	Estimate doses to workers and public from spent fuel transport	Risks due to radiological effects from transportation accidents are small	AEC (1972)
Transportation EIS	Estimate radiological effects from land, water, and air transport of radioactive materials	Impacts of normal transport and accidents are sufficiently small to allow continued shipments of radioactive materials by all modes	USNRC (1977)
Modal study	Improve understanding of spent fuel package performance under severe accident conditions	Risks from severe accidents involving spent fuel were lower by at least a factor of three than estimated in the transportation EIS	Fischer et al. (1987)
Reexamination study	Update the modal study using improved models and data	Risks from severe accidents are comparable to or lower than modal study estimates	Sprung et al. (2000)
Final Yucca Mountain EIS	Estimate spent fuel and high-level waste shipping risks for transportation of spent fuel and high-level waste to Yucca Mountain, Nevada	Expected transportation impacts are small given the size of the transport program	DOE (2002a)

WASH-1238 Study (1972)

The first analytical study of the health effects of spent fuel transportation in the United States was undertaken by the AEC in 1972 (AEC, 1972). This study is known as the "WASH-1238" study after its report identification number. This study estimated doses to workers and the general public from nuclear fuel and solid radioactive waste transport both under normal transport conditions and for severe accidents. The study's main conclusion was (AEC, 1972, p. 2) that "[w]hen both the probability of occurrence and extent of the consequences are taken into account, the risk to the environment due to radiological effects from transportation accidents is small."

The study has limited applicability to modern-day spent fuel transport programs: It examined highway transport of spent fuel along routes with very different population densities than present-day routes using transportation packages that do meet current regulatory requirements.

Transportation Environmental Impact Statement (1977)

A transportation Environmental Impact Statement (EIS) was undertaken by the USNRC to evaluate the effectiveness of its regulations for the transport of radioactive materials by air and other modes (USNRC, 1977). This EIS provided a more complete analysis of the radiological consequences for land, water, and air transport of radioactive materials than WASH-1238 and has become the "baseline" analysis for assessing radioactive materials transportation risk in the United States.

The 1977 transportation EIS characterized environmental impacts in terms of fatalities, expressed as an annual probability of occurrence for two types of transport: incident-free transport, where the main health impact is expected to be cancer fatalities due to exposure of workers and the general public to small doses of radiation from the shipping containers; and accidents that produce either conventional traffic fatalities or, for more severe conditions, latent cancer fatalities resulting from the release of radioactive materials from a damaged transport package.

Sandia National Laboratories performed this study. A computer code (RADTRAN 1; Sidebar 3.4) was developed to estimate radiation doses and latent cancer risks of transporting (including temporary storage and modal transfers) 25 different radioactive materials by plane, truck, train, ship, or barge. One of the 25 categories of materials considered was spent power reactor fuel. The study estimated risks to workers involved in shipping the materials and to members of the general public who lived near or traveled on the transportation routes. Latent cancers during incident-free transport were assumed to arise solely from external radiation doses from the transport packages. Latent cancers during accidents were assumed to arise from both internal and external exposure pathways.

The study estimated risks for transport of spent fuel on a generic highway route and a generic rail route. The study provided only a limited consideration of accidents in highly urbanized areas—one analysis was carried out for an accident in New York City. Additional analyses of releases in highly urbanized areas were undertaken in a subsequent series of studies known as the "urban studies" (DuCharme et al., 1978; Finley et al., 1980; Sandoval et al., 1983; Sandoval, 1987). Public circulation of all but one of the unclassified versions of these reports (Sandoval et al., 1983) was restricted after the September 11, 2001, terrorist attacks; consequently, these reports were not available for review by the committee.

SIDEBAR 3.4 RADTRAN

RADTRAN is a computer code that can be used to estimate radiological exposures and consequences under both incident-free and accident conditions. It can provide estimates of collective dose as well as doses to maximally exposed individuals. The code was first developed by Sandia National Laboratories for use in the 1977 transportation EIS (USNRC, 1977). The code has been expanded and refined several times and is now in version 5 (RADTRAN 5). This code, which is written in FORTRAN 77, is available from Sandia National Laboratories. It has become a worldwide standard for assessing incident-free radiological transportation risks.

The program allows risks to be estimated for seven different transport modes (two highway modes, rail, barge, ship, and two air modes) using a series of models that account for the following:

- Sources and isotopic contents of transport packages,
- Transportation routes and stops,
- Population distributions of workers, residents who live along transportation routes, and vehicle occupants on transportation routes,
- Number and severity of accidents,
- Releases of radioactivity from packages in accidents,
- Dispersion of released radioactivity in the environment,
- Radiation exposure pathways for inhalation, ingestion, resuspension, cloudshine, and ground-shine exposures,
- Radiological fatalities using a dose-response model,
- Nonradiological fatalities, including vehicular fatalities and fatalities from vehicle emissions.

Like all computer codes, the validity of the results is only as good as the information used as input to the various models and embedded assumptions in the models themselves. Of particular importance in this regard is the input information for population densities, accident numbers and severities, radiation releases, and dispersion in the environment.

INTERTRAN 2 is a development of RADTRAN for international application. It was developed by the Swedish Nuclear Power Inspectorate (SKI) for the IAEA.

The 1977 transportation EIS concluded that the average radiation doses to at-risk populations from radioactive materials transportation were a small fraction of the limits recommended for the general public from all anthropogenic sources of radiation other than medical sources. The USNRC determined that the "environmental impacts of normal transportation of radioactive material and the risks attendant to accidents involving radioactive material shipments are sufficiently small to allow continued shipments by all modes" (USNRC, 1977, p. viii). The Commission concluded that "present regulations are adequate to protect the public against unreasonable risk from the transport of radioactive materials" (46 FR 21629, April 13, 1981).

The applicability of the 1977 transportation EIS for estimating risks for current and potential future transport of spent nuclear fuel is limited, owing mainly to the simple models and limited data used in the analysis. Since the issuance of this EIS, the USNRC has sponsored two additional studies to improve its understanding of the risks from commercial spent nuclear fuel transportation by road and rail. Those studies are discussed in the following sections.

Modal Study (1987)

The modal study (Fischer et al., 1987) was undertaken to improve understanding of spent fuel shipping package performance under severe accident conditions. The study examined the response of generic truck and rail spent fuel packages to both severe impact and fire conditions. As part of the analysis, historical data on real accidents were compiled from government and private databases to develop accident scenarios and their probabilities (see Figure 2.3). The scenarios are displayed as "event trees." These trees provide a graphic illustration of the sequence of events leading to an accident along with the probability of each event. Each branch of the tree depicts a sequence of events that leads to the accident outcome depicted at the end of the branch. The probability of an accident is equal to the product of the probabilities of each segment along the branch.

Historical data also were used to estimate the magnitudes of impacts and fire loads associated with each scenario. These were calculated from records of accident speeds and angles, the hardness of the objects involved in the impacts, and the frequency and duration of accident-associated fires. A total of 31 truck accident scenarios and 24 train accident scenarios were developed for the analysis. The 1977 transportation EIS, in contrast, defined only eight accident severity categories based on expert judgment.

Analyses were carried out to assess the effects of these accident scenarios on generic rail package and truck packages. The design of these packages was based on analyses of the features of rail packages and truck packages in service at the time of the study and accounted for typical built-in safety margins (see Sidebar 2.4).

The analysis involved a two-stage screening process. Phase 1 screening used dynamic linear and standard transient heat-transfer models to identify those accident scenarios in which the impact and fire conditions would not exceed the regulatory requirements in 10 CFR Part 71 (see Sidebar 2.1). For these scenarios, any releases from the packages are presumed to be below regulatory limits. Approximately 99.4 percent of truck accidents and

99.7 percent of the rail accident scenarios analyzed fell into this category (Fischer et al., 1987, p. 9-2).

Phase 2 screening involved more sophisticated analyses of package responses and radiological releases for those accident scenarios that exceeded the 10 CFR Part 71 limits. This screening employed dynamic nonlinear models to estimate package deformation and transient thermal models that took into account the phase change that accompanied the melting of lead shielding at high temperatures. The analyses assumed that the packages contained five-year-cooled pressurized water reactor fuel having a burn-up of 33,000 megawatt days per metric ton, typical of spent fuel for that day.[24] The analyses considered breaches of the spent fuel cladding due to both impact and thermal creep (see Sidebar 2.5). The radiological effects considered included releases of radioactive materials from the package as well as increased radiation doses from a loss of package shielding.

Based on the Phase 2 screening, the authors concluded that roughly 0.39 percent of severe accidents involving truck or rail packages would result in radioactive material releases or doses that approached or slightly exceeded the regulatory limits in 10 CFR Part 71. The report concluded that fewer than 0.001 percent of the truck and 0.012 percent of the rail accident scenarios could actually produce hazards that would likely exceed regulatory limits.[25] No attempt was made to model the radioactive releases for the most severe accidents, because these conditions pushed the capabilities of modeling programs. Instead, estimates of the releases for these very severe accidents were simply extrapolated from the release behavior during less severe accidents (see Fischer et al., 1987, Table 8.3).

The 1987 study did not include "consequence calculations" to estimate risks to workers and the public from exposure to radiation as was done in the 1977 transportation EIS. However, a comparison of the frequencies and magnitudes of radiological releases from the two studies led the authors of the 1987 study to conclude that their risk estimates for both truck and rail were at least three times lower that those documented in the 1977 transportation EIS (Fischer et al., 1987, p. 9-11). There are several possible reasons

[24]Present-day fuel burn-ups are typically between 50,000 and 60,000 megawatt-days per metric ton. High-burn-up fuel produces more decay heat than low-burn-up fuel of the same age, which has implications for internal package heating in an accident involving fires. However, the generic packages used in the modal study analysis probably would not be suitable for transporting high-burn-up fuel.

[25]To place these numbers in perspective, consider that the planned transportation program to a federal repository at Yucca Mountain by the "mostly rail" scenario (described later in this chapter) would, according to DOE, involve about 9600 rail shipments and 1100 truck shipments (see Table 3.8). Multiplying these shipment numbers by the number of modal study scenarios that result in radiation releases above regulatory limits results in about 0.01 truck release and 1.1 rail releases during the life of the repository transportation program.

for this difference, including reductions in accident rates in the decade between this study and the 1977 transportation EIS.

Resnikoff (1994, unpublished paper) criticized several aspects of the modal study. These included the transport package designs used in the analyses; the failure to model the end closures of the packages; and the methods and data used to estimate impact and fire conditions. He provides an analysis of 38 severe accidents that he claims shows that the 1977 transportation EIS and 1987 modal study underestimate severe impact and fire conditions and probabilities. He provides his own estimates of releases based on a reanalysis using the RADTRAN 4 code. He concludes that these releases are many times higher than estimated by the 1977 transportation EIS.[26] Resnikoff's analysis prompted Sandia researchers to perform additional historical accident reconstructions (described in Chapter 2), for which they concluded that spent fuel packages would likely maintain their containment integrity in all but possibly the most severe accidents involving long-duration fires.

Reexamination of Spent Fuel Shipment Risk Estimates (2000)

In 2000, Sandia National Laboratories published a reexamination of spent fuel shipment risk estimates (Sprung et al., 2000) using updated models and data. Since this is the most current generic study and has been used as the basis for a more recent analysis of transportation to a Yucca Mountain repository, as discussed later in this chapter, it is described in some detail in this section.

The 2000 reexamination study utilized an updated version of RADTRAN (RADTRAN 5; see Sidebar 3.4) to estimate population doses to workers and the general public during both incident-free transport and severe accidents involving releases of radioactivity or loss of shielding. Estimates were obtained for five potential exposure pathways (direct inhalation, resuspension inhalation, ingestion, cloud shine, and ground shine; see glossary in Appendix D) versus the single direct inhalation pathway considered in the 1977 transportation EIS.

The 2000 study considered transport of spent fuel along 741 truck and 741 rail routes: 249 truck routes and 249 rail routes developed in a previous routing study (Cashwell et al., 1986), as well as 492 truck and 492 rail routes developed specifically for this study. The latter routes connect the 79 spent fuel storage sites in existence when the study was initiated with six hypothetical interim storage sites located in different regions of the United States (474 routes), and those interim storage sites to three hypothetical

[26]The committee did not undertake a detailed review of Resnikoff's claims.

permanent repository sites (18 routes). One of the permanent repository sites considered was Yucca Mountain.

To make the analyses tractable, these routes were sampled to obtain a smaller set of routes having representative characteristics. The samples were generated as follows. First, distributions were constructed of route lengths; fractions of those lengths that contain urban, suburban, and rural population densities; and the actual population densities for each of those length fractions for the 741 truck and 741 rail routes. Next, these distributions were sampled using Monte Carlo methods to generate two sets of "representative" routes: 200 highway routes and 200 rail routes. These representative routes were used in the RADTRAN calculations.

The 2000 reexamination study used slightly modified versions of the accident event trees from the 1987 modal study (Figure 3.1). An additional branch was added to the rail event tree to account for accidents involving fires that did not result from either collisions or derailments. Additionally, new estimates of route wayside hardness were developed based on surveys of selected transportation routes and Department of Agriculture data on near-surface locations of coherent rock formations. This information was used to estimate how "yielding" these surfaces would be in accidents involving impacts of truck and rail transportation packages.

Truck and train accident rates used in the study were estimated separately from state-level data for the 48 contiguous U.S. states. These data represent heavy-truck accidents on interstate highways and train accidents on mainline rail routes. The heavy-truck data were detailed enough to support the development of accident rate distributions for suburban and rural routes as well as a single average accident rate for urban routes. The rail data were only detailed enough to support the development of a single accident rate distribution by combining all of the state-level data.

The study modeled the performance of four generic package types whose physical specifications were based on a review of data on packages in use at the time of the study: steel-lead and steel-depleted uranium truck packages and steel-lead and monolithic steel rail packages. The packages were assumed to contain total activities equivalent to three-year-cooled pressurized water reactor (PWR) fuel with a burn-up of 60,000 megawatt-days per metric ton, or three-year-cooled boiling water reactor (BWR) fuel with a burn-up of 50,000 megawatt-days per metric ton. These estimates are characterized in the study as conservative.[27] The rail package was

[27]The study characterizes the total activities used in the RADTRAN calculations as being conservative by about a factor of 4. While these high burn-up levels are now achieved routinely in U.S. power reactors, most stored spent power reactor fuel has much lower burn-ups. Moreover, spent fuel is generally expected to be stored for at least five years before being moved from pools to dry casks for storage or transport, and some of it will have been stored for several decades.

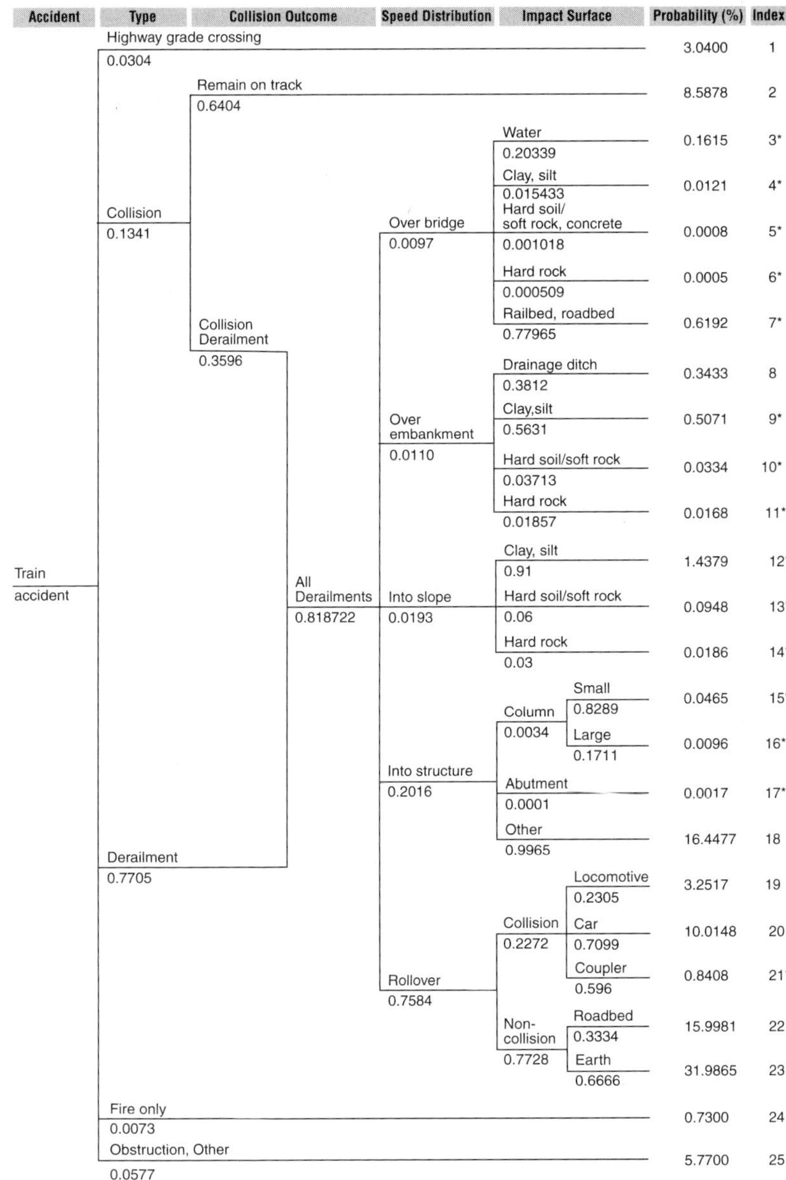

FIGURE 3.1 Accident event trees for rail accidents from the 2000 reexamination study, slightly modified from the modal study (see Figure 2.3). The numbers shown at each branch are probabilities for the accident branch based on an analysis of historical data. The accident scenarios that are marked with an asterisk were determined to produce consequences that would approach or exceed regulatory limits. SOURCE. Sprung et al. (2000, Figure 7.4, p. 7-12).

assumed to contain 24 PWR or 52 BWR assemblies. The truck package was assumed to contain between 1 and 3 PWR or 2 and 7 BWR assemblies.

External package surface dose rates were also important inputs to the model. These were estimated for commercial spent fuel discharged from reactors in the United States. These dose rates depend on fuel burn-up, time since discharge (or cooling time), and package shielding. For conservatism, the dose rate distribution estimates were rescaled upward so that their upper limits were equal to the regulatory dose limit of 0.1 mSv (10 mrem) at 2 meters (about 6.5 feet) from the package surface.

Analyses of package behavior in severe accidents were carried out using a standard finite element code (PRONTO 3D). This computer code is commonly used to model high strain rates in nonlinear materials. Although material failure is not included explicitly in this code, such failure can be estimated based on the calculated deformation of package components such as lid bolts. This code was used to estimate package deformations resulting from impacts onto unyielding surfaces (see Sidebar 2.2) at speeds of 30, 60, 90, and 120 miles per hour (48, 96, 144, and 192 kilometers per hour). To make the results conservative, the impact limiters were assumed to be attached to the package but fully crushed before impact.

The analyses indicate that even at the highest of these impact speeds, strains were well below the levels needed to fail or penetrate the package body or lid. The analyses also indicate that the truck package seals would not fail in any impact orientation at any impact speed. Nevertheless, it was arbitrarily assumed that seal leaks having a cross section of 1 square millimeter would result from impacts of these packages onto unyielding surfaces at speeds of 120 miles per hour (193 kilometers per hour). The analyses also suggest that for rail packages, some seal leakage could occur for some impact orientations at impact speeds onto unyielding surfaces as low as 60 miles per hour (97 kilometers per hour) and possibly at all orientations at speeds of 120 miles per hour (193 kilometers per hour).

Surfaces along transportation routes (e.g., soils, concrete structures) are likely to be partially yielding and will absorb some impact energy in severe accidents. Consequently, the finite element results for impacts onto unyielding surfaces must be adjusted to account for this energy loss. This was done by calculating the impact speeds onto three types of yielding surfaces (soil, concrete, hard rock) that would result in the same peak contact forces on the package as the equivalent impact onto an unyielding surface. This calculation took into account the energy-absorbing effects of the package impact limiters.

If the package seal was determined to fail in an accident, the release of radioactive material (noble gases, volatiles, and particulates), collectively referred to as the *accident source term*, was calculated. These calculations took into account several phenomena: rod cladding failures; radionuclide inventory releases from the failed rods; partial deposition of radionuclides on internal surfaces in the package; pressurization of the package interior from failed rods or package heating; and leakage of radionuclides through package seal failures driven by package depressurization (see Sidebar 2.5). The estimates of particulate releases and behavior were based on actual experiments performed on spent fuel rods subjected to burst failure.

For accidents involving fires, package heating calculations were carried out using a commercial code (PATRAN/PThermal), which can be used to model heat convection, conduction, and radiation transport processes. Calculations were made for each of the four generic packages for a fully engulfing, optically dense hydrocarbon fire that would heat the package sufficiently to cause the pressurized rods to fail by burst rupture. The calculations included the effects of internal package heating from radioactive decay of three-year-cooled spent fuel.[28] Package seal leakage and accident source terms were estimated in a manner similar to that described previously for severe accidents.

Several sets of RADTRAN calculations were performed in this study:

- Calculations for the 200 truck and 200 rail routes obtained by Monte Carlo sampling as described previously;
- Calculations for 5 truck and 5 rail routes selected from the 1977 transportation EIS or from the 474 routes that connect spent fuel storage sites to the locations of the hypothetical interim storage facilities considered in this study—the latter calculations were carried out to demonstrate that results for real routes would fall within the envelope of results for the representative 200 rail and 200 truck routes;
- Calculations comparing the consequences and risks for RADTRAN 1 with RADTRAN 5 for a single transportation route; and
- Calculations comparing the risks and consequences using the package inventories and assumptions about radionuclide releases developed for the 1977 transportation EIS, 1987 modal study, and this study.

The study provided RADTRAN calculations for both incident-free and severe accident scenarios. The conservative assumptions used in these analyses (e.g., the packages contain three-year-old spent fuel with high burn-ups; the external package dose rate distribution estimates were rescaled upward so that their upper limits were equal to the regulatory dose limits; package

[28]The use of three-year-cooled spent fuel in the calculations yielded external steady-state package surface temperatures as high as 194°C (Sprung et al., 2000, Table 6.4). This would have exceeded the regulatory limits in 10 CFR 71.43(g), which restricts external surface temperatures to 85°C for exclusive-use shipments. The authors described these temperatures as "conservative" for the purposes of the analysis.

deformation estimates were made using fully crushed impact limiters; truck package seals were assumed to have small leaks at high-impact speeds even though the analyses did not indicate seal failure) are reasonable for producing bounding estimates of accident consequences or radiological exposures. However, the 200 rail and 200 truck routes selected through Monte Carlo techniques for use in the analyses were based on realistic, not bounding, characteristics. Consequently, the overall results of the Sandia analyses are likely to be neither realistic nor bounding and probably overestimate the transport risks.[29]

One result is discussed below for the sake of illustration: Population risk estimates for severe accidents involving rail transport of PWR spent fuel in a steel-lead rail package. This example was selected because PWR fuel is the most common fuel used in U.S. power reactors and because train transport is the preferred mode for shipping to a federal repository and to Private Fuel Storage (see Chapter 2).

The population risk estimates for each of the 200 route calculations for the rail package are displayed as *complementary cumulative distribution functions* (CCDFs), which are also sometimes referred to as "risk curves" (Figure 3.2). The horizontal (x-axis) is referred to as the accident consequence value. Simply put, this is the collective dose that would be received by the population defined in the model as a result of the assumed accident scenario calculated by the model. The vertical (y-axis) is the probability that the collective dose will exceed that accident value on a per-shipment basis. This probability is given in dimensionless units ranging from 0 to 1. The right vertical axis is the expected number of years between accidents exceeding the accident consequence value when 100 shipments per year are assumed.

The total set of 200 CCDFs would produce a plot like that shown in the inset in Figure 3.2. To improve the visual utility of such plots, *compound CCDFs* can be constructed that represent certain statistical properties of the 200 individual CCDFs. For the compound CCDFs shown in Figure 3.2,

- The mean compound CCDF is computed by averaging the 200 CCDFs;
- The 50th percentile compound CCDF represents the median value of the 200 CCDFs; and
- The 95th and 5th percentile compound CCDFs represent the 190th highest and 10th lowest of the 200 CCDFs.

[29]The committee hedges this statement with the word "probably" because there are a great many other uncertainties in the input data to the calculations, especially with respect to local accident rates and route wayside conditions, that could affect the realism of the calculations.

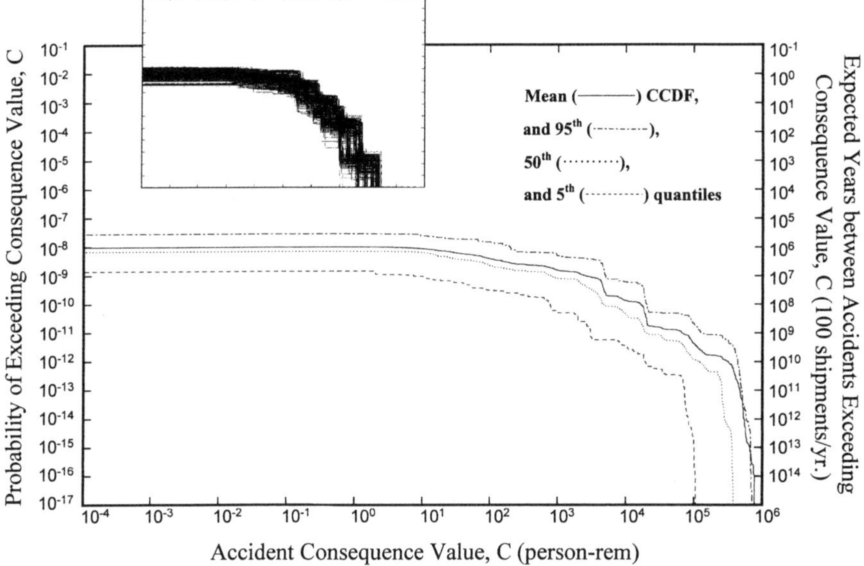

FIGURE 3.2 Compound CCDF (complementary cumulative distribution function) for train transport of PWR fuel using steel/lead packages. The left vertical axis is the probability expressed on a per-shipment basis; the right vertical axis represents the expected years between accidents assuming 100 shipments per year. Inset: Plot showing individual CCDFs for 200 routing calculations of the type used to derive the compound CCDFs shown in the main figure. SOURCE: Sprung et al. (2000, Figure 8.7).

Two general observations can be made from this plot. First, the estimated risk of exposure from an accident that is severe enough to compromise fuel rod and package seal integrity is very small on a per-shipment basis. For example, the expected (mean) probability of receiving a population dose of 1 person-rem (i.e., 10^0 person-rem in the figure) is about 1 in 100 million (10^{-8}) per package shipment. Assuming a shipment frequency of 100 packages per year, the expected mean time between such accidents is estimated to be about 1 million years.

Second, the spread of probability (left vertical axis) values of the compound CCDFs at a given consequence value (horizontal axis), which is represented by the vertical distance between the 5th and 95th percentile compound CCDFs, reflects the sensitivity of the calculations to route characteristics (e.g., route length, wayside hardness, traffic conditions). The variability is about an order of magnitude at low consequence values and increases to more than 5 orders of magnitude at high consequence values.

The collective dose risk[30] for all of the RADTRAN 5 calculations displayed in the figure is equal to 9.4×10^{-6} person-rem on a per-shipment basis. If an accident has a 1 in one million probability of occurrence (i.e., 1×10^{-6}), the mean collective dose received by the population (the size of which must be specified) would be about 9 rem per accident. The mean collective dose risks are most useful as a tool to compare different accident risks as shown in Table 3.7.

The population (collective) dose risks for all of the packages modeled in the 2000 reexamination study are shown in Table 3.7. The authors of this study also calculated the collective dose risks using the source terms from the 1977 transportation EIS and the1987 modal study. Two observations are particularly noteworthy. First, estimated population dose risks for the reexamination study are on the order of 10^{-6} to 10^{-7} person-rem per shipment for all of the packages and scenarios examined in the study. Given the uncertainties in the parameters used in the calculations, these values are essentially identical. Second, estimated population dose risks from the 2000 reexamination study are about three orders of magnitude lower than estimates calculated using the source terms in the1977 transportation EIS and the 1987 modal study. These differences are probably significant and may reflect a lack of realism in some of the assumptions used in the earlier analyses, especially with respect to package release behavior in a severe accident. As noted previously, this behavior was based largely on expert judgment in the 1977 transportation EIS. Release behavior was modeled explicitly in the 2000 reexamination study.

Final Yucca Mountain EIS (2002)

DOE has prepared an EIS (DOE, 2002a) as part of a larger effort to site and construct a federal repository for spent fuel and high-level waste at Yucca Mountain, Nevada. This EIS provides estimates of spent fuel and high-level waste transportation risks. The EIS considers two scenarios for transporting spent nuclear fuel and high-level radioactive waste from 72 commercial and 5 defense sites to the proposed repository at Yucca Mountain, Nevada: a *mostly truck scenario* that would involve transporting most of the spent fuel and high-level waste by legal-weight truck across the

[30]The mean collective dose risk is the collective dose that is received by the population from an accident times the probability of occurrence of that accident. This follows from the general risk equation, risk = probability x consequences, where consequences = collective dose. The collective dose is conditional because it will be received only if the accident occurs. In that case, the collective dose can be calculated by dividing the risk by the probability that the accident will occur, or consequences = risk ÷ probability. The mean collective dose risk is most useful as a comparative tool.

TABLE 3.7 Population (Collective) Dose Risks for Severe Accidents from the 2000 Reexamination Study

	Population Dose Risk (person-rem)[a]			
	Truck		Rail	
Package Type	PWR	BWR	PWR	BWR
2000 reexamination study				
Steel-lead	8.0×10^{-7}	3.3×10^{-7}	9.4×10^{-6}	9.2×10^{-6}
Steel-DU	2.3×10^{-6}	1.1×10^{-6}		
Monolithic steel			2.0×10^{-6}	1.5×10^{-6}
Calculated using 1977 transportation EIS and 1987 modal study sources terms	1.3×10^{-2} to 1.3×10^{-4}		1.9×10^{-2} to 4.9×10^{-4}	

NOTE: Numbers are rounded to two significant figures; BWR = boiling water reactor; DU = depleted uranium; PWR = pressurized water reactor.

[a]Expected values per shipment.

SOURCE: Sprung et al. (2000, Tables 8.4, 8.5, E.2).

nation's highways, and alternatively, a *mostly rail scenario* that would involve transporting most of this material using commercial railroads and a railroad spur to be constructed in Nevada. The numbers of truck and rail shipments under each scenario are shown in Table 3.8.

DOE considered the impacts of two repository scenarios in this EIS. The first assumes that Yucca Mountain would receive the legislatively mandated limit of 70,000 metric tons of spent fuel and high-level waste over a period of 24 years. The second assumes that the repository would operate for 38 years and would receive between 119,000 and 125,000 metric tons of spent fuel, high-level radioactive waste, and other special requirements waste (e.g., greater-than-class-C waste).

DOE's analysis of transportation impacts examined several classes of hazards for workers and the general public, two of which are of particular interest in this discussion:

1. Incident-free transportation in which populations in proximity to transportation routes would receive small radiation doses during the routine transport of spent fuel and high-level waste. These doses would be the result of radiation shine from the transport packages.
2. Accidents that involve a loss of transport package shielding, which could result in more severe radiological exposures. The EIS analyzed a

TABLE 3.8 Transportation Scenarios, Collective Doses, and Radiological Impacts from the 2002 Yucca Mountain Repository EIS for Routine Transport and Accidents

	Mostly Truck Scenario	Mostly Rail Scenario	Comment
Scenario Definitions			
Operational period (years)	24		
Repository capacity (MTHM)	70,000		
Number of legal-weight truck shipments			
Commercial SNF	41,000	1100	
DOE SNF	3500		
DOE HLW	8300		
Number of rail shipments			
Commercial SNF	300	7200	
DOE SNF		770	
DOE HLW		1700	
Radiological Impacts, Routine Transport[a]			
Worker collective dose (person-rem)	29,000	7900–8800	Collective dose received over 24 years assuming specified crew sizes for loading, transport, and inspections; total numbers of workers not specified
Dose to maximally exposed worker (rem)	48	48	Assumes that worker receives the DOE occupational administrative dose limit of 2 rem per year for 24 years[b]
Estimated number of worker latent cancer fatalities	12	3.2–3.5	

continues

TABLE 3.8 Continued

	Mostly Truck Scenario	Mostly Rail Scenario	Comment
Dose to maximally exposed member of the public (rem)	2.4	0.29	Maximally exposed person for mostly truck is a service station worker; mitigation would be required to keep doses below 0.1 mSv (100 mrem) per year. Maximally exposed person for mostly rail is resident near a rail stop
Public collective dose (person-rem)	5000	1200–1600	Distributed across 10.4 million people for mostly truck scenario and 16.4 million people for mostly rail scenario over 24 years
Estimated number of public latent cancer fatalities	2.5	0.6–0.8	
Total estimated number of latent cancer fatalities	14	3.8–4.3	
Radiological Impacts, Maximally Reasonably Foreseeable Accident (MRFA)[c]			
Accident scenario	Long-duration fire that leads to breach of a package and dispersal of a portion of its contents	Long-duration fire that leads to breach of a package and dispersal of a portion of its contents	
Annual probability that the accident will occur	2.3 in 10 million	2.8 in 10 million	During each year of the 24-year shipping campaign
Dose to maximally exposed individual assuming the accident does occur (rem)	3	29	Maximally exposed individuals are located downwind of the package and receive dose from a cooling plume of radioactive particles released from the package

continues

TABLE 3.8 Continued

	Mostly Truck Scenario	Mostly Rail Scenario	Comment
Collective dose assuming the accident does occur (person-rem)	1100	9900	Analysis assumes that the accident occurs in an urban area, and that populations up to 50 miles (80 kilometers) from the release point could receive a dose. Population densities used in the calculations were based on 1990 census data extrapolated to 2035 for 21 large urban centers in the United States. Accident was assumed to occur at the center of the population zone
Total number of latent cancer fatalities assuming accident does occur	0.6	5	Calculated by multiplying the collective dose by the nominal probability coefficient
Annual collective dose risk (rem)	2.5×10^{-4}	2.8×10^{-3}	See footnote 30
Nonradiological Impacts			
Total fatalities from vehicular collisions, industrial accidents, and air emissions	6.8	3.1–4.2	

NOTE: Numbers are rounded to two significant figures.

[a]The dose estimates shown in this section of the table have a probability of occurrence of 1; that is, it is certain that these doses would be received by workers and members of the public if the Yucca Mountain transportation program were carried out as described in Appendix J of the final Yucca Mountain EIS.

[b]The final Yucca Mountain EIS also notes that if a lower administrative dose limit is imposed on transportation workers in the future, maximally exposed worker doses would be correspondingly lower.

[c]Dose estimates shown in this section of the table are conditional upon the actual occurrence of the accident, which has a very low probability of occurrence.

SOURCES: DOE (2002a, Table 6-1, Table J-1, Table J-16, table on p. S-69, table on p. S-80).

"maximum reasonably foreseeable accident"[31] to provide some perspective on the largest expected transportation impacts on populations that live along potential transportation routes. The EIS analyzed the consequences of accidents that are expected to occur with a frequency greater than 10^{-7} (i.e., with a likelihood greater than 1 in 10 million times per year).

The EIS estimated the consequences of these hazards to several groups of individuals: workers involved in loading, transporting, inspecting, and escorting the shipments; members of the public in vehicles that share transportation routes with the shipments; members of the public who live in proximity to transportation routes; and members of the public who are exposed while shipments are stopped en route to Yucca Mountain.

Appendix J of the final Yucca Mountain EIS provides a detailed discussion of the models, data, and assumptions that were used to produce these estimates. In developing these analyses, DOE used the RADTRAN 5 code developed by Sandia National Laboratories (see Sidebar 3.4) for estimating collective radiological doses under both incident-free and accident conditions. DOE relied heavily on the accident scenarios and the transport package release mechanisms developed in the 2000 reexamination study. The RISKIND[32] computer code was used to calculate radiological doses to

[31]According to DOE, maximally reasonably foreseeable accidents are characterized by extremes of mechanical and thermal forces, and other conditions not specified, that lead to the "highest reasonably foreseeable consequences" (DOE, 2002a, p. 6-45). The thermomechanical forces in these accidents would exceed regulatory limits and would be applied to a package in such a way as to cause the greatest damage and would lead to the release of radioactive materials. DOE defines any accident that has the chance of occurring more than 1 in 10 million times per year as being reasonably foreseeable. The Final Yucca Mountain EIS analyses of maximally reasonably foreseeable accidents were based on an examination of the accident scenarios presented in the reexamination study (Sprung et al., 2000). The analyses are described in Appendix J of the EIS. The scenario determined to be most severe for both rail and truck packages is a long-duration fire.

[32]The RISKIND code was developed in 1993 by Argonne National Laboratory to estimate local, scenario-specific radiological doses to maximally exposed individuals. This code performs similar calculations for incident-free exposures as the RADTRAN code. However, the two codes use different mathematical representations for external dose rate as a function of the distance between the source and receptor (Steinman and Kearfoot, 2000). Steinman et al. (2002) compared the estimates from these models against experimental measurements on moving conveyances containing radioactive materials. They found that both the RADTRAN and the RISKIND models predict doses to within an order of magnitude of the experimentally measured values. The RISKIND model estimates agreed more closely with the measured values at short distances (within a few meters) from the package, whereas RADTRAN provided better agreement at distances that the authors characterized as more typical of residential populations alongside of roads.

maximally exposed individuals for incident-free transportation and to populations and maximally exposed individuals for maximally reasonably foreseeable accident conditions.

To make incident-free radiological impact assessments, DOE assumed that each transportation package would have the maximum external dose rate allowed under DOT transport regulation 49 CFR 173.441(b) (as well as USNRC regulation 10 CFR 71.47(b)(3)): 0.1 mSv per hour (10 mrem per hour) at 2 meters (6.5 feet) from the lateral surfaces of the transport vehicles. This is a conservative assumption when the packages contain aged spent fuel.[33]

Estimates of the number of shipments were based on information on current inventories of spent fuel at reactor sites as well as projections of future inventories based on industry trends. The analysis took into account factors such as package handling capabilities[34] at each site and package capacities to meet heat generation and criticality requirements. In some cases, these requirements would necessitate the shipment of partial packages. The analysis used 31 shipping package configurations: 9 for legal-weight truck and 22 for rail.

Highway routing selections were made using actual highway data and DOT rules for Highway Route-Controlled Quantities of Radioactive Materials in 49 CFR 397.101. Population densities within 800 meters (0.5 mile) of the routes were used to calculate incident-free doses.[35] These densities used data derived from the 1990 and 2000 census data and were extrapolated to the year 2035 based on Bureau of the Census forecasts.

Rail routing selections were made using rules based on the historical routing practices of U.S. railroads from a database of 94 rail networks representing current railroad conditions. Rail routes were determined by minimizing shipping "impedance," which is accomplished by reducing travel distances and the number of railroad companies involved and by using main line (i.e., generally better maintained) tracks. Population densities within 800 meters (0.5 mile) of the routes were used to calculate incident-free doses. These densities used data derived in the same manner as for highway shipments. Route selections were made for all but six sites that do not have the capacity to load or handle rail packages.

---

[33]However, as discussed in Chapter 5, the first fuel shipped to Yucca Mountain might not be thermally and radiologically cool.

[34]For the mostly rail scenario, the analysis assumed that sites that had insufficient crane capacity to handle rail packages would be upgraded after the plant shut down.

[35]Calculating doses out to 800 meters is a very conservative approach. If the dose rate at 2 meters (6.5 feet) is 0.1 mSv per hour (10 mrem per hour), then the dose rate at 800 meters (0.5 mile) is 6.25×10^{-7} mSv per hour (6.25×10^{-5} mrem per hour). This is a negligible exposure and becomes trivial when the conveyance is moving.

For the mostly truck scenario, DOE assumed that all shipments would be made by legal-weight truck except for naval spent fuel, which would be shipped by rail. For the mostly rail scenario, DOE assumed that all sites would ship by rail except for the six commercial sites that do not have the capability to load rail packages. DOE assumed that these sites would ship by legal-weight truck until the sites shut down. They would then be upgraded to load rail packages and would ship by direct rail (or heavy-haul truck or barge). Another 24 commercial sites that do not have rail access would ship by heavy-haul truck or barge to railheads.

Table 3.8 provides a summary of the EIS analyses for Yucca Mountain for a 24-year transportation program involving the movement of 70,000 metric tons of spent fuel and high-level waste to the repository. The table provides several types of consequence estimates:

- Estimates of radiation exposures during incident-free transport. Two types of exposures are estimated: the collective dose (see Sidebar 3.3) to workers and to members of the public during incident-free transport, and doses to the maximally exposed worker and member of the public.
- Estimates of the annual probabilities of the maximally reasonable foreseeable accident.
- Estimates of collective doses and the maximally exposed individual if the maximally reasonably foreseeable accident does occur.
- Estimates of latent cancer fatalities from these exposures calculated as described in Sidebar 3.3.
- Estimates of the annual collective dose risk.

For comparison purposes, the estimated number of fatalities from nonradiological exposures are given at the bottom of the table. These fatalities are estimated to arise from vehicular collisions, industrial accidents during loading and handling of the transport packages, and air emissions from the transport vehicles.

Several observations from Table 3.8 are noteworthy. First, and perhaps most important, a Yucca Mountain transportation program will not be risk free. Workers and members of the public who are exposed to radiation from the transportation packages could have an elevated risk of developing fatal cancer. However, the absolute risk, as measured by the total number of fatalities, will be very small for either rail or truck transport for both incident-free and accident scenarios.

It is important to recognize that these risk estimates are based on a large number of parameter estimates having varying degrees of uncertainty. In view of these uncertainties, it is unclear whether the estimates of radiological fatalities for either the mostly truck or the mostly rail scenarios are significantly different. Also, the estimates of fatalities represent averages for all shipments over all routes. There may be individual routes, shipments or persons that could have risks that are significantly higher or lower than those shown in Table 3.8.

The maximally foreseeable reasonable accident probabilities are similar for both truck and rail accidents (on the order of 10^{-7} occurrences per year [i.e., 1 occurrence every 10 million years]). This is a very low recurrence rate compared to other kinds of transportation accidents that result in the release of hazardous materials to the environment. The collective dose estimates shown in Table 3.8 are also based on some conservative assumptions: for example, the accidental releases are assumed to occur at the center of a large urban area having a population density extrapolated to 2035. Also, the collective doses are calculated out to a distance of 50 miles (80 kilometers) from the release point and would include large numbers of people who receive very small doses. The resulting collective dose risks shown in the table (2.8×10^{-3} to 2.5×10^{-4}) reflect these conservative assumptions.

It is worth emphasizing the differences between the dose estimates shown in Table 3.8 for incident-free transport and the maximally reasonably foreseeable accident. The incident-free estimates have a probability of occurrence of 1 (i.e., the hazard is always present during transport) if the transportation is carried out as described in Appendix J of the final Yucca Mountain EIS (DOE, 2002a). The dose estimates for the maximally reasonable foreseeable accident, on the other hand, are conditional on the occurrence of that accident. The probability of occurrence of such an accident is estimated to be very low, on the order of 1 chance in 10 million each year the program is in operation. This is a very low probability, so these doses are very unlikely ever to be received by workers or the public. This low probability is reflected in the small collective dose risk estimates shown in the table.

The State of Nevada provided extensive commentary on the Yucca Mountain draft EIS estimates of transportation risk (Nevada, 2000). The state believes that incident-free transportation risks based on truck transport have been underestimated and that the models and scenarios used for the accident consequence estimations are also unrealistic, echoing earlier criticisms by Resnikoff (1994). As noted previously, the final EIS for Yucca Mountain relied heavily on the data and modeling approaches used in the 2000 reexamination study. The State of Nevada has criticized this approach because the 2000 study was performed by a contractor and was not subjected to the public review and comment process used for the 1977 transportation EIS.

Some participants at the committee's information-gathering meetings suggested that the accident statistics used in these analytical studies needed to be reanalyzed in light of increased truck traffic and vehicular speed limits on interstate highways. Participants also raised concerns about whether these studies have appropriately analyzed the consequences of the very low frequency but high-magnitude accident scenarios that could result in releases from spent fuel packages.

The committee has not performed an in-depth analysis of the methods used in the final Yucca Mountain EIS to estimate the radiological impacts shown in the table. The calculation of maximum incident-free impacts can be made if reliable data on shipments, routes, and populations can be obtained. The State of Nevada's specific concerns about incident-free exposures for the mostly truck scenario would have limited relevance for a mostly rail scenario, which DOE has announced as its preferred scenario for shipments to a federal repository. Many fewer total shipments would be required under this scenario (see Table 3.8), and in general there are likely to be greater distances between packages and members of the public along main line railways. To the extent that truck shipments are made under this scenario, however, the likelihood of exposure to radiation will depend to a great extent on the routing of these shipments through populated areas. Since routing decisions have not yet been announced, the committee cannot evaluate these potential impacts.

3.2 SOCIAL RISKS

As defined by the committee, *social risks* arise from both social processes and human perceptions (see footnotes 1 and 2 in this chapter). They can arise during the construction of transportation facilities, during routine transportation operations, and as a result of transportation accidents. Social risks are associated with the two types of impacts shown in the rightmost column of Table 3.1: direct social and economic (i.e., socioeconomic) impacts,[36] and perception-based impacts. These two types of impacts can be difficult to separate in practice because they can have similar manifestations, as described below.

A number of direct socioeconomic impacts can result from the transport of spent fuel and high-level waste. Routine transport operations, for example, might result in increased visual impacts (i.e., increased numbers of visually conspicuous shipments of spent fuel through communities), especially from large-quantity shipping programs at the "funnel end" of a transportation system where large numbers of conveyances would be expected to travel along a single route. These activities may have direct impacts on

[36]These kinds of "standard" socioeconomic impacts are included as factors that must be considered in environmental impact statements under the National Environmental Policy Act: for example, see Chapter 8 of the final EIS for Yucca Mountain (DOE, 2002a).

quality of life, property values, and/or business activities, especially if they persist over extended periods of time. Severe accidents involving loaded transportation packages might lead to the temporary loss of use of a transportation route, which could result in business disruptions and other inconveniences with economic and quality-of-life impacts.

These direct socioeconomic impacts arise from generally well understood social processes. For example, most people prefer to live in neighborhoods with roads that carry low volumes of mostly local vehicular traffic. Such neighborhoods tend to be quieter and safer for unsupervised children. The preference for such neighborhoods is reflected by their higher property values compared with nearby neighborhoods along major roads. Similarly, most people prefer to shop at stores that offer easy access by foot, public transportation, or (in most suburban areas) automobile. People will tend to avoid stores along highly congested routes if comparable but more easily accessible alternatives are available. These preferences are examples of social processes in action.

Perception-based impacts arise from people's beliefs and values concerning the consequences of transportation activities on their well-being and that of their communities (Sidebar 3.5). Such perceptions can shape

SIDEBAR 3.5 Well-being and Social Risk

The general proposition that peoples' well-being changes systematically as events interact with beliefs and values has been central to economic and social theories for generations. Broadly speaking, these processes take place around us (and to us) continuously. Except for truly isolated hermits, people's well-being is inevitably affected by the behavior of others; this is why being treated respectfully so often matters, cell phone use is unwelcome in concert halls, and social ostracism hurts.

Social rules governing behaviors that affect people's well-being are often informal and subject to change over time (this explains why Ann Landers, an expert in informal rules governing behaviors that impinge on the well-being of others, had such a hugely successful career). When there is broad agreement that a behavior damages the well-being of a sufficient number of others, it is often proscribed by formal rules: Noise ordinances in some cities preclude the too-enthusiastic sharing of music, zoning rules forbid certain kinds of property use, nudity is illegal in most public places, and smoking is heavily regulated. Many arguments can be (and are) made about the nature and extent of the losses associated with these behaviors, but in common they reflect a sense of lost well-being on the part of affected people. The bottom line remains that in a representative political system, if enough people perceive a loss of well-being, rules regulating the offending behavior are likely to be forthcoming.

TABLE 3.9 Examples of Impacts of Social Risks

Social Risk Impacts	Reference Examples
Increased stress and anxiety (psychosocial risks)	NCRP (2001); Slovic (2001a,b); Tuler (2002); Webler (2002)
Loss of sense of security and safety	NCRP (2001); Slovic (2001a,b); Williams et al. (1999)
Loss of trust and confidence in government and government agencies	Freudenberg (1993); Slovic (1993); Satterfield and Levin (2002); Tuler (2002); Rosa and Clark (1999)
Reduced desirability as a place to live	Hunsperger (2001); Slovic et al. (2001)
Reduced economic activity (e.g., tourism and other business activities)	Easterling (1997); Easterling and Kunreuther (1993); Hunsperger (2001); Slovic et al. (2001); UER (2001, 2002)
Reduced property values	UER (2000); Gawande and Jenkins-Smith (2001); Hunsperger (2001)

peoples' behaviors and can materially affect both individual and community welfare.[37] The impacts of social risks may be manifested in different ways, ranging from stress or depression to more direct socioeconomic impacts including losses in jobs or wealth and to sociopolitical impacts that include loss of institutional trust, but in common they result from the ways in which people understand and then respond to the effects of a hazard on their well-being.

While these perception-based impacts can produce a systematic reduction in people's sense of well-being (or utility), the mechanism of loss can vary. Table 3.9 provides examples of some of the potential consequences resulting from actual or feared exposures to radioactive materials or beliefs about the ways such materials are managed. Previous research suggests that while such consequences *may* result from concerns about radiation, these kinds of consequences are not guaranteed to result; the social dynamics of how perceptions and consequences emerge is complex and incompletely understood.

Perception-based impacts arise in many different contexts. With respect to a spent fuel and high-level waste transportation program, risks might arise as follows. The advent of a program for transporting spent fuel and high-level waste—perhaps even in the planning stages—might produce im-

[37]Another term used to describe these effects is "special impacts." Appendix N of the final EIS for Yucca Mountain (DOE, 2002a) includes a section that evaluates these potential impacts.

agery and messages (i.e., signals) about the program's impacts on local communities (Slovic et al., 1991). The signals originate from an array of sources, including the program's implementing organization, opposition groups, government agencies, and others. These signals are typically transmitted to individuals though the news media (Kasperson et al., 1988), but also through public meetings and government reports.

These signals are *discerned* (filtered, interpreted, and evaluated) by members of affected communities in light of their prior beliefs and values (Jenkins-Smith and Smith, 1994). Given that signals and images of nuclear waste are usually very negative (Slovic, 1987; Slovic et al., 1991), it is not surprising that the prospect or advent of a transport program through a community may widely be perceived to be threatening.

The discernment of a threat posed by spent fuel and high-level waste shipments—regardless of whether it is consistent with technical estimates of risk—has real implications for affected individuals. The threat can diminish individuals' sense of well-being, sometimes in an acute manner, as the understood threat undermines health expectations and increases emotional and physical stress (MacGregor and Flemming, 1996; but see Renn, 1997). The sense of loss can be exacerbated by a sense that the imposition of the risk is unjust or inequitable. When a community—for example, a low-income or minority community—has historically been subjected disproportionately to harms emanating from industrial and other undesirable activities, is less endowed with the resources needed to manage the risks, or holds values that are unusually susceptible to infringement by additional discerned threats (e.g., cultural or spiritual beliefs attached to a place), the loss is likely to be seen as unjust.

This "substantive" injustice can also be matched by "procedural injustice": that is, a sense of injustice stemming from the belief that the *process* by which the threat was imposed is unfair (Gusterson, 2000). The losses potentially associated with a pervasive sense of injustice are numerous and may include loss of trust in government institutions, reduced faith that citizen involvement can result in appropriate public policy (i.e., a sense of disempowerment), and a reluctance to participate in planning processes (e.g., public meetings) (Kasperson, 1983; Fischer, 2001; Bradbury et al., 2003).

Federal (e.g., EO, 1994; CEQ, 1997; EPA, 1998a; DOE, 1995a; DOT, 2002), some state (American Bar Association, 2004) and city governments, and a growing number of other organizations (e.g., NCHRP, 2004) have recognized the importance of environmental justice[38] impacts, and some

[38]Environmental justice is the fair treatment and meaningful involvement of people regardless of race, gender, national origin, or level of attained education in the development of laws, regulations, and policies that affect them (see IOM, 1999).

have enacted laws, regulations, and policies to address them. Additionally, a rich literature on environmental justice has developed over the past 20 years (e.g., GAO, 1983; United Church of Christ, 1987; Bullard, 1990a,b; Bryant and Mohai, 1992). It has many thrusts, such as identifying and characterizing the causes of disproportionate harms (Adeola, 1994), often through community-based participatory research (Shepard et al., 2002); investigating the effects of scale of measurement (e.g., census tract, block group) on disproportionate impacts on communities (Eady, 2003); and developing mitigating tools, such as good neighbor and community benefits agreements.

Should the perceived threat become broadly associated with a place or community, it could have a potentially lasting stigmatizing[39] effect (Kasperson et al., 2001; Slovic et al., 2001). For spent fuel and high-level waste shipments, concern about stigma would be associated chiefly with severe accidents, but it could also result from frequent and widely publicized shipments or minor vehicular accidents involving spent fuel or high-level waste shipments.

Publicity about transportation incidents, even minor incidents, can result in the *social amplification of risk* in which "the consequences of risk and risk events . . . often exceed the direct physical harm to human beings and the ecosystems to include more indirect effects on the economy, social institutions, and well-being associated with the amplification-driven impacts" (Kasperson et al., 2001, p. 18; see also, Slovic et al., 1991). Discerned risks from hazards such as spent fuel and high-level waste shipments also can change behaviors. A perceived threat to health may modify the way people use residential properties (Berrens et al., 2002) or change the attractiveness of areas for residency, vacations, and conferences (Easterling and Kunreuther, 1993).

Changed perceptions of places resulting from discerned threats may result in changes in perceived values of residential and commercial properties (Ketkar, 1992; Mendelsohn et al., 1992; Kiel, 1995; Hunsperger, 2001). These perceived losses may translate into reductions in market values as sellers become more eager to leave and buyers more wary of the affected locale.

Because perception-based impacts derive from the manner in which individuals recognize and understand the hazard, social risks are sometimes mistakenly treated by technical experts and policy makers as imaginary, or less real, than the health and safety risks discussed in Section 3.1. The general difficulty in quantifying social risks no doubt contributes to this

[39]Stigma marks a person, place, product, or technology as deviant, flawed, or undesirable. When the particular stigmatizing characteristic is observed, the person, place, product, or technology may be denigrated or avoided.

view. Nevertheless, it is clear from social science research, some of which is described in the next section of this chapter, that social risks are as real and important to many people as the associated health and safety risks. They can impact individuals and communities in ways other than injury or death. In addition, social risks may exacerbate concerns about the likelihood of future, unanticipated health and safety risks. For example, an erosion of trust in a program or the agency overseeing such a program can arise from frequent minor problems; these continuing problems may lead people to conclude that the agency lacks the capacity to effectively manage the program over the long term. Left unaddressed, these risks could diminish the ability of implementers to mobilize the necessary resources for managing the health and safety risks of transportation systems.

Technical experts and policy makers sometimes attribute the concerns about social risks expressed by others as based on misinformation about or ignorance of the "real" (i.e., health and safety) risks. This attribution is frequently coupled with calls for better public education about risk, with the unspoken implication that such education would encourage the public to behave more rationally (i.e., more like technical experts). Although there is no doubt that the public would benefit from more accurate information about transportation risks, one should not expect that such information would result in a widespread change of public behavior. Such "information deficit" approaches to behavior change have largely been discredited (e.g., Kollmuss and Agyeman, 2002).

In fact, people may be acting rationally if they oppose spent fuel and high-level waste transportation on health and safety grounds even if they agree with the experts that the estimated health and safety risks are low. Most people recognize that transportation programs are run by fallible institutions and that institutional and human errors play a large role in determining transportation risks. There are many examples of technological systems where the experts were wrong or overly optimistic (Schrader-Frechette, 1995; Perrow, 1999; Freudenburg, 2003). They also recognize that the risk of an accidental release from a spent fuel shipment, while low, is not zero and, moreover, that such a release can have a range of consequences: health, safety, and social. Rational people care about all consequences that can impact their lives and communities, not just health and safety consequences that are the main concern of technical experts (NRC, 1996).

3.2.1 Research on Social Risk

There is a large body of research on social risk that focuses on the perceptions of and responses to nuclear power, nuclear weapons, and radioactive waste disposal. This research has shown that perceptions of risk can have gender, racial, and cultural predispositions. A brief review of this research is provided below.

Studies have found that men tend to "rate a wide range of hazards as lower in risk than do women" including nuclear technologies (e.g., Finucane et al., 2000, p. 169). Flynn et al. (1994, p. 1101) found that white men have been found on average to perceive risks, including technologies such as nuclear power and waste "as much smaller and much more acceptable" than other people (also see Finucane et al., 2000). Vaughan (1995; see also Vaughan and Nordenstam, 1991) found that African Americans, when compared with others, tend to perceive risks as higher and support stricter regulatory actions for issues involving nuclear power plants or the disposal of radioactive waste. These predispositions may be related to the sociocultural contexts within which people live, including their beliefs about the trustworthiness of risk management institutions. Vaughn (1995, p. 175) notes:

> Because prior beliefs about risk, perceptions about the trustworthiness of various government agencies and beliefs about risk management process evolve within sociocultural contexts, they likely are not independent of broader social experiences that bound and structure perspectives and worldviews. . . . Different patterns of belief and value systems relevant to risk management are likely to be observed across diverse ethnic and socioeconomic communities to the degree to which these communities' social and cultural contexts have varied.

These sociocultural predispositions have been the topic of numerous studies (e.g., Dake, 1991, 1992; Jenkins-Smith and Smith, 1994; and Peters and Slovic, 1996).

There is also quantitative and qualitative evidence (e.g., United Church of Christ, 1987; GAO, 1983; Zimmerman, 1993; Goldman and Fitton, 1994) that not all people and communities bear the burden of environmental problems equally, including those arising from transportation of hazardous materials. This disproportionate burden has been associated particularly with minority and low-income communities (Bryant and Mohai, 1992). The imposition of preexisting risks on these communities may affect their conceptualization and framing of risk problems and make them even more vulnerable to risks from new activities (Sidebar 3.6).

A large body of published work has examined public perceptions concerning the proposed federal repository at Yucca Mountain, Nevada; much of the initial work was supported by the State of Nevada (e.g., Kasperson et al., 2001; see also, NRC, 1984; Slovic et al., 1991, 1994; Gregory et al., 1995). Consistent with the Nevada-sponsored studies, a National Research Council committee that examined the development of geologic repositories noted that the "[g]eneral public in almost every nation where data have

SIDEBAR 3.6 Publics, Affected Communities, and Vulnerable Communities Defined

There is a temptation to talk of "the public" or "the community" when thinking of risk estimates or tasks such as communication or the provision of information. However, the situation is far more complex than this. There are many *publics* (Jacobson, 1999) differentiated by both demographic (ethnicity, income) and interest-based criteria, and many groups or *communities* differentiated by numerous criteria of which demographics is just one.

In addition, for any project with potentially adverse consequences, such as those that might result from the transportation of spent nuclear fuel and high-level waste, there are *affected* groups and *vulnerable* groups. Affected groups are all communities within an influence zone of a transportation project. This is a hypothetical zone, which, for spent fuel and high-level waste transportation, would extend out some distance on either side of designated shipping routes. Within these affected communities will be vulnerable groups who, because of disproportionate exposures to other health-affecting substances, or because of ethnic, linguistic, or socioeconomic issues, may be less able to read or understand information from the authorities, to act in a first-responder role, to exit the area in a timely manner in an emergency, or to otherwise cope with an emergency. The 2005 Hurricane Katrina disaster in New Orleans is an unfortunate illustration of this point: many poor, mostly black, New Orleans residents had no means to evacuate the city and were stranded in their homes or in shelters of last resort. These vulnerable groups may have to be identified and given special consideration by the authorities, including—but not limited to—translated materials, emergency warnings in different languages, and appropriate first-responder training.

been collected perceives nuclear technologies and radioactive wastes as the riskiest of all hazards and expresses great concern about them" (NRC, 2003, p. 56).

There is a smaller body of research that specifically examines the potential social risks of transporting radioactive waste. This research has generally followed one of two approaches. The first and most common approach relies on survey interviews taken from systematic samples of people (adults) from the affected populations. The second approach, often referred to as "hedonic studies," relies on direct measures of behaviors that reflect responses to the transportation of nuclear waste.

Survey research has the immediate advantage of obtaining individuals' responses to specific questions about their own risk perceptions, beliefs, preferences, and anticipated behaviors. Of course, anticipated behaviors do not always match real behaviors. Surveys can be targeted to respondents of particular relevance to the transportation program. Quite a number of surveys have focused on nuclear waste transport, many of which are accessible on the Nevada Nuclear Waste Project Office Web site (*http://www.state.nv.us/nucwaste/*). Some of the most directly relevant survey research is summarized in the following paragraphs.

The University of New Mexico (UNMIPP, 2004) reported on a decade-long series of surveys on public perceptions of the risks associated with the Waste Isolation Pilot Plant among New Mexico residents. One survey item asked respondents to indicate whether they believed the facility was safe to open or was slightly, somewhat, or very risky. Over time, support gradually increased, and it appeared to increase significantly once the first shipments of transuranic waste reached the facility without incident.

A DOE-sponsored telephone-based survey of South Carolina residents living near the Savannah River Site examined residents risk perceptions of radioactive waste cleanup and disposal activities at the site, including the transport of radioactive waste (not explicitly including spent fuel) to and from the site. The study authors (Williams et al., 1999, p. 1028) noted that respondents were also more than four times more likely to believe that transport of waste from the site posed a fair to certain chance of harm than a small or no chance of harm. Truck was seen as being the most risky mode of transport. The study found that heightened risk perceptions among these residents were based upon their expectation of economic loss, their financial security, proximity to the site, and their trust in Savannah River Site officials (Williams et al., 1999, p. 1033).

The State of Nevada is located at the end of the transportation funnel for shipments to the planned Yucca Mountain repository. Consequently, larger numbers of spent fuel and high-level waste shipments will pass through communities in that state than anywhere else in the country if a federal repository at Yucca Mountain is opened. Two surveys—one of residents, the other of bankers and appraisers—were undertaken by the Urban Environmental Institute, LLC (UER) in 2000 to assess potential transportation-related impacts in Clark County, Nevada, which includes the city of Las Vegas. More than 70 percent of the respondents to UER's residential survey indicated that they would not consider purchasing residential property near a highway designated for spent fuel transport. The UER survey of appraisers and bankers indicates that under routine transport conditions, residences 1 mile from transportation routes may see property value decreases of 2.0 to 3.5 percent, while commercial-office properties and industrial properties might decline in value from 0.5 to 3.0 percent. The study found that significant and adverse impacts on property values are likely to extend up to at least 3 miles from planned transportation routes (UER, 2000, p. 71).

UER also has studied the potential impacts of spent fuel transport near Moapa tribal lands in Nevada (UER, 2001). These lands are located near

Interstate 15, a route identified for possible truck shipments to Yucca Mountain. Tribal lands are located within 5 miles of I-15, and the major revenue source for the tribe is a gaming center located near the interstate. The UER concluded that an accident near Moapa lands would put the "tribe in an extremely vulnerable position in terms of economic well-being since that enterprise generates 90 percent of the tribe's revenues" (UER, 2001, p. 40). The tribe has plans for future economic developments along I-15, including a truck stop and sales of agricultural produce. The study concluded that an accident involving a spent fuel shipment near the Moapa exit along I-15 may cause declining property values and lost revenues, resulting in potentially severe adverse economic impacts on the tribe.

The UER has also conducted a survey to identify the potential impacts of spent fuel transportation on the Las Vegas and Moapa Paiutes (UER, 2002; see also, Nevada, 2002). Survey respondents noted that they were concerned not only with economic impacts, but also with what they termed the "moral" issue of transporting nuclear waste through Indian communities that have already experienced exposure to radioactivity from atomic bomb tests at the Nevada Test Site.

Intertech Services Corporation completed a study for Lincoln County, Nevada, on the potential adverse impacts to Caliente, Nevada, from a Yucca Mountain transportation program (Intertech Services Corporation, 2001). Caliente is the planned site for a rail spur junction to Yucca Mountain. The development of this junction is supported by city and county leaders. The study concluded that the transportation program would have "negative impacts on community cohesion, population driven effects, emergency management, highway accident risk, radiation exposure risk, and impacts from stigma that may reduce the desirability of Lincoln County as a place to live and as a destination for tourists" (Intertech Services Corporation, 2001, ES-2). The report notes (p. 51) that although "scientific estimates of risk for an accident resulting in a release of radiation into the environment may be quite low, the risk is not zero." The report also notes (p. 51–52), "To the degree that media amplifies the incident, even when there is no radiation release, the economic and fiscal consequences can be expected to be much greater than from a similar accident without nuclear waste."

Transportation-related surveys taken in other parts of the country are broadly consistent with these results. A national survey of 972 people, which was part of the University of Maryland 1997 Omnibus Survey (Flynn et al., 1998), examined the perceived risks and property value losses resulting from the transportation of spent fuel. The survey found that nearly two-thirds of the respondents thought the value of properties located along spent fuel transportation routes would be lowered. Seventy percent of respondents thought that terrorist groups could successfully attack spent fuel shipments. Most respondents stated an unwillingness to live near a spent fuel transportation route, and a majority of the respondents reported that they considered the transport of spent fuel to be more risky than the transport of industrial chemicals or gasoline.

There are fewer hedonic studies of people's reactions to nuclear waste transport, in part because these studies are more expensive and more difficult to carry out. Only one study, which is described below, was available for review by the committee.

A DOE-sponsored study of real estate transactions in South Carolina (Gawande and Jenkins-Smith, 2001) measured the effects of a series of highly publicized shipments of foreign spent fuel to the Savannah River Site (research reactor spent fuel shipments to the Savannah River Site are described in more detail in Chapter 4). The authors found no correlation between the spent fuel shipments and property values in rural Aiken County, where there is a long experience with nuclear materials management. In urban and populous Charleston County, however, "the net gain in value associated with being 5 miles away from the route relative to a property on the route was nearly 3 percent of the average home value" (Gawande and Jenkins-Smith, 2001, p. 229) once the shipments were under way. The authors caution about making generalizations concerning the effects of hazardous material shipments based on this survey, however, because of data limitations.

During the scoping process for the draft EIS for Yucca Mountain, DOE received many public comments about the need to address risk perceptions and the potential stigmatization of Nevada, in particular the Las Vegas area, due to the transportation and disposal of spent fuel and high-level waste. In response, DOE sponsored a review of relevant perception-based impacts and stigma effects literature, in which qualitative assessments were made of the likelihood of these impacts. The literature reviewed included studies sponsored by DOE, Nevada, and others. The report resulting from this review concluded (O'Connor, 2001, p. 2):

> . . . absent accidents, there is no reason to expect impacts to property owners in areas beyond the transportation corridors. Even absent accidents, however, two studies report that, at least temporarily, a decline in residential property values of approximately 3 percent may be expected in transportation corridors in urban areas. . . . More research on whether property values have fluctuated with the transportation of radioactive materials would be beneficial, although the research would not allow analysts to know with certainty whether there would be any impacts from perceptions of shipments of spent nuclear fuel and high-level waste to a Yucca Mountain Repository, or how long such impacts would persist.

The final EIS for Yucca Mountain reached a similar conclusion (DOE, 2002a, p. N21):

> There is a consensus among social scientists that a quantitative assessment of the potential impacts from risk perceptions of the proposed repository and the transportation of spent nuclear fuel and high-level radioactive waste is impossible at this time and probably unlikely even after extensive additional research. The implication is not that impacts would probably be large, but simply difficult to quantify.

In summary, scientific research has generally shown the following:

- The public generally perceives nuclear-related activities to carry a higher risk than non-nuclear activities.
- These risks are not perceived in isolation, but in a broader context of social experiences and risk management processes—for example, in the context of proposals for increasing reliance on nuclear energy that would also require transportation of additional spent fuel.
- Social processes have the potential to amplify or attenuate social risks arising from the transport of spent fuel and high-level waste.
- Risk perceptions may have gender, cultural, and ethnic predispositions.
- Social risks are difficult to quantify and there are no universally agreed-upon metrics for comparing them.
- Trust and confidence can play important roles in modulating these risks.

The last point about trust and confidence suggests that risk perceptions can be modulated (i.e., amplified or dampened) by the actions of organizations involved in implementing or opposing a transportation program. On the one hand, the publics' trust of groups that seek to mobilize opposition to a transportation program may increase the perceived risks of transportation. On the other hand, public trust and confidence in government agencies that manage and regulate transport (e.g., DOE, USNRC, and their contractors) may amplify or dampen these effects. In other words, *trust and confidence* serves to amplify or dampen the publics' response to signals sent by those who make claims and counterclaims about the risks of transportation (Freudenberg, 2003; Frewer, 2003).

Responses to programs for transporting spent fuel and high-level waste among members of the public will depend to a great extent on perceptions about the need for such transport as well as the risks involved. Those responsible for implementing transport programs can take several steps to inform public understanding of needs, options, risks, and benefits. They can also benefit from a better understanding of the reasons for public responses, whether in the form of support or opposition to a proposed program. Improved understandings of these issues by the public and transportation implementers can support better planning and operational decisions, improve confidence in transportation management, improve safety, and potentially reduce conflict.

3.3 COMPARATIVE RISK

The statement of task for this study (Sidebar 1.1) directs the committee to provide a comparison between the principal risks for transporting spent fuel and high-level waste and other risks that confront members of society. In this section, the committee provides a comparison of the health and safety risks of spent fuel and high-level waste transportation with other types of risks to address this charge.

The committee made no attempt to compare social risks because of the difficulties in quantifying such risks. The committee has followed the common approach for risk comparisons by comparing risks on dimensions for which the most information currently exists. The committee is well aware that these dimensions may not represent the outcomes that are most important to some people. However, data are lacking to make meaningful quantitative or qualitative social risk comparisons, and there are no agreed-upon metrics for making such comparisons.

The committee's objective in presenting this comparison is to inform readers' understanding about the risks of spent fuel and high-level waste transportation, not to persuade readers that such risks are—or are not—acceptable. The committee recognizes that acceptability is a normative judgment; that is, there is no basis in science for judging the acceptability of transportation risks. Societal acceptability of risk is ultimately a public policy decision and may vary over time and among different societies. Individual acceptability is based on personal judgments.

There is a rich literature on risk comparisons that informed the committee's work (Sidebar 3.7). An important finding from research on risk communication is that different audiences frequently find different information and comparisons more (or less) informative and relevant; some may be critical of a particular comparison, while others will find it helpful for understanding particular risks. At best, comparisons can help to improve people's understandings about the risks of a given activity, which may or may not change their views about its acceptability. At worst, such comparisons can be seen as trying to manipulate public opinion or to "sell" a particular technology or approach. People can reasonably disagree about the "best" comparisons to use because each comparison will privilege some aspects of the risk context at the expense of others (Vaughan and Seifert, 1992; NRC, 1996). In short, there is no single "right" comparison that will satisfy and be understood by all audiences or that will convey all of the relevant information associated with a complex risk.

The committee was guided by two principles in developing compari-

SIDEBAR 3.7 Comparative Risks

A growing body of research has been used to inform the practice of risk comparisons (e.g., NRC, 1989, 1996; Roth et al., 1990; MacGregor et al., 2002a, 2002b; Johnson, 2003, 2004; Johnson and Chess, 2003). Key insights from this research include the following:

- Comparisons can have subtle effects on judgments about relative risks and their significance.
- Comparisons on multiple dimensions can be more helpful than comparisons on a single dimension.
- Comparisons of different risks on a single dimension can be problematic when the risks are viewed as qualitatively different (e.g., voluntary vs. involuntary, familiar vs. unfamiliar).
- Use of point estimates can be misleading when uncertainties are large.
- Risk denominators (e.g., periods of exposure, population base) to the risk should be defined.
- Risks should be meaningful to those presented with the comparisons.

Choices about which risk comparisons to make involve personal judgments about which risk *outcomes* are important (Crouch and Wilson, 1982; Vaughan and Seifert, 1992; NRC, 1996): for example, human fatalities, human morbidity, ecological impacts, economic costs, procedural fairness, distributional equity, intergenerational effects, personal rights, effects on institutional trust and risk management regimes. It may also be important to consider benefits alongside such outcomes to fully inform understandings and decisions. Choices about which outcomes to use in the comparison are not only technical or scientific; rather, they also reflect values—implicitly or explicitly—about the characteristics of the risks that are important to people. They are embedded in and reflect social values.

There are a potentially large number of competing outcomes and measures that may be relevant in a risk comparison. Meaningful comparison of risks need not consider all possible outcomes or measures. Instead, decisions must be made about what will be most useful and relevant. A previous National Research Council committee offers some cautionary advice in this regard (NRC, 1989, p. 172).

> Risk comparisons can be helpful, but they should be presented with caution. Risk comparisons must be seen as one of several inputs to risk decisions, not as determinants of decisions. There are proven pitfalls when risks of diverse character are compared, especially when the intent of the comparison can be seen as that of minimizing a risk (by equating it to a seemingly trivial risk). More useful are comparisons of risks that help convey the magnitude of a particular risk estimate, that occur in the same decision context (e.g., risks from flying and driving to a given destination), and that have a similar outcome. Multiple comparisons may avoid some of the worst pitfalls.

There is a frequent tendency to compare risks on dimensions that can be quantified (e.g., estimates of fatalities). These dimensions do not always represent the outcomes that may be most important to people, but they may be the outcomes for which most information exists. Other outcomes—for example, those for the social risks described in Section 3.2—may be equally important, but data may be lacking to make meaningful comparisons.

sons of health and safety risks. First, the committee considered it important to compare risks associated with *like physical causes*. Radiation is the primary hazard of concern in spent fuel and high-level waste transportation. Thus, the committee's comparisons focus on risks associated with radiological exposures.

Second, the committee considered it important to compare risks associated with similar outcomes. Human exposure to radiation can lead to undesirable health outcomes. The primary health effect of concern from exposure to ionizing radiation is cancer.[40] Exposure to low and moderate levels of radiation can lead to the induction of cancer, usually years to decades after the exposures occur; some of these induced cancers will be fatal. Exposure to high levels of radiation can result in radiation sickness and death in a much shorter period of time. Thus, the committee separates comparisons of *doses* associated with routine radiological transport risks, which have the potential to provide chronic exposures, and severe accident risks, which have the potential to provide acute exposures.

Risk estimates for cancer incidence and mortality from exposure to radiation and radionuclides are available (e.g., EPA, 1999; NRC, 2005a). Some individuals may find cancer incidence to be a more meaningful factor for risk comparisons given the dread that is often associated with this disease. Nobody welcomes a cancer diagnosis, even if the cancer is treatable. Cancer incidence is also used as the basis for U.S. compensation programs for workers, veterans, and members of the public exposed to radiation from national defense activities (NRC, 2005b). Moreover, the incidence of health effects (including cancer) is commonly used in risk assessments for chronic exposures to hazardous chemicals.

The numerical relationship between cancer incidence and cancer mortality varies with cancer site within the human body.[41] Average lifetime lung cancer mortality in males, for example, is about 100 percent of cancer incidence. On the other hand, average lifetime prostate cancer mortality in males is about 22 percent, and average lifetime breast cancer mortality in females is about 25 percent. The average lifetime mortality for all solid cancers[42] in the U.S. population is about 48 percent of cancer incidence. For leukemia, the average lifetime mortality rate is closer to 85–90 percent.

---

[40]Radiation exposure may have other health effects besides cancer. For example, recent research suggests that such exposures can contribute to the development of cardiovascular disease.

[41]Average lifetime risk estimates for non-radiation-induced cancer incidence and mortality for several cancer sites are available for the U.S. population. See NRC (2005a, Table 12-4) for example.

[42]Solid cancers (cancers manifested by the formation of tumors) constitute more than 98 percent of all human cancers in the U.S. population. Leukemia (cancer of the blood or blood-forming organs) constitutes the remaining cancers.

Cancer incidence is most readily used for comparing chronic radiation exposures during routine transport to chronic radiation exposures from other activities. Alternatively, one can compare the magnitudes of the chronic exposures for different activities directly. People who are concerned about exposures to any anthropomorphic radiation, regardless of its health effects, may find this sort of comparison useful. For those interested in how these exposures relate to cancer incidence and mortality, the appropriate multiplicative conversion factors (see Sidebar 3.3) can be applied. The committee uses exposures in its comparisons for routine transport but also provides examples of how these exposures relate to cancer incidence and mortality.

The committee uses mortality in its comparative assessments for transportation accidents. This allowed the committee to compare cancer mortality from exposures to ionizing radiation in a spent fuel transportation accident with other types of hazardous material accidents—for example, deaths from an accident involving releases of chlorine from a rail tanker car.

The committee has used a wide range of comparison factors, some of which are more directly comparable than others. As noted previously, different individuals will find different measures to have more or less meaning. The committee hopes that its presentation of a large range of factors will provide most individuals with comparisons that are helpful.

3.3.1 Risks for Normal Transport

As described elsewhere in this chapter, transportation packages do not completely shield the radiation emitted by the spent fuel or high-level waste contained within them. Consequently, individuals who travel, work, and live along the routes used for shipping spent fuel and high-level waste might receive small radiation doses when loaded packages are transported in their vicinity. The dose received by given individuals will vary as the inverse square of their distances from the packages[43] and directly in proportion to their exposure times. Individuals closest to the packages will receive comparatively larger doses for a given exposure time. Doses will drop off quickly, however, as the distance between the individual and the package increases.

The doses received by any one individual may be very small, but a large number of individuals may receive radiation doses over the life of a transportation program. The *collective dose* to the population of people exposed

[43]That is, doubling the distance results in one-fourth of the exposure. The inverse square law assumes that the radiation emanates from a point source. A transportation package is not a point source, however, so the inverse square law is only a rough approximation at close distances to it.

to radiation, which is obtained by summing the doses received by all individuals within the population, can be used to estimate radiological risks from normal transport (Sidebar 3.3). The outcome of principal concern for routine exposures is cancer incidence and cancer mortality.

Quantitative risk assessment models that describe the association between radiation dose and cancer incidence or cancer mortality have been deduced from epidemiological and biological studies. A linear no-threshold association between dose and cancer risk from exposure to ionizing radiation such as X-rays and gamma rays[44] is consistent with current epidemiological and biophysical understanding (NRC, 2005a). That is, the risk of radiation-induced cancer rises linearly with dose with no threshold below which the risk falls to zero. The relationship between radiation dose and radiation-induced cancer can be expressed as a straight-line function that passes through zero risk at zero dose.[45] The slope of the line, referred to as the nominal probability coefficient, is between about 4×10^{-2} and 5×10^{-2} fatal cancer per sievert (4×10^{-4} to 5×10^{-4} fatal cancer per rem; ICRP, 1991). These estimates have high uncertainties at the low doses typical of normal transport conditions (NRC, 2005a).

This risk model is a probabilistic function. It expresses the average number of fatal cancers that would be expected to occur in a population of individuals having a typical age and gender distribution for workers or the U.S. population for a given level of radiation exposure. Such risk models are used for setting standards for radiation exposure in the United States and many other countries. These models are also used for estimating risks to populations from specific activities involving the use of radiation. For example, such models were used to estimate latent cancer fatalities in the final Yucca Mountain EIS (DOE, 2002a; see Table 3.8) for the transport of spent fuel and high-level waste.

The routine radiological risks to a given population are scenario specific. That is, the risks depend on factors such as the number of packages transported; package inventories; shipping modes and routes; and population densities along shipping routes. The most complete estimate of scenario-specific routine radiological risks for spent fuel and high-level waste transportation is provided by the planned Yucca Mountain transportation program. Those estimates, which are provided in the final Yucca Mountain EIS and described in Section 3.1.2, were used by the committee for some of the comparisons in this section.

---

[44]This radiation is sometimes referred to as low energy transfer (LET) radiation.

[45]While the risk of developing a radiation-induced cancer is zero at zero dose according to the linear no-threshold model, the risk of developing cancer from other causes is much greater than zero. As noted elsewhere in this chapter, about 42 percent of the U.S. population will develop some form of cancer in their lifetimes due to causes other than radiation exposure.

The final Yucca Mountain EIS (DOE, 2002a) provides estimates of routine radiological risks for both the mostly truck and mostly rail transport scenarios (see Section 3.1.2 and Table 3.8). While DOE has announced its preference for the mostly rail scenario, the committee provides estimates of radiological risks for both scenarios to help readers put these risks in perspective.

The estimated collective dose for the mostly rail scenario is 1200 to 1600 person-rem (Table 3.8), which applies to the public. Multiplying this dose by the nominal probability coefficient for fatal cancers produces an estimated average of about one latent cancer fatality among the 16.4 million people estimated to be exposed to radiation during the 24-year operational life of the transportation program. The estimated collective dose for the mostly truck scenario is 5000 person-rem, which would be expected to produce on average about three latent cancer fatalities out of the 10.4 million people that are estimated to be exposed to this radiation during the 24-year transport program. Of course, this comparison does not address some associated issues that people may care about, such as the voluntariness of the exposures.

To put these numbers in perspective, it is instructive to compare them to average cancer incidence and mortality in the U.S. population. Approximately 42 out of 100 people in the United States will be diagnosed with solid cancers during their lifetimes, and about 20 of those cancers will be fatal (NRC, 2005a, Table 12-4). Thus, of the 10.4 million to 16.4 million people who are estimated to be exposed to radiation (in almost all cases at very small levels) from transport of spent fuel and high-level waste to Yucca Mountain, approximately 4 million to 6 million would be expected to be diagnosed with solid cancers during their lifetimes for causes unrelated to the transportation program; about 2 million to 3.3 million of those cancers would be expected to be fatal. The estimated cancer fatalities from exposure to radiation during incident-free transport to Yucca Mountain—one for the mostly rail scenario and about three for the mostly truck scenario—would not be detectable in this much larger population of fatal cancers.

Other comparisons based directly on radiation dose[46] are also possible. In Table 3.10 and Figure 3.3, the committee compares the estimated doses for maximally exposed workers and members of the public to radiation from the Yucca Mountain transportation program (DOE, 2002a, Chapter 6 and Appendix J), to three other types radiation exposures:

1. Permissible maximum doses to workers and the public under current radiation standards and regulations

[46]Dose and risk are interchangeable in an arithmetic sense by multiplying or dividing by the nominal probability coefficient.

2. Doses received by members of the U.S. public from natural background radiation

3. Doses received from selected medical diagnostic procedures that utilize radiation

Radiation standards and regulations establish ceilings for the maximum permissible radiation doses that workers and members of the public are allowed to receive from anthropogenic activities involving ionizing radiation. International standards are developed by international groups of radiation experts and health practitioners. These standards do not have the force of law but are frequently used as starting points by U.S. authorities (and authorities in other nations) for establishing national regulations.

U.S. regulations have been developed by the federal government through an elaborate administrative procedure[47] that provides opportunities for public input. As such, these regulations represent a kind of social contract between the government and its citizens to protect worker and public health and safety. These standards set limits on what are often involuntary exposures to radiation, especially for members of the public. The exposures from spent fuel transport to a Yucca Mountain repository also will be largely involuntary for the individuals who receive them.

Natural background radiation consists of cosmic and solar radiation, external radiation exposure from radioactive materials present in rocks and soil, and radioactivity that is inhaled or ingested (see Sidebar 3.2). The committee presents four different estimates of natural background radiation in Table 3.10 (see also Figure 3.3): (1) the annual natural radiation background dose in Florida, the state with the lowest estimated annual natural background dose; (2) the annual natural radiation background dose in South Dakota, the state with the highest estimated annual natural background dose; (3) the average annual natural radiation background dose in the United States; and (4) the galactic cosmic background radiation dose received in a single round-trip airline flight between New York and Tokyo and also between St. Louis and Tampa.[48]

Natural background radiation is usually viewed as an involuntary and

[47]This procedure is specified by the Administrative Procedures Act: United States Code, Title 5, Part I, Chapter 5, SubChapter II.

[48]Airline travel subjects passengers to elevated doses of cosmic radiation originating from stars and galaxies. Radiation exposure increases with altitude and latitude and can also increase significantly during solar disturbances. The Federal Aviation Administration has developed a computer program (CARI-6) that calculates the effective dose of cosmic radiation received by individuals flying in aircraft on great circle routes between two airports. See *http://www.faa.gov/education_ research/research/med_humanfacs/aeromedical/radiobiology/cari6/index.cfm.*

TABLE 3.10 Radiation Dose Comparisons

Estimated Radiation Doses Received by Yucca Mountain Transportation Workers and the Public for Routine Transportation Operations[a]	Maximum Radiation Doses Allowed by International Standards and U.S. Regulations[b]
	DOE annual occupational dose limit established in 10 CFR Part 835
Approximate annual dose to maximally exposed transport worker, mostly truck and mostly rail scenarios (DOE, 2002a, Tables 6-9, 6-12)	ICRP recommended annual occupational dose limit (ICRP, 1991, Table 6)
	Current DOE annual occupational administrative dose limit (DOE, 1999)
	All-pathways annual dose limit to reasonably maximally exposed individual near Yucca Mountain at time periods greater than 10,000 years after repository closure (70 FR 49014, August 22, 2005)
Approximate annual dose to maximally exposed service station worker, mostly truck scenario (DOE, 2002a, Table 6-9)	ICRP recommended annual public dose limit (ICRP, 1991, Table 6)

Examples of Natural and Anthropogenic Radiation Exposures	Radiation Dose Limits or Exposures, mSv (mrem)	Notes
	50 (5000)	
	20 (2000)	ICRP standards are for doses averaged over defined periods of 5 years, not to exceed 50 mSv (5000 mrem) in any one year
	20 (2000)	Maximally exposed transport worker is assumed to receive the maximum allowable DOE occupational administrative dose
Single whole-body CT scan (Brenner and Elliston, 2004)	12 (1200)	Weighted average dose to major organs
Approximate annual natural background radiation dose in South Dakota (Mauro and Briggs, 2005)	9.6 (960)	Includes doses from exposure to radon
	3.5 (350)	EPA draft standard (40 CFR Part 197)
Average U.S. annual natural background radiation dose (NCRP, 1987)	3 (300)	Includes doses from exposure to radon
Approximate annual natural background radiation dose in Florida (Mauro and Briggs, 2005)	1.3 (130)	Includes doses from exposure to radon
	1 (100)	Maximally exposed worker is assumed to receive the maximum allowable dose under ICRP guidelines and 10 CFR Part 20 ICRP guideline is for doses from all sources except natural, medical, and accidental exposures

continues

TABLE 3.10 Continued

Estimated Radiation Doses Received by Yucca Mountain Transportation Workers and the Public for Routine Transportation Operations[a]	Maximum Radiation Doses Allowed by International Standards and U.S. Regulations[b]
	All-pathways annual dose limits for release of radiation to the environment from land disposal facilities (10 CFR Part 61)
	All-pathways annual dose limit to reasonably maximally exposed individual near Yucca Mountain for first 10,000 years after repository closure
Approximate annual dose to maximally exposed resident near rail stop, mostly rail scenario (DOE, 2002a, Table 6-12)	
	Maximum hourly dose allowed at 2 meters (about 6.5 feet) from the lateral surfaces of a transport vehicle carrying spent fuel or high-level waste (49 CFR 173.441(b) and 10 CFR 71.47(b)(3))
Approximate annual dose to maximally exposed resident along rail route, mostly rail scenario (DOE, 2002a, Table 6-12)	

NOTE: EPA = Environmental Protection Agency; ICRP = International Commission on Radiological Protection.

[a]Annual doses were calculated by dividing the total estimated dose given in DOE (2002a) by the 24-year length of the transportation program.

Examples of Natural and Anthropogenic Radiation Exposures	Radiation Dose Limits or Exposures, mSv (mrem)	Notes
X-ray of human hip joint (Ngutter et al., 2001)	0.60 (60)	
	0.25 (25)	
	0.15 (15)	EPA standard 40 CFR Part 197
Round-trip airline flight between New York and Tokyo (Friedberg and Copeland, 2003)	0.145 (14.5)	45-year average (1958–2002) dose calculated using the CARI-6 computer program, which estimates the effective dose of galactic cosmic radiation
	0.12 (12)	Applies to maximally exposed residents who live near rail yards and crew change stops
Single chest X-ray (NRC, 2005a)	0.1 (10)	
Single X-ray of a human extremity (Mettler et al., 2000)	0.01 (1)	
Round-trip airline flight between St. Louis, Mo. and Tampa, Fla. (Friedberg and Copeland, 2003)	0.009 (0.9)	45-year average (1958–2002) dose calculated using CARI-6 computer program, which estimates the effective dose of galactic cosmic radiation
	0.0007 (0.07)	

[b]Radiation protection standards and regulations also include an ALARA (as low as reasonably achievable) requirement that usually results in doses to workers and the public that are well below the limits in this table. Moreover, constraints are sometimes placed on individual sources of radiation or practices involving the use of radiation to limit worker and public exposures.

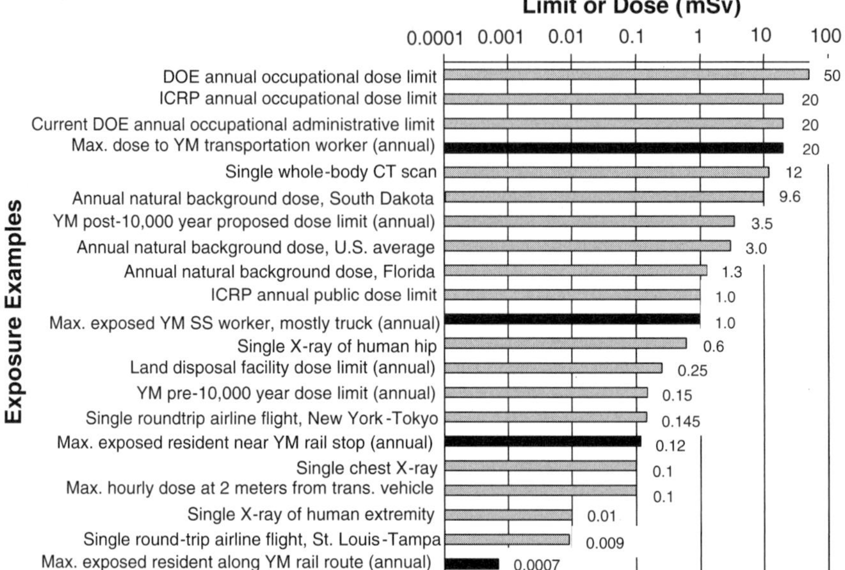

FIGURE 3.3 Graphical illustration of the radiation dose comparisons shown in Table 3.10. The dose data are plotted on a logarithmic scale to better illustrate the spread of values. Doses shown in the figure are annual limits (for standards and regulations) or exposures except for medical procedures and round-trip airline flights, which are one-time exposures. Black bars depict doses to workers or residents for the Yucca Mountain (YM) transportation program (SS = service station).

uncontrollable risk. In principle, people do have some control over the background dose they receive based on where they choose to live. In practice, however, the great majority of people probably do not explicitly include radiation dose considerations in decisions about where to live. Most people probably do not even know the average background radiation dose in their current location of residence.

Similarly, the radiation received by members of the public from a Yucca Mountain transportation program is also frequently viewed as involuntary and uncontrollable. Again, this is not completely true in principle; people have some control based on where they choose to live. In practice, however, it would be very difficult to make such a choice, given that transportation routes and schedules have not been established by DOE (see Chapter 5).

There is also a qualitative difference between natural background radiation and radiation from a Yucca Mountain transportation program. The former is natural, whereas the latter is the result of human activities. Although exposures to radiation from these two sources have identical biological effects on living organisms, the committee recognizes that some people may view the acceptability of these exposures differently given their different origins.

Under the linear no-threshold risk model, background doses would be expected to elevate the risk of a fatal cancer.[49] The use of background radiation in risk comparisons has been criticized by some (e.g., MacGregor et al., 2002a,b) because it implicitly suggests that anthropogenic exposures of the same magnitude as background radiation are acceptable to society and it does not address the possible effects of cumulative exposures. The committee uses natural background radiation to give interested readers an established benchmark for making comparisons and makes no value judgments about the acceptability of doses at background levels.

Table 3.10 and Figure 3.3 also present the radiation doses received from a small number of common medical treatments. Comparisons using medical diagnostic procedures might be viewed by some people as inappropriate because the circumstances under which they are received are qualitatively different from spent fuel and high-level waste transportation: medical diagnostic procedures are voluntary and familiar and are widely perceived as having positive health benefits. The committee acknowledges these differences, but nevertheless decided to use medical procedures in its comparisons precisely because they would be familiar to many readers. The committee selected medical diagnostic procedures that represent relatively high, medium, and low radiation exposures to aid readers in making comparisons.

The entries in Table 3.10 and Figure 3.3 are arranged from high to low dose to provide a visual comparative ranking. This "risk ladder" is commonly used to display comparative information (Covello et al., 1989). Several noteworthy observations can be made. First, according to analyses presented in the final Yucca Mountain EIS (DOE, 2002a), maximally exposed workers (primarily transportation crews, escorts, and inspectors) for the Yucca Mountain transportation program (first column in Table 3.10) are assumed to receive annual doses at the limits of the current international standards and DOE administrative limits shown in the second column of the table. The final EIS (DOE, 2002a, p. 6-43) notes that "individual crew members who operated legal weight trucks and escorts for rail shipments could be exposed to as much as 48 rem over 24 years of operation (maximum exposure of 2 rem each year)." In practice, this probably means that DOE will monitor worker doses to ensure that they do not exceed these limits. These exposures will be about seven times higher than average annual natural background radiation doses and about twice as high as the dose received in a whole-body CT (computed tomography) scan.

[49]However, epidemiologic studies have not observed an association between background exposures and latent cancer incidence or fatalities.

According to the final EIS, under the mostly truck scenario, the maximally exposed service station worker could receive a dose of about 1.3 mSv (130 mrem) per year, which would exceed current allowable dose limits to members of the public of 1.0 mSv (100 mrem) per year from all anthropogenic, nonmedical sources. The final Yucca Mountain EIS (DOE, 2002a, p. 6-40) notes that measures would be taken to keep this dose at or below the 1.0 mSv (100 mrem) limit. It should be noted that this worker dose estimate is very conservative: it assumes that every spent fuel and high-level waste truck shipment bound for Yucca Mountain stops at the service station during the 1800 hours the worker is on duty each year for 24 years.

Maximally exposed members of the public are estimated to receive substantially lower annual radiation doses from a Yucca Mountain transportation program as shown by the two bottom-most entries in the first column of the table. The maximally exposed resident near a rail stop (for the mostly rail scenario) would receive an annual dose of about 0.12 mSv (12 mrem). This is roughly equivalent to the dose from about one chest X-ray or one round-trip airline flight between New York and Tokyo. The maximally exposed resident near a rail route (again for the mostly rail scenario) would receive about 0.0007 mSv (0.07 mrem), which is about 6 percent of the dose received in an X-ray to a human extremity (e.g., hand or foot).

3.3.2 Transport Accident Risks

Given the robust construction of spent fuel transportation packages and the rigorous regulatory requirements for transporting them (Chapter 2), significant releases of radioactive material are very unlikely except possibly in extreme accidents, as indicated by the studies in Section 3.1.2. The final Yucca Mountain EIS estimates that the probability of such accidents is very low: 2.3 in 10 million per year for trucks to 2.8 in 10 million per year for trains (Table 3.8). This EIS also estimates exposures from releases in such accidents (Table 3.8): Estimated exposures in a maximally reasonably foreseeable accident would range from about 1100 person-rem for truck accidents to 9900 person-rem for train accidents. The maximally exposed individual is estimated to receive between 3 and 29 rem of radiation, which would be insufficient to cause acute radiation sickness or death. This exposure is estimated to produce between 0.5 and 5 latent cancer fatalities.

The committee provides a comparison of the potential consequences of extreme accidents involving spent fuel transportation packages with those for other types of hazardous materials transport using cumulative complementary distribution functions. The construction of these functions for accidents involving a loaded spent fuel package is described in Section 3.1.2.

For this comparison, the committee used the mean CCDF for accidents involving rail transport of PWR spent fuel that was analyzed in the reexamination study (Sprung et al., 2000) and is shown in Figure 3.2. The committee compared this mean CCDF to those for accidents involving rail transport of three other kinds of hazardous materials: a flammable liquid (methanol), a flammable gas (propane), and a toxic gas (chlorine). These materials were selected because they behave differently under accident conditions and produce a wide range of consequences.

The CCDFs for these hazardous materials were estimated using a computer model that was designed to study the risks of hazardous materials shipments by rail.[50] The model was a joint effort of the Chemical Manufacturers Association (now the American Chemistry Council), the Association of American Railroads, and the Railway Progress Institute. The model was designed to be used by the participating associations and their member companies to evaluate changes to rail hazardous material transportation equipment, routings, and operating practices and to evaluate the effectiveness of options for reducing the risk of accident-caused hazardous material releases from tank cars through such changes. The model has not been published in the open literature, but it was peer reviewed during its development.

The overall model has two main components:

1. A frequency submodel that provides an estimate of the probability of occurrence and size of a release as a function of railroad operating factors (e.g., speed, track class) and tank car type
2. A consequence submodel that provides estimates of the consequences of a release of a defined volume of a specific chemical for human and/or environmental impact[51]

The model also has an extensive database that contains accident rates, benefits of risk reduction options, release probabilities, ignition probabilities, spill size distributions, basic sets of weather conditions, chemical properties for eight preselected materials, and other information needed to run the model. The eight materials included in the model are acetaldehyde, ammonia, chlorine, ethylene oxide, methanol, propane, sodium hydroxide, and styrene.

---

[50]Inter-Industry Rail Safety Task Force's Detailed Rail Model, Version 2.0, 1996.

[51]SuperChems™ is the consequence modeling package used within the model to generate the hazard zones for potential human impacts.

The model is designed to run "projects." A project is defined by

- A hazardous material of interest,
- A designated type of railcar,
- A specified number of trips (trains) of interest, and
- A route, which includes information on length, track class, train speed, train length, number of hazardous material cars of interest per train, population density (for human impact) and/or soil type (for environmental impact).

Each route is generally subdivided into pieces, called segments, within which the variables listed above are essentially constant.

This model was used to calculate CCDFs for a single railcar carrying 20,000 gallons of three types of hazardous materials (chlorine, propane, and methanol) being transported in a general train with typical train speeds, track conditions, and train lengths. The number of shipments (100), the route lengths (about 1600 miles), and the population densities along the shipping routes were approximately the same for these three types of hazardous material shipments and the spent fuel shipments. The results should not be taken as exact estimates, but they are useful for comparison purposes.

The results of the calculations are plotted in Figure 3.4. The horizontal axis represents the number of expected fatalities from an accident having an annual frequency shown on the vertical axis. The CCDF for spent fuel shown on the figure was plotted by multiplying the mean CCDF curve shown in Figure 3.2 by a nominal probability coefficient of 5.75×10^{-2} fatal cancer per sievert (5.75×10^{-4} fatal cancer per rem)[52] (EPA, 1998b, Table 7.3).

Several features of this plot are noteworthy: First, the mean CCDF for chlorine has a relatively flat shape and has the highest accident frequencies and fatalities of the four cases examined. Chlorine gas is highly toxic and can be fatal if inhaled. Once released, gas can be dispersed widely by wind and can have adverse consequences even at relatively low concentrations. Thus, accidental releases can have adverse consequences even in lightly populated areas and can produce many casualties in densely populated areas.

Accidental releases of flammable gases such as propane can have similar consequences to toxic gas releases, but the expected number of fatalities is lower. The primary consequence of concern in a flammable gas release is an explosion or large fire. This requires an ignition source to be present when the proper fuel-air mixture is attained following the accident. The

[52]For low-dose, low linear energy transfer radiation, assuming uniform irradiation of the body. This nominal probability coefficient is age and gender averaged.

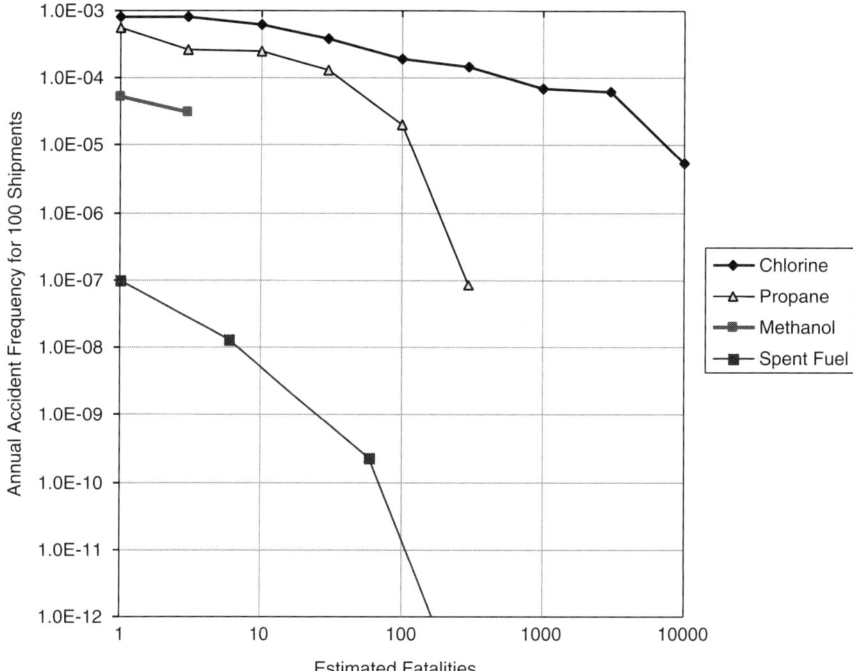

FIGURE 3.4 Complementary cumulative distribution functions showing expected fatalities from hypothesized accidents during transport of three types of hazardous materials and spent fuel. Calculations are explained in the text.

consequences of an explosion or fire will be more localized than a toxic gas release; hence, the expected fatalities from accidental releases are lower.

Accidental releases of flammable liquids such as methanol would be expected to have even fewer consequences. These consequences will generally be more localized to the area of the spill, and not all releases will yield fatalities. The mean CCDF is truncated at about three fatalities due to these localized effects.

The mean CCDF for accidental releases of radioactive material from spent fuel packages has the same general shape as the CCDF for propane. However, the frequency of accidents that lead to such releases is expected to be four to five orders of magnitude lower because of the robust construction of the transportation packages. Figure 3.4 shows that on a comparative basis, the likelihood of extreme accidents that would lead to fatalities is several orders of magnitude lower for spent fuel than for the other hazardous materials shown in the figure. In Section 3.1.2, the committee notes that the risk estimates provided in the reexamination study (Sprung et al., 2000) from which this CCDF was taken are likely to be neither realistic nor bounding and may overestimate the risks. In other words, the risk estimates for accidents involving spent fuel shown in Figure 3.4 may be higher than is actually the case.

3.4 TRANSPORTATION RISKS: FINDINGS AND RECOMMENDATIONS

The committee concludes this chapter with findings and recommendations in response to the first three charges of its original statement of task shown in Sidebar 1.1:

FINDING: There are two potential sources of radiological exposures from transporting spent fuel and high-level waste: (1) radiation shine from spent fuel and high-level waste transport packages under normal transport conditions; and (2) potential increases in radiation shine and release of radioactive materials from transport packages under accident conditions that are severe enough to compromise fuel element and package integrity. The radiological risks associated with the transportation of spent fuel and high-level waste are well understood and are generally low, with the possible exception of risks from releases in extreme accidents involving very long duration, fully engulfing fires. While the likelihood of such extreme accidents appears to be very small, their occurrence cannot be ruled out based on historical accident data for other types of hazardous material shipments. However, the likelihood of occurrence and consequences can be reduced further through relatively simple operational controls and restrictions and route-specific analyses to identify and mitigate hazards that could lead to such accidents.

RECOMMENDATION: Transportation planners and managers should undertake detailed surveys of transportation routes to identify potential hazards that could lead to or exacerbate extreme accidents involving very long duration, fully engulfing fires. Planners and managers should also take steps to avoid or mitigate such hazards before the commencement of shipments or shipping campaigns. (See also the recommendation to transportation regulators in Chapter 2 on operational controls and restrictions on spent fuel and high-level waste shipments to reduce the chances that such hazards might be encountered in actual service.)

The finding that "radiological risks . . . are well understood and are generally low" is based on a large set of observational data and studies described in this chapter and in Chapter 2. These include the following:

- Rigorous international standards and U.S. regulations for the design, construction, testing, and quality assurance of spent fuel packages, including the built-in safety margin requirements for package designs (see Chapter 2); and more than four decades of worldwide experience in transporting spent fuel (Section 3.1). Although there have been accidents and incidents, to the committee's knowledge there has never been a large-scale release of radioactive materials reported from the failure of a spent fuel package during an accident. The broad sharing of information on experiences and best practices by transportation planners, implementers, and regulators through organizations such as the IAEA promotes the continued maintenance of this safety record.
- Full-scale crash testing of transport packages under severe accident conditions (Section 2.3). These tests show that properly constructed spent fuel packages can withstand severe accidents without a loss of containment that would result in releases of radioactive material that exceed regulatory limits. These tests also illustrate that the regulatory requirements for spent fuel packages (e.g., free-drop tests) produce in many cases more severe tests of package integrity than do severe accidents.
- A series of increasingly sophisticated analytical and computer modeling studies of spent fuel transport package performance (Section 3.1.2). The most recent of these studies (Sprung et al., 2000; DOE, 2002a) have attempted to estimate risks using actual spent fuel transport package, fuel, route, and severe accident characteristics and generally conservative assumptions and models.
- Other studies that examine the mechanical and thermal loading conditions from severe accidents that did not involve spent fuel transport (Section 2.2.3). These studies have shown that with the possible exception of very long duration fires, the loading conditions from these accidents would not have exceeded regulatory limits.

Of course, spent fuel transportation is not risk-free, and past experience is not necessarily a useful predictor of future performance. The fact that spent fuel transportation risks have been low in the past does not necessarily mean that risks will also be low in the future. Future risks depend on a number of factors including the quantities and ages of spent fuel transported, associated scaling issues related to the overall size of the transport program, transport modes, and the care taken in fabricating and maintaining transport packages and executing transportation operations. Ongoing vigilance by regulators and shippers will be essential for maintaining low-risk programs in the future, especially for the scale-up and operation of large-quantity shipping programs. Any accident or terrorist attack that results in the large-scale release of radioactive material into the environment would likely have worldwide implications and could result in a temporary or even permanent halt to ongoing transportation programs for spent fuel in the United States.

The recommendation calls for transportation implementers to survey the routes they plan to use for spent fuel and high-level waste to identify hazards that could lead to very long duration fires. This recommendation arises from the finding in Chapter 2 that very long duration, fully engulfing fires might produce thermal loading conditions sufficient to compromise package containment effectiveness. The recommended survey would involve traveling the route in advance of a shipment (or shipping campaign if several shipments are planned) to identify

- Facilities close to the route that use or store large quantities of flammable materials (e.g., refineries, petroleum and gas storage tanks);
- Large-volume flammable hazardous material shipments along the routes to be used; and
- Other route conditions (e.g., the presence of multitrack tunnels, bridges, rail yards, and sidings, as well as remote locations) that could make it difficult to deploy an effective firefighting capability.

Once these conditions have been identified, implementers can take steps to avoid or mitigate these hazards. For example, routes can be altered to avoid multitrack train tunnels, and time spent in rail yards and sidings, where packages could be exposed to other trains carrying large amounts of flammable materials, can be minimized. Where such hazards cannot be avoided completely, shipments can be scheduled to minimize encounters with other hazardous materials trains, or emergency response preparedness can be improved along specific route segments of concern.

The committee judges that none of these recommended survey and mitigation actions would be difficult or expensive to implement. Transportation implementers and regulatory authorities now routinely survey routes to identify other safety and security concerns prior to shipping spent fuel. This recommended action simply represents an expansion of an activity that many implementers already carry out on a routine basis.

FINDING: The *social risks* for spent fuel and high-level waste transportation pose important challenges to the successful implementation of programs for transporting spent fuel and high-level waste in the United States. Such risks, which can result in lower property values along transportation routes, reductions in tourism, and increased anxiety, have received substantially less attention than health and safety risks, and some are difficult to characterize. Current research and practice suggest that transportation planners and managers can take early proactive steps to characterize, communicate, and manage the social risks that arise from their operations. Such

steps may have additional benefits: they may increase the openness and transparency of transportation planning and programs; build community capacity to mitigate these risks; and possibly increase trust and confidence in transportation programs.

RECOMMENDATION: Transportation implementers should take early and proactive steps to establish formal mechanisms for gathering high-quality and diverse advice about social risks and their management on an ongoing basis. The committee makes two recommendations for the establishment of such mechanisms for the Department of Energy's program to transport spent fuel and high-level waste to a federal repository at Yucca Mountain: (1) expand the membership and scope of an existing advisory group (Transportation External Coordination [TEC] Working Group; see Chapter 5) to obtain outside advice on social risk, including impacts and management; and (2) establish a *transportation risk advisory group* that is explicitly designed to provide advice on characterizing, communicating, and mitigating the social, security, and health and safety risks that arise from the transportation of spent fuel and high-level waste to a federal repository or interim storage. This group should be comprised of risk experts and practitioners drawn from the relevant technical and social science disciplines and should be convened under the Federal Advisory Committee Act or a similar arrangement to enhance the openness of its operations. Its members should receive security clearances to facilitate access to appropriate transportation security information. The existing federal Nuclear Waste Technical Review Board, which will cease operations no later than one year after the Department of Energy begins disposal of spent fuel or high-level waste in a repository, could be broadened to serve this function.

This finding and recommendation spring from several factors: Social risk is a poorly understood phenomenon; expert opinion frequently differs; DOE does not, to the committee's knowledge, have any precedent to guide its understanding and management of social risks; and most transportation program staff are not likely to be well acquainted with either theory or practice on this issue. Consequently, the committee concluded that broad input and advice on social risks will be essential to the establishment and ultimate success of programs to transport spent fuel and high-level waste to a federal repository or interim storage.

The recommendation represents pragmatic steps that transportation implementers can take immediately and at relatively low cost to better understand and (working with affected communities) manage the social risks from their programs. These groups are not intended to undertake research on risk. Instead, the committee intends that they have a practical, problem-solving focus and be committed to working closely with program staff to help it become more effective in carrying out the program's mission. One of the most important functions of these advisory groups would be to foster continuous learning and improvement.

The recommendation to expand the scope and membership of the TEC Working Group builds on and complements existing public participation and communication activities within DOE's transportation program for Yucca Mountain. The TEC Working Group is now comprised of state, tribal, local, and industry representatives, and it provides a conduit for communication and advice on topics such as emergency response, inspection and enforcement, training, and public information. The committee recommends that the membership of TEC be expanded to include social risk experts and representative stakeholders from affected communities to provide information on social risks of DOE's transportation operations and their management.

The committee also recommends the establishment of a separate transportation risk advisory group that would advise DOE on characterizing, communicating, and mitigating the social, security, and health and safety risks to communities near transportation routes. The suggestion that the Nuclear Waste Technical Review Board could be broadened to serve this function is intended to take advantage of an established capability within the federal government. This group is independent of DOE and its membership is drawn from the scientific and technical communities. The procedures for nominating and appointing members to this board (i.e., presidential appointments based on nominations by the National Academy of Sciences) are designed to ensure that it is balanced and credible to carry out its mission.

Finally, although this recommendation is focused primarily on DOE, it also applies to any large-quantity shipping program, including the program to ship commercial spent fuel to centralized interim storage (e.g., Private Fuel Storage, LLC, in Utah).

4

Transport of Research Reactor Spent Fuel to Interim Storage

Since the 1950s, the Department of Energy (DOE), other federal civilian agencies, and U.S. universities have regularly transported spent nuclear fuel from research reactors[1] to DOE facilities. These shipments have passed through many regions of the country. These transportation programs have at times been controversial and have led to conflicts between DOE and state governments, and DOE has been compelled to revise and improve its practices regarding evaluation, planning, and consultation with states and tribes. This chapter responds to the U.S. Department of Transportation's (DOT's) request, as directed by congressional study charge (Sidebar 1.2), for the committee to examine the procedures that are followed in selecting routes for these shipments. The routing of research reactor spent fuel is also a good example of a "current concern" identified in the original statement of task for this study (Sidebar 1.1).

The first section of this chapter describes DOE's involvement in managing spent fuel from research reactors. The second section summarizes provisions of federal regulations governing transportation of spent nuclear fuel that are particularly relevant to the congressional study charge. DOE routing practices for research reactor spent fuel shipments are described in the

---

[1]As the term is used here, *research reactors* are small nuclear reactors used primarily to conduct research, to develop theoretical practices, and for education or medical purposes. Their output is typically a fraction of a percent of the output of a commercial electric utility reactor. They serve as sources of neutrons for spectrographic and radiographic applications and for the manufacture of isotopes for medical and other uses.

third section, in the order presented in the study charge. The final section presents the committee's findings and recommendations.

4.1 DOE MANAGEMENT OF RESEARCH REACTOR SPENT FUEL

DOE has responsibilities for managing spent nuclear fuel from three categories of research reactors (Table 4.1):

1. Research reactors located at DOE facilities: there are two operating reactors, one at Oak Ridge National Laboratory in Tennessee (the High Flux Isotope Reactor) and one at Idaho National Laboratory (the Advanced Test Reactor).
2. Foreign research reactors located in 41 countries that use fuel manufactured in the United States from fissionable material provided by the U.S. government under the Atoms for Peace Program.
3. Research reactors operated by U.S. universities, U.S. government agencies other than DOE, and private-sector firms. All such reactors are required to be licensed by the U.S. Nuclear Regulatory Commission (USNRC). As of July 2005, there were 33 operating research reactors and 11 in the process of decommissioning (USNRC, 2005c) (Figure 4.1).

TABLE 4.1 DOE Research Reactor Spent Fuel Management Activities

Activity	Points of Origin[a]	Packages Shipped, 1996–2004	USNRC Approval of Shipment Route Required?	DOT Highway Routing Regulations Apply?
FRR spent fuel acceptance	2[b]	168	yes, by DOE policy	yes
Non-DOE U.S. research reactors	36	45[c]	yes	yes
DOE research reactors	2	93[b]	no	yes

NOTE: FRR = foreign research reactors. USNRC = U.S. Nuclear Regulatory Commission.

[a]Number of places within the United States where shipments of research reactor spent fuel originated.
[b]Charleston Naval Weapons Station and DOE Savannah River Site. Future shipments from Canada are also possible.
[c]Through 2002.

SOURCE: DOE (2004c); DOE Office of Environmental Management, written communication.

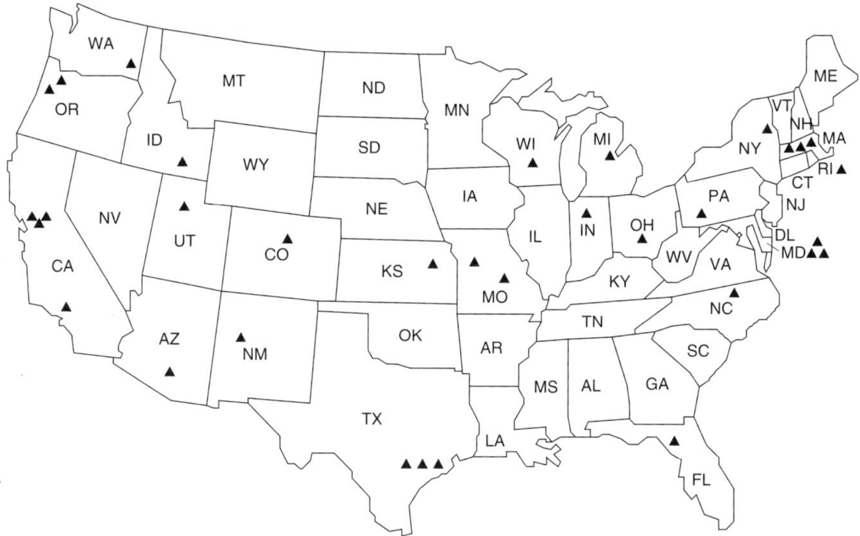

FIGURE 4.1 Sites of U.S. domestic non-DOE research reactors. SOURCE: Modified from USNRC (2003).

DOE provides interim storage for the spent nuclear fuel it receives from these reactors and is responsible for preparing that fuel for eventual shipment to a federal repository for disposal.

Research reactor fuel includes highly enriched uranium (HEU), fuel enriched in the fissile isotope uranium-235 to concentrations of 20 percent or greater, as well as low enriched uranium (LEU). By the 1970s, most research reactors in the United States and abroad were using HEU.[2] HEU fuel contains material that is potentially usable in nuclear weapons and is therefore a nuclear proliferation concern. Also in the 1970s, the United States began programs to promote the conversion of research reactors to LEU and to return all U.S.-origin HEU to the United States, with the goal of eliminating HEU in civilian applications worldwide (GAO, 1994, 2004a, pp. 10–11). Today, DOE transports and stores both HEU and LEU spent fuel from some foreign and U.S. research reactors (GAO, 2004a, pp. 11–22; 2004b, p. 28).

[2]HEU was used for applications that were thought not to be possible with LEU reactors. Use of HEU also allowed some economies, in part because less frequent refueling was required (GAO, 2004a, p. 10).

Research reactor spent fuel is received and stored at two DOE facilities: the Savannah River Site in South Carolina and the Idaho National Laboratory. Savannah River stores research reactor spent fuel containing aluminum-uranium matrices and aluminum cladding. Idaho National Laboratory stores other types of spent fuel, for example, stainless steel-clad fuel with uranium-zirconium matrices.

Since 1996, between about 20 and 60 packages containing research reactor spent fuel have been shipped annually to Savannah River or Idaho National Laboratory. Foreign research reactor spent fuel accounts for about 55 percent of the packages shipped during this period (Figure 4.2). Domes-

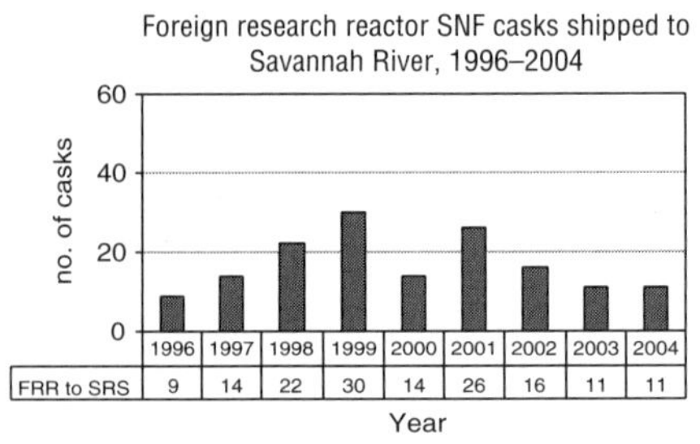

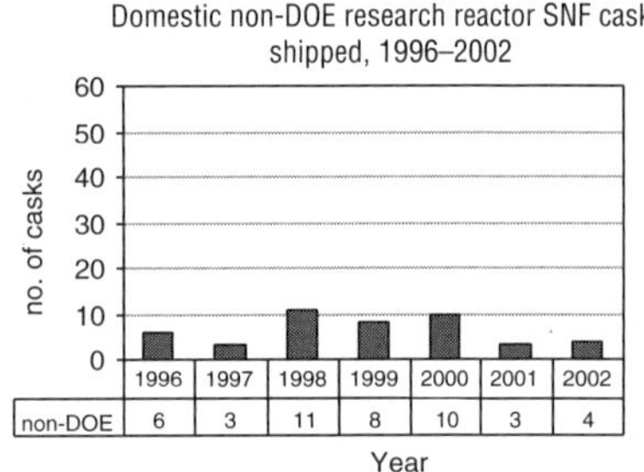

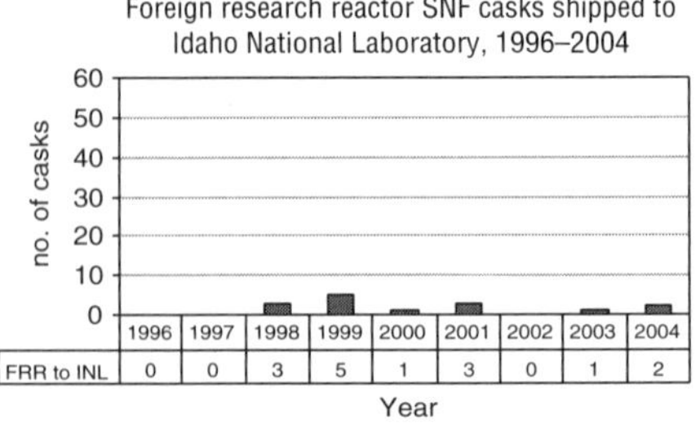

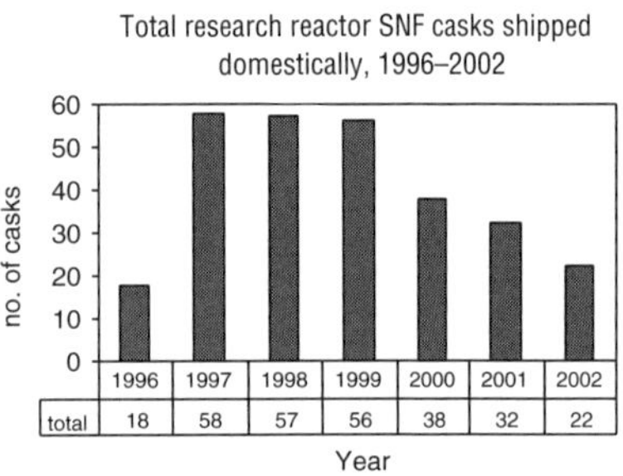

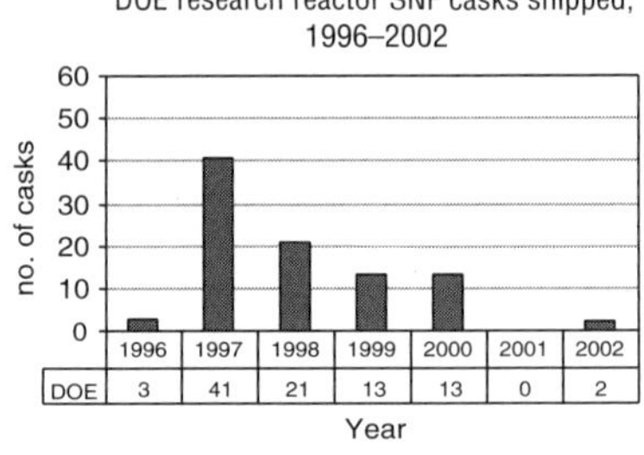

FIGURE 4.2 Numbers of research reactor spent fuel casks shipped domestically, 1996–2004. NOTE: SNF = spent nuclear fuel. SOURCES: DOE (2004c); DOE Office of Environmental Management, written communication.

tic research reactor spent fuel is transported to Savannah River or Idaho by truck (Figure 4.3). Foreign research reactor spent fuel arrives from overseas at the Charleston Naval Weapons Station in South Carolina and is transported to Savannah River (141 packages by rail and 9 packages by truck during 1996–2004). Foreign research reactor spent fuel to be stored at Idaho National Laboratory is transported from Savannah River to Idaho National Laboratory (12 packages during 1996–2004, all by truck). In addition, three packages of foreign research reactor spent fuel that were landed at Concord Naval Weapons Station in California in 1998 were shipped to Idaho National Laboratory by rail. Also, three packages have been shipped from Canada to Savannah River by truck.

DOE has prepared Environmental Impact Statements (EISs) evaluating its management of research reactor spent fuel (DOE, 1995b, 1996a). These EISs and the Records of Decision (RODs) that followed (DOE, 1995c, 1996b) described the anticipated scope of DOE's research reactor spent fuel transportation activities (Table 4.2). Projections from the EISs indicate the

FIGURE 4.3 The package pictured above shows the GE Model-2000 package, which is used to transport research reactor spent fuel and other byproduct, source, or special nuclear materials. This package has a specially fabricated liner and basket for shipping High Flux Isotope Reactor spent fuel from Oak Ridge to the Savannah River Site. SOURCE: DOE (2001e).

TABLE 4.2 EIS Projections of Quantities of Research Reactor Spent Fuel to be Shipped

	To be Shipped During Program[a]		
Origin of Spent Fuel	MTHM	Packages	Program Duration
Foreign research reactors			
Stored at SRS	19	815	1996–2019[b]
Stored at INL	1	162	1996–2019[b]
DOE domestic	1.8		1996–2035[c]
Non-DOE domestic	5.5		1996–2035[c]

NOTE: INL = Idaho National Laboratory; MTHM = metric tons of heavy metal; SRS = Savannah River Site.

[a]The quantities of actual shipments probably will differ considerably from these projections, which were prepared in 1995 and 1996.

[b]DOE originally scheduled the foreign research reactor program to end in 2009 (DOE, 1995b), but in 2004 DOE extended it to 2019 (DOE, 2004a).

[c]2035 is the planning horizon of the EIS but not necessarily the end of the programs.

SOURCES: DOE (1995b, Table 1.1; 1996a, p. S-21).

order of magnitude of expected shipments, but actual quantities will differ because of changes in the utilization of reactors and because some countries that received research reactor fuel from the United States have decided not to ship it back. By 2004, 29 percent of the quantity (measured in numbers of fuel assemblies) of foreign research reactor spent fuel receipts anticipated in the 1996 EIS had been shipped, but presently scheduled shipments would bring the total over the life of the foreign research reactor acceptance program to only about half the quantity projected in the EIS (DOE, 2005a).

Because research reactors are small compared to commercial power reactors, the sizes of shipments of research reactor spent fuel are also relatively small. Most types of transportation packages used for research reactor fuel have loaded weights of 40,000 pounds (about 18,200 kilograms) or less, small enough to be carried by a legal-weight truck (DOE, 2004b).

DOE, employing commercial carriers, is directly responsible for the transportation of spent nuclear fuel from its own research reactors and for transportation from Savannah River to Idaho National Laboratory. DOE oversees all aspects of the planning and conduct of shipments from Charleston Naval Weapons Station to Savannah River. However, in some cases, the shipper makes arrangements for contracting with a commercial carrier to transport the fuel. DOE does not arrange the shipment.

Shipment of spent nuclear fuel from U.S. university and other domestic research reactors, including selection of routes to comply with DOT regulations and submission of routes to the USNRC for approval as required by USNRC regulations (see Table 1.3), is the responsibility of reactor licensees and their commercial carriers. Because DOE plays no role in route selection for these shipments, this transportation activity is not within the scope of the congressional charge for the present study (Sidebar 1.2). However, the committee examined practices for these shipments because this experience is relevant to the problem of route selection and to the committee's original task statement (Sidebar 1.1) to identify technical and societal issues concerning spent fuel transport. The management of transportation of spent fuel from the two sources (foreign research reactors and U.S. university reactors) demonstrates two alternative organizational approaches: assigning responsibility to DOE versus leaving responsibility with the private owners of the fuel.

4.1.1 Controversies Regarding Shipment of Research Reactor Spent Fuel

The questions about DOE's routing practices that are embodied in the congressional charge arose as a result of past DOE shipments of research reactor spent fuel. A review of the history of some of these controversies is helpful in understanding the intent of the study charge.

Research reactor spent fuel was shipped in the United States for many decades in quantities equal to or exceeding present shipment rates with relatively little public attention. The first shipment of spent fuel from a foreign reactor under the Atoms for Peace Program occurred in 1958. However, in the 1980s, DOE was challenged in court by environmental organizations, states, and others for failing to comply with National Environmental Policy Act (NEPA) requirements for evaluation of potential impacts of its transportation activities. To justify its activities, DOE had relied on earlier EISs (USNRC, 1977; DOE, 1980) and other evaluations that had concluded that transportation of spent nuclear fuel is generically a safe activity with negligible environmental impacts. Complainants argued that DOE was required to perform analyses of the actual conditions for specific planned shipments and to evaluate alternative shipping routes. Some of the subsequent court rulings found that DOE analyses had been inadequate (DOE, 2002a, pp. 16–19). DOE suspended acceptance of foreign research reactor spent fuel from 1988 to 1994 while new analyses were prepared. In 1995 and 1996, DOE published EISs evaluating its management of research reactor spent fuel and other materials (DOE, 1995b, 1996a) and RODs defining new DOE policies for transporting and storing these materials (DOE, 1995c, 1996b). The contents of these EISs and RODs that are relevant to the committee's study charge are summarized below.

Before DOE's EIS studies were completed, the State of South Carolina challenged DOE plans to store foreign research reactor spent fuel at the

Savannah River Site, which is located in the state. The state's primary concerns centered on the safety of indefinite storage of the spent fuel in South Carolina, rather than on transportation (Schill, 1996, 1997, 1998). The state filed three state lawsuits in federal court that were unsuccessful.

Following the publication of the foreign research reactor EIS (DOE, 1996a) and ROD (DOE, 1996b), the California State government criticized DOE's designation of the Concord, California, Naval Weapons Station as one of the ports of entry for foreign research reactor spent fuel. The state argued that DOE had failed to take into account analyses showing that alternative routes (using ports in Washington or Oregon) were safer than routes using Concord. The state had been one of the parties challenging DOE spent fuel transportation activities in the 1980s. However, the state decided not to take legal action in 1996, so the legal challenge of a California city and county to DOE's transportation plans failed (California, 1998).

Three packages of research reactor spent fuel from Korea were shipped through Concord and then onto Idaho National Laboratory by rail in July 1998. This shipment was made following discussions between DOE and the affected states and tribes concerning routes and other procedures and after extensive preparation of emergency responders along the route. Since 1998, all shipments of research reactor spent fuel from East Asia have arrived at Charleston Naval Weapons Station.

In 2001, DOE decided to send a shipment of three packages of foreign research reactor spent fuel from Savannah River to Idaho National Laboratory via Interstate 70 through Missouri. The state objected, as California had earlier, that DOE was not basing its routing decisions on comparisons of the safety of alternative routes. Missouri had not acquiesced to DOE's plan, announced in 1998, identifying three potential highway routes for shipments from Savannah River to Idaho National Laboratory. Cross-country shipments in 1999 and 2000 had avoided the state, traveling over the alternative route through Illinois and Iowa instead (see Figure 4.4). The development of the Savannah River-to-Idaho National Laboratory highway routes is described below.

DOE staff reported to the committee that for the 2001 shipment, DOE selected the route through Missouri at the insistence of the USNRC, because it was shorter than the Illinois-Iowa alternative route. The USNRC was insistent that travel time be minimized because one of the packages in the 2001 shipment was nearing the end of its design life. After DOE announced its intention to use the Missouri route in 2001, the state attempted to block use of the route by declaring that Interstate 70 was unsuitable for spent fuel shipments because of high accident rates and because of construction. The state argued that the route through Illinois and Iowa would be safer and questioned the basis for DOE's decision to switch to the Missouri route from the route used for the 1999 and 2000 shipments. A negotiated resolu-

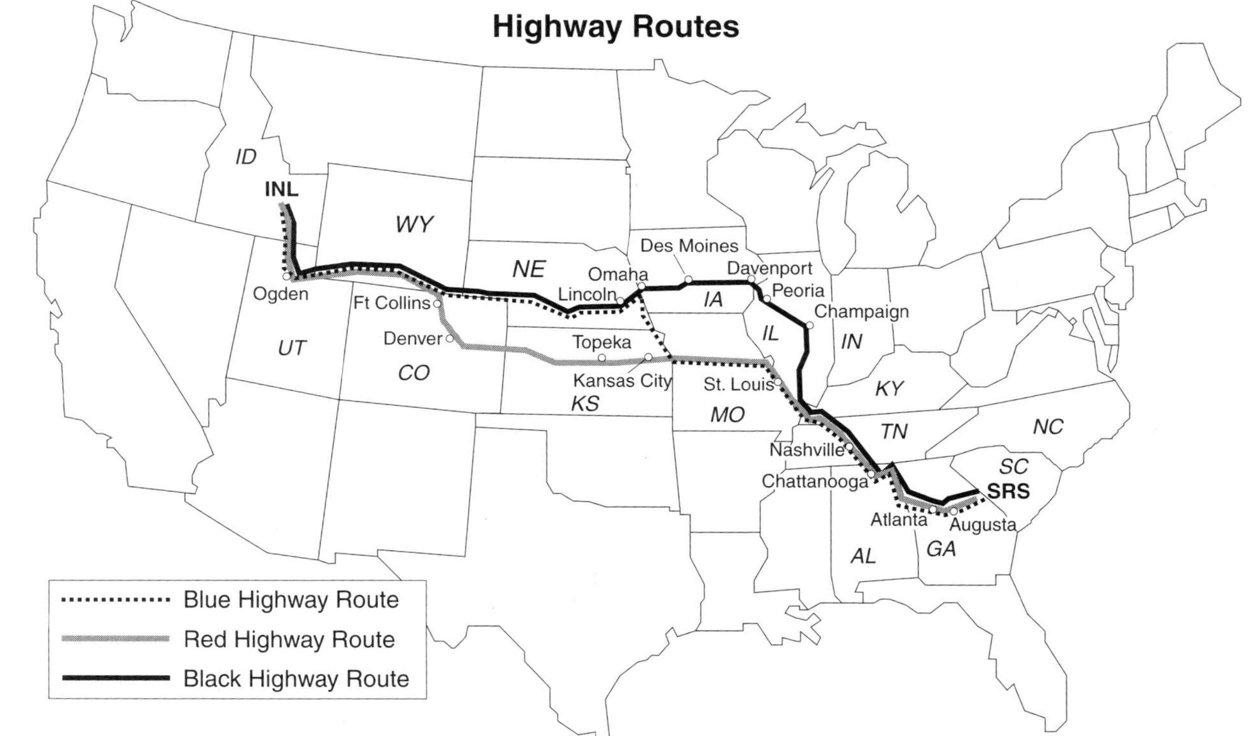

FIGURE 4.4 Highway routes for transportation of foreign research reactor spent nuclear fuel from the Savannah River Site to Idaho National Laboratory. SOURCE: DOE (2003a).

tion was reached that allowed the shipment to proceed as originally planned with state cooperation. This agreement provided for special safety measures in Missouri, including additional vehicle inspections and escorts, training of state personnel, and scheduling to avoid city rush hours. The state later complained that DOE had not abided by all of these commitments, including the schedule for notifying the state, in its management of this shipment (Bell, 2001; DOE, 2001b,c; Holden, 2001; Shields, 2001).

In summary, the history of these controversies shows that states and others have challenged DOE to justify its selection of specific routes for its shipping campaigns. As the description of DOE route selection practices below indicates, DOE generally has not based route selection on quantitative comparisons of risks of alternative routes, but rather on application of the routing rules contained in DOT regulations, taking into account advice from states and tribes along potential routes.

4.2 REGULATIONS GOVERNING SELECTION OF ROUTES FOR SHIPPING SPENT FUEL

DOT and USNRC regulations that affect the selection of routes for domestic shipments of spent nuclear fuel are described briefly in Section 1.2.3. Certain provisions of these regulations, relating specifically to the route selection practices relevant to this chapter, are summarized below.

The DOT administers regulations governing the routing of highway shipments of spent nuclear fuel (49 CFR 397.101 and 397.103) that apply to any shipment of a quantity of radioactive material meeting the regulatory definition of "highway route controlled quantity." These require the following:

- The carrier must operate the vehicle containing the material only over "preferred routes." Preferred routes include all Interstate System highways (see Figure 1.1), except where the state has designated an alternative route to a particular Interstate System highway segment, and other state-designated routes.
- A state may designate a preferred route for transport of radioactive materials after it conducts a risk analysis and consults with neighboring states, and must notify DOT of the designation.
- Among preferred routes, the carrier must select the route that minimizes time in transit, except that the carrier must select Interstate System bypasses around cities unless the state has designated an alternative.
- The carrier may deviate from preferred routes only for security reasons (as specified in a required security plan or at the direction of the USNRC), for pickup and delivery of shipments, for rest stops, and for emergencies.

- In traveling between the origin and a preferred route and from a preferred route to the destination, the carrier may select either the shortest-distance route or another route that minimizes radiological risk considering accident rates, transit times, population density, activities, time of day, and day of week, provided this route does not exceed the length of the shortest-distance route by more than 25 miles and is less than five times the length of the shortest-distance route.

Literal application of these regulations would practically dictate a unique highway route in many circumstances. The regulation does not explicitly require the carrier to consider risks of individual routes, except in the circumstance that the shortest-distance route is not chosen for travel to a designated highway from the shipment origin or from a designated highway to the destination.[3] Rather, the primary responsibility of the carrier is to keep to the designated preferred route system, and a burden is placed on the states to make whatever adjustments are necessary in the designated route system to ensure safety.

In a Notice of Proposed Rulemaking, DOT explained its justification for the routing regulation as follows (DOT, 1980, p. 7144): "In view of statistics showing lower accident rates and reduced travel times in travel on Interstate highways, this proposal favors use of the Interstate System. [DOT] believes that in most cases this policy will produce the most significant transportation safety impact reduction and it offers a clear standard for compliance and enforcement purposes." For quantitative support, DOT cites the USNRC's transportation EIS (USNRC, 1977; this EIS is described in Chapter 3 of this report), which, DOT states, determined that restricting carriers of large quantities of radioactive materials to Interstate System highways would be a cost-effective measure (DOE, 1980, p. 7149).

For rail shipments of spent fuel, there are no federal regulations governing route selection analogous to the highway routing regulations. As the following section on DOE's practices for shipping research reactor spent fuel describes, the absence of regulation has not meant in practice that railroads have selected routes for these shipments without government oversight. Historically, DOE has specified rail routes in its contracts with the

---

[3]The regulation states: "Except as provided in paragraph (b) of this section [which requires the carrier to operate on preferred routes and to minimize travel time] . . . , a carrier . . . operating a motor vehicle that contains [a highway route controlled quantity of radioactive material] . . . shall: (1) Ensure that the motor vehicle is operated on routes that minimize radiological risk; (2) Consider available information on accident rates, transit time, population density and activities, and the time of day and the day of week during which transportation will occur to determine the level of radiological risk . . . ." Therefore, following preferred routes will always comply with the regulation, although the carrier apparently is required in addition to consider time of day and activities along the route in planning the shipment.

railroads. The states lack the authority over rail routes that federal regulations give them over highway routes, although DOE has consulted the states on rail route selections.

USNRC regulations (10 CFR 73.37) require that any licensee shipping spent nuclear fuel exceeding a threshold quantity obtain the USNRC's approval of shipment routes. Normally, the transportation services contractor arranging a spent fuel shipment submits the planned route for USNRC review. USNRC approval of a route is valid only for shipments by the party submitting the application and for a term of two years. Rail as well as highway routes must be submitted for review.

The USNRC route review is among the regulatory requirements intended to "minimize the possibilities for radiological sabotage . . . [and] . . . [f]acilitate the location and recovery of spent fuel shipments that may have come under the control of unauthorized persons" (10 CFR 73.37 (a)). The regulation does not specify the security factors that the USNRC takes into account in its review. USNRC staff reported to the committee that proposed routes are examined in detail and that the review considers travel time and distance and, for highway shipments, adequacy of provisions for safe havens, that is, preplanned locations where the vehicle may stop in case of an emergency and receive protection by police or other security forces.

Shipments of spent fuel made directly by DOE or by contractors to DOE are not legally subject to USNRC regulations because DOE is not a licensee. The regulations do apply to shipments performed by non-DOE operators of research reactors, including universities, private firms, and other government agencies. These operators are USNRC licensees. Certain domestic shipments of foreign research reactor spent fuel are carried out by contractors to the foreign operators of the research reactors, and these are subject to USNRC regulations as well. For shipments of foreign research reactor spent fuel carried out by DOE contractors, DOE as a matter of policy seeks USNRC approval of the routes (DOE, 2003a, p. 6; DOE, 2003b, p. 4).

4.3 SELECTION OF ROUTES FOR SHIPPING RESEARCH REACTOR SPENT FUEL

As specified by the Congress, the charge in this task (Sidebar 1.2) is to analyze how DOE

- Selects potential routes for shipment of research reactor spent fuel to DOE facilities;[4]

[4]The congressional study charge refers to shipments "between or among" DOE facilities. This language could be interpreted as limiting the charge to shipments from Savannah River

- Selects a route for a specific shipment;
- Assesses risks of such a route; and
- Considers proximity to population, current traffic and accident data, road quality, emergency response capabilities, and proximity to gathering places.

The questions about routing of shipments of research reactor spent fuel, and also the points of contention in the controversies about specific shipments that led to the congressional request for this study, focus on methods of comparing risks of the selected routes with risks of alternatives and the use of such comparisons in routing decisions.

The steps that were followed by DOE to select routes for shipment of foreign research reactor spent fuel are the same steps that DOE expects to follow in its program to ship commercial spent nuclear fuel to a federal repository: first, development of a plan for the program that identifies sets of potential or candidate routes and, then, selection of a specific route at the time of each shipment, with each step guided by, among other considerations, an assessment of the risks of the favored routes versus alternative routes and modes. These steps are dictated by DOE's general policies regarding transportation of spent nuclear fuel (DOE, 2002b) and are consistent with DOE statements concerning its plans for selecting routes for shipping spent fuel to a federal repository (see Chapter 5). DOE's experience with research reactor spent fuel shipments therefore has general relevance to commercial spent fuel transportation, although the relevance is limited because the scales of the two activities differ greatly. The foreign research reactor spent fuel transport program is a small-quantity shipping program, whereas the transport program to the federal repository will be a large-quantity shipping program (see glossary, Appendix D).

The subsections below address the questions posed to the committee in the congressional study charge: The first is about procedures for selecting potential routes; the second, about selecting routes for specific shipments; and the third, about consideration of population, traffic, accident, and emergency response capabilities in route selection. DOE's methods of assessing risk are described in each of these subsections. The descriptions refer primarily to procedures for selecting routes for shipping spent fuel from foreign research reactors, which is the largest category of shipments for which DOE has direct responsibility (see Table 4.1 and 4.2). The final

---

to Idaho National Laboratory or from Oak Ridge National Laboratory to Savannah River or Idaho National Laboratory. Such shipments constitute only a minority of all domestic shipments of research reactor spent fuel. The committee decided to consider routing practices for all domestic research reactor spent fuel shipments as explained previously.

subsection identifies differences in practices for transport of spent fuel from domestic research reactors.

4.3.1 Potential Routes: Foreign Research Reactor Spent Fuel

DOE has identified the following potential routes for shipments of foreign research reactor spent fuel from Charleston Naval Weapons Station to the Savannah River Site, from Savannah River to Idaho National Laboratory, and from Concord Naval Weapons Station to Idaho National Laboratory:

- Three alternative highway routes from Savannah River to Idaho National Laboratory (Figure 4.4), named the blue, red, and black highway routes and published in DOE's transportation plan for foreign research reactor spent fuel shipments from Savannah River to Idaho National Laboratory (DOE, 2003a, Appendix 8.1). The plan specifies each road segment and junction on the routes.
- Preferred and alternate rail routes and one highway route from Charleston Naval Weapons Station to Savannah River published in the DOE transportation plan for foreign research reactor spent fuel shipments between these points (Figure 4.5) (DOE, 2003b, Appendix 8.1).
- Preferred and alternate rail routes and one highway route from the Concord Naval Weapons Station in California to Idaho National Laboratory announced in advance of the shipment that departed from Concord in 1998 (Figure 4.6) (Nevada, 1997; California, 1998).

These routes were evaluated and published well in advance of shipments, with the intent that routes for specific shipments occurring over a period of years would be selected from among these potential routes. In addition to shipments on these routes, highway shipments of research reactor spent fuel from Canada to Savannah River occurred in 1996 (one package) and 2000 (two packages). No future shipments from Canada are planned at this time (DOE, 2004c, 2005a).

The milestones in development of these potential routes were as follows:

- The EIS on foreign research reactor spent fuel (DOE, 1996a), which identified and evaluated representative truck and rail routes between 10 potential ports of entry and five potential DOE storage sites.
- The ROD on the foreign research reactor spent fuel management program (DOE, 1996b), which identified two ports of entry and two storage sites and declared that the preferred transport mode would be rail.
- Discussions of potential routes with states and tribes through two working groups formed in 1996: the Cross Country Transportation Work-

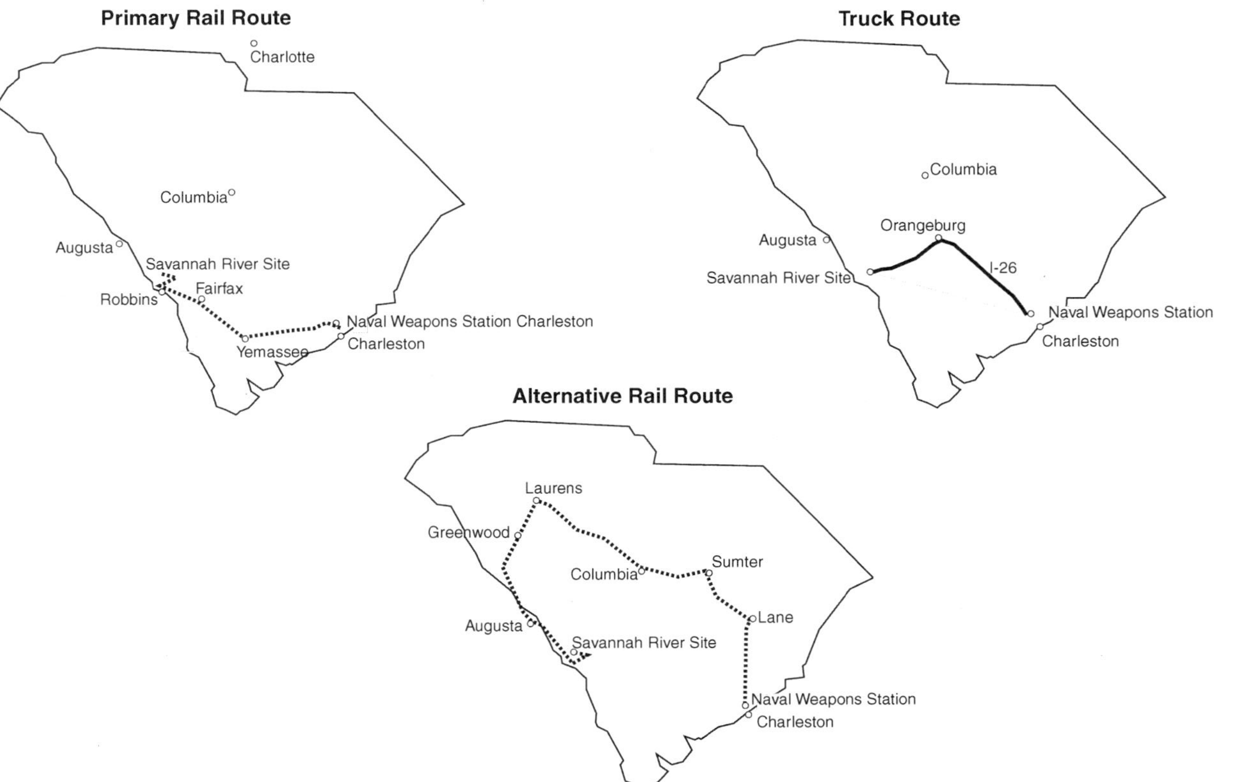

FIGURE 4.5 Highway and rail routes in South Carolina for transportation of foreign research reactor spent nuclear fuel from Charleston Naval Weapons Station to the Savannah River Site. SOURCE: Modified from DOE (2003b).

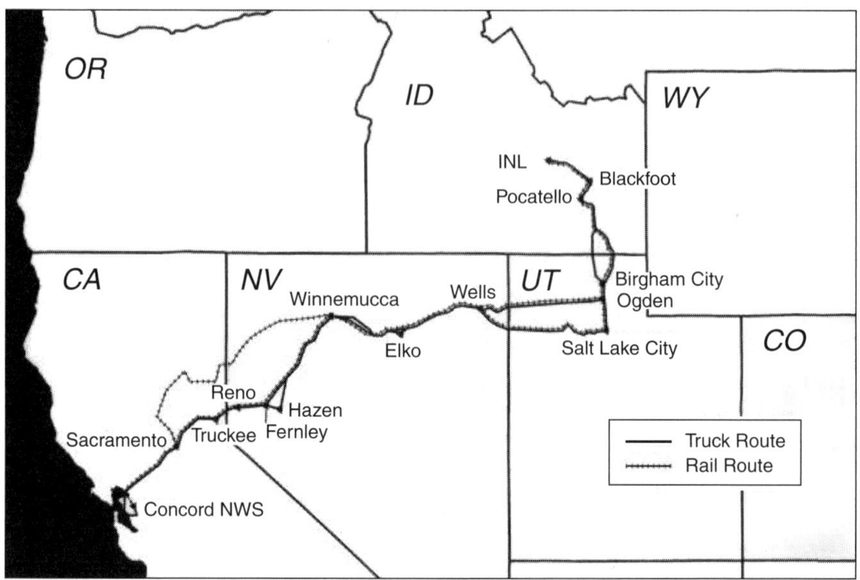

FIGURE 4.6 Highway and rail routes for transportation of foreign research reactor spent nuclear fuel from Concord Naval Weapons Station to Idaho National Laboratory. SOURCE: Nevada (1997).

ing Group, dealing with shipments from Savannah River-to-Idaho National Laboratory, and the Foreign Research Reactor Spent Nuclear Fuel Transportation Working Group, dealing with shipments from Charleston to Savannah River. Similar discussions took place between DOE and the affected states and tribes concerning the Concord-to-Idaho National Laboratory route.

- Publication of transportation plans for shipments from Charleston to Savannah River in 1996, with updates in 1998 and 2003 (DOE, 2003b), and publication of plans for shipments from Savannah River to Idaho National Laboratory, most recently updated in 2003 (DOE, 2003a), specifying the potential routes as well as procedures for scheduling, advance information, and safety and security en route.

Route Assessments in the EIS and ROD

The 1996 EIS evaluated the risks of transportation of foreign research reactor spent fuel over "representative routes," by truck and rail, from

Charleston to Savannah River, Savannah River to Idaho National Laboratory, and Concord to Idaho National Laboratory, as well as other pairs of origins and destinations that were later eliminated from consideration (DOE, 1996a, Appendix E). The risk assessment evaluated specific highway and rail routes, but the selection of routes for evaluation was not intended to indicate that the evaluated routes would necessarily be used for shipments. The EIS states that "specific routes cannot be identified in advance because the route would not be finalized until it had been reviewed and approved by the [US]NRC. The selection of the actual route would be responsive to environmental and other conditions . . . at the time of shipment" (DOE, 1996a, p. E-21).

The routes evaluated were chosen using two models, HIGHWAY and INTERLINE. These employ databases characterizing the highway and rail networks, respectively, with information on speed and distance for each link in the networks and population densities adjacent to each link. The models can be used to search for routes that minimize time or distance traveled (or some function of distance and time) between a specified origin and destination. Highway routes are constrained to comply with DOT regulations regarding use of the Interstate System highways or state-designated alternate preferred routes, use of Interstate System bypasses around cities, and minimizing the distance traveled between origin and destination points and the Interstate System highway (DOE, 1996a, p. E-33).

The risk of fatalities during transport of foreign research reactor spent fuel was estimated in the EIS using RADTRAN (described in Section 3.1.2 and Sidebar 3.4). The model, which was developed by Sandia National Laboratories, can be used to estimate the expected health consequences of human exposures to radiation during routine transport and in severe accidents. The model also can be used to estimate the expected number of conventional fatalities (i.e., vehicular and pedestrian fatalities) during transportation.

RADTRAN was used to estimate the expected fatalities for the entire foreign research reactor shipping program, based on the quantities of materials and numbers of shipments projected in the EIS, for a scenario in which all shipments are transported by truck and with fuel destined for Idaho National Laboratory or Savannah River according to fuel type (DOE, 1996a, Tables E-14 and E-17). The estimates are as follows:

- Fatalities from routine transport: 0.1 to 0.2 LCF (latent cancer fatality)
- Fatalities from releases in severe accidents: 10^{-5} to 10^{-4} LCF
- Conventional fatalities: 0.05 to 0.1 expected fatality

On the basis of these estimates, DOE concluded that "the Final EIS demonstrates that the spent fuel and target material could be safely transported overland within the United States by either truck or rail . . ." (DOE, 1996b, Sec. VII).

The EIS risk analysis did not seek minimum-risk routes or compare the risks of alternative routes or modes between origin-destination pairs. Truck-versus-rail comparisons may be derived from the results presented, but DOE evidently did not regard the differences in risks as significant. DOE states in its ROD that it "will generally seek to use rail" because "there appears to be a strong preference by some members of the public in the port areas for the use of rail" (DOE, 1996b, Sec. I, IX), rather than on the basis of a comparison of truck and rail risks. In response to public comments about the absence of route specifications in the EIS, DOE explains that "conditions could well change [by] the time the shipments would be made. . . . Selection of the actual route would be accomplished in consultation with the affected States, Tribes, local officials, and the carrier . . ." (DOE, 1996b, Sec. VI).

State Working Groups and Transportation Plans

After publication of the ROD, DOE convened two state working groups to serve as forums for consultation with the states and tribes along potential foreign research reactor shipping routes from Charleston to Savannah River and from Savannah River to Idaho National Laboratory. The consultations were to cover road conditions, emergency response capabilities and needs, and any specific state highway route designations. This approach followed the model of consultation on radioactive waste transportation that had been developed for earlier DOE activities, especially the Waste Isolation Pilot Plant (WIPP) program (see Section 5.2.2).

The Foreign Research Reactor Spent Nuclear Fuel Transportation Working Group addressed plans for transportation from Charleston to Savannah River and involved South Carolina officials from health, law enforcement, and emergency response agencies. It was organized with the cooperation of the Southern States Energy Board, an organization formed by an interstate compact of 16 southeastern states providing for cooperative energy and environmental programs.

For highway transportation in South Carolina, DOE originally proposed to the working group the predominantly Interstate System route dictated by DOT routing regulations. This route had been evaluated in the EIS (DOE, 1996a, p. E4) and cited in the Web version of the ROD (DOE, 1996c). The state rejected this route because it is indirect, passes near the urban areas of Columbia and Augusta, and includes an interchange with a high accident frequency. The state proposed another highway route, which was adopted in the Charleston-Savannah River transportation plan (DOE,

2003b, Appendix 8.1). The alternate route follows mostly roads other than Interstate System highways but is shorter than the Interstate System route, avoids the cities of Columbia and Augusta, and avoids the high-accident interchange. DOT has not published this route as a state-designated preferred route. Selection of the alternate route appears to have been based primarily on the local knowledge and professional judgment of the officials involved, although state officials report that a comparison of accident rates on the routes was carried out.

Because the 1996 ROD had declared that rail would be the preferred mode, the DOE-South Carolina working group began by developing procedures for rail shipments. Whereas the states have authority under federal law to regulate highway routes of spent fuel shipments, they have no legal control over rail routes. Nonetheless, primary and alternate rail routes for Charleston-Savannah River shipments (Figure 4.4), as well as safety procedures, were defined by DOE after consultation with the state and discussions involving the railroad and the Federal Railroad Administration. Rail shipments have been by dedicated train, with routes specified in DOE contracts with the carrier. It is part of DOE's arrangement with the state that rail will be used for all shipments from Charleston to Savannah River except that truck may be used when four or fewer packages are awaiting transport. Since 1996, 20 of the 23 shipments from Charleston to Savannah River (most comprising multiple packages) have been by rail.

To consult with the states and tribes on Savannah River-to-Idaho National Laboratory shipments, DOE convened the Cross-Country Transportation Working Group, with support from the Southern States Energy Board and the Council of State Governments-Midwestern Office, an association of midwestern states that coordinates those states' interactions with DOE on radioactive materials transportation. The membership of the working group included representatives of 17 states, two tribal nations, the state regional organizations, and federal agencies (Huizenga et al., 1999). At DOE's request, the state members were gubernatorial appointees.

DOE initially proposed four highway routes to the group, developed with the HIGHWAY model and similar to the "representative routes" in the EIS. These routes were modified according to recommendations of working group members, and one of the four (the green route, departing South Carolina to the north, through North Carolina) was eliminated because of the group's concerns about weather and terrain. Some of the state recommendations were based on more detailed examinations of the physical characteristics of the routes than DOE had carried out. For example, South Carolina recommended an improved access route from Savannah River to the Interstate System. The states also favored routes that had been used earlier for radioactive waste shipments, because emergency responders along these routes had already received training.

During this consultation process, DOE analyzed the transportation risks of the alternative routes using RADTRAN (Weiner and Mills, 1999) and presented the results to the working group. These show small differences among the routes in risks of radiation exposure (ranging from 1 LCF in 10 million trips for the route with the lowest radiation risk to 1 LCF in 6 million trips for the route with the highest risk) and of truck accidents (ranging from 1 fatal truck accident in 16,000 trips to 1 truck accident in 13,500 trips). There is no indication that these estimates had any influence on the initial specification of the routes or on subsequent selection of routes for individual shipments. The green route, the route dropped from consideration because of objections from several states, appears in the RADTRAN estimates to have the lowest risk of radiation exposure fatalities, because it has the lowest total population in proximity to the route, and the second lowest risk of fatalities from truck crashes. It should be noted that many of the working group objections to DOE route proposals arose from particular local conditions that are not taken into account in the RADTRAN estimates. For example, RADTRAN uses state-level average truck accident rates rather than rates specific to individual highway sections.

The understandings that DOE reached with the states and tribes in the working groups regarding routes and transportation procedures for foreign research reactor shipments to Savannah River and Idaho National Laboratory were documented in the two transportation plans (DOE, 2003a, 2003b). These include the following:

- Maps specifying the highway and (for shipments from Charleston to Savannah River) rail routes to be used
- Definitions of the responsibilities of all federal and state agencies involved and of commercial carriers
- Specification of advance notification and shipment tracking practices
- Specification of additional safety practices, including state-by-state vehicle inspection procedures and use of dedicated trains for rail shipments
- A public communications plan
- An emergency response plan that specifies the responsibilities of the parties in the event of an incident during transport
- In the plan for shipments from Savannah River to Idaho National Laboratory, a list of special events and of urban areas with rush hours that the states and tribes asked DOE to avoid in scheduling shipments; DOE agreed to minimize conflicts and to notify the state if a conflict were to arise

4.3.2 Routes for Specific Shipments: Foreign Research Reactor Spent Fuel

All recent and currently planned domestic foreign research reactor spent fuel shipments have been from Charleston to Savannah River or from Savannah River to Idaho National Laboratory. Packages destined for Idaho National Laboratory normally would arrive at Charleston by sea, be shipped by rail or truck to Savannah River, and remain at Savannah River no more than a few days before being shipped by truck to Idaho National Laboratory (DOE, 2003a, p. 9). Selection of a route for a specific shipment therefore entails deciding which of the potential routes published in the two DOE plans for these movements (DOE, 2003a,b) will be used, and checking to determine if any immediate circumstances require modifying the route.

A transportation services contractor organizes transport and all related activities, with oversight from DOE. For shipments originating in high-income economy countries,[5] the foreign reactor operator hires the contractor. For shipments from other than high-income economy countries,[6] DOE hires the contractor. The contract specifies that transportation must comply with the provisions of the DOE transportation plan (DOE, 2003a, pp. 5–9). The contractor is responsible for obtaining approval of the intended route from the USNRC, according to the regulatory requirements of 10 CFR 73.37.

DOE officials reported to the committee that in selection of routes for specific shipments, the following factors are considered:

- DOT highway route selection regulations
- State and tribal advice regarding
  —Road conditions and construction zones
  —Planned events (e.g., sporting events or festivals)
  —Emergency response and radiological training needs
  —Shipment and truck inspection requirements
  —Rush hour periods through cities
- RADTRAN accident analysis
- Shipment schedule, particularly the season of the year

[5]DOE (1996b) identifies the following countries as high-income economy countries: Australia, Austria, Belgium, Canada, Denmark, Finland, France, Germany, Israel, Italy, Japan, Netherlands, Spain, Sweden, Switzerland, Taiwan, and United Kingdom.

[6]DOE (1996b) identifies the following as other than high-income economy countries: Argentina, Bangladesh, Brazil, Chile, Colombia, Greece, Indonesia, Iran, Jamaica, Malaysia, Mexico, Pakistan, Peru, Philippines, Portugal, Romania, Slovenia, South Africa, South Korea, Thailand, Turkey, Uruguay, Venezuela, and Zaire.

- Number and type of packages to be shipped
- Possibility of coordination of the foreign research reactor shipment with shipments from DOE or university facilities
- Any other factor that could affect shipment transit time

DOE reported that the route selection process leads to a recommendation to the Assistant Secretary for Environmental Management for approval.

4.3.3 Consideration of Risk Factors: Foreign Research Reactor Spent Fuel

The study charge asks how route selection takes into account proximity to populations, traffic and accident data, road conditions, emergency response capabilities, and proximity to gatherings. These characteristics of individual trips all affect risk. They depend on the schedule of the trip (time of day, day of the week, and season) as well as the route.

DOE's practice, in dealing with states and tribes through working groups devoted to foreign research reactor spent fuel transportation, has been to place responsibility for detailed review of highway routes on the states and tribes. In particular, DOE has relied on the states' and tribes' local knowledge of accident rates, road and traffic conditions, and public events. Assigning this responsibility to the states is consistent with DOT highway routing regulations, which give states authority to designate preferred routes.

For rail routing, some general guidelines in DOE's *Radioactive Material Transportation Practices Manual* (DOE, 2002c, p. 16) state that DOE is to consider track quality (including guidance from the Association of American Railroads concerning rail lines suitable for carrying spent fuel and other hazardous materials) and "operational input from carriers," and to consult with states and tribes on rail routes. DOE would primarily be dependent on carriers for information on line conditions that would affect safety.

DOE's quantitative risk analysis of representative routes in the 1996 EIS takes into account population density along the routes, but the risk estimation procedure does not employ data for specific road segments about traffic condition, accident rates, road quality, or places of public gatherings or about analogous factors for specific rail lines. These factors have been explicitly addressed in DOE's consultation with the states and tribes. The foreign research reactor transportation plans (DOE, 2003a,b), developed cooperatively with the states and tribes, place responsibility on the states and tribes to identify particular conditions that would affect the safety of a route proposed for an individual shipment. During development of the plans, state officials recommended adjustments to the potential routes based on their knowledge of accident histories and traffic conditions at specific locations. The plan for truck shipments to Idaho National Laboratory lists events, public gatherings, and other circumstances along the routes and stipulates that shipments will avoid "major" special events (DOE, 2003a, p. 11). According to the descriptions of the consultative process presented to the committee by the involved parties, decisions on adjustments to routes and schedules arising from consideration of these factors have sometimes been made without benefit of supporting quantitative analysis.

4.3.4 Domestic Research Reactor Spent Fuel

Foreign research reactor spent fuel transportation is an ongoing program managed directly by DOE. The program conforms to detailed published plans and has received considerable scrutiny in the press, from interest groups, and from the state-federal working groups. Transportation of spent fuel from university and other domestic research reactors, in contrast, is a decentralized activity, managed by individual reactor operators. Each operator is responsible for arranging for transportation from its site to a DOE facility and for ensuring that DOT and (for shipments from non-DOE domestic reactors) USNRC regulations are complied with. Perhaps because of this structure, spent fuel transport from domestic research reactors seems rarely to have been a focus of controversy. Thus, for example, while a single shipment of foreign research reactor spent fuel across Missouri in 2001 led to a federal-state confrontation, the University of Missouri research reactor has shipped spent fuel to Savannah River several times a year for many years without comparable notice.

A university that plans to ship spent fuel from a research reactor that it operates will usually contract with a transportation services firm to arrange all aspects of the shipment. The firms that undertake this work are the same firms that contract with DOE to handle its shipments, because the work requires special equipment and expertise. The contractor selects the route for shipment and submits it to the USNRC for approval. The USNRC review checks for compliance with DOT routing regulations (all shipments in recent years from domestic research reactors have been by truck) and with USNRC's own security requirements. USNRC publishes approved routes. To satisfy USNRC security and notice requirements as well as state procedures, the contractor must coordinate with state public safety officials to arrange for inspections, escorts, permits, and any other special state requirements. Some universities have acted as their own prime contractor, selecting routes themselves and dealing directly with jurisdictions along the routes, and have employed a contractor solely for transportation.

For domestic university research reactor shipments, there is no published plan analogous to the plans DOE prepared for shipments from

Charleston to Savannah River and Savannah River to Idaho National Laboratory (DOE, 2003a, 2003b) and no formal standing multistate working groups have been organized. DOE did consider representative routes between university reactors and DOE facilities in its 1995 EIS on spent fuel management (DOE, 1995b), although that analysis did not compare risks for alternative routes.

DOE has shipped research reactor spent fuel from its facilities at Oak Ridge National Laboratory in Tennessee to Idaho National Laboratory and to Savannah River. Transportation plans have been prepared for these shipments following a format similar to that of the foreign research reactor spent fuel transportation plans. The plans were submitted to the states for comment. DOE procedures for preparation of these plans, including the outline of the plans' contents and provision for state review, are specified in DOE's *Radioactive Material Transportation Practices Manual* (DOE, 2002c).

4.3.5 Discussion

In the committee's judgment, DOE's procedure for selecting transportation routes in the foreign research reactor spent fuel program appears on the whole to be adequate and reasonable. The elements of this procedure have been the following:

- A quantitative risk analysis of representative routes and alternative modes is conducted as part of an EIS, to judge whether the transportation activity meets a threshold standard of acceptable safety (but not for the purpose of choosing among alternative routes). In the case of research reactor spent fuel, these evaluations have always concluded that risks are very low.
- Potential routes are identified following DOT regulations (for highways) in consultation with states and tribes and in discussions with railroads, states, and tribes for rail routes.
- The actual route used for an individual shipment is selected from among the potential routes, again in consultation with states and tribes, after a review of immediate circumstances (e.g., special events, road construction). Actual route choices also have reflected DOE's desire to avoid conflicts and minimize delays.

This procedure reflects DOE's position (which is consistent with DOT regulations) that the states and tribes are competent and responsible for selecting highway routes and, in particular, for having detailed and current local knowledge about accident rates, road and traffic conditions, and events. The route selection process may be described as risk-informed; that is, quantitative estimates of risks are considered alongside other factors, including costs, administrative feasibility, local preferences, and inevitable political considerations. Route selection is not determined wholly by a risk assessment because DOE recognizes that these other valid factors must also be considered.

Experience with research reactor spent fuel shipments indicates that selection of highway routes by complying with DOT highway routing regulations is a reasonable substitute for a process that selects routes through quantitative risk assessments that explicitly compare alternative routes, provided that the shipper actively and systematically consults with the states and tribes along potential routes and that states comply with DOT route designation regulations. Analyses indicate that differences in risk among routes attributable to the factors that current models represent adequately (e.g., distance and population density) are relatively small. Up-to-date, comprehensive, and detailed data on accident rates and other risk factors that would be required for more refined quantitative comparisons of alternative routes do not exist. Factors that cannot readily be incorporated into a quantitative assessment (e.g., emergency response capabilities, schedules of road construction and other transient events) may be predominant influences on differences in risk among alternative routes and must therefore be considered alongside the quantitative risk estimates.

Information on local transport conditions supplied by states and tribes is an essential element in route selection decisions. Detailed state reviews allow for the identification of high-accident-rate segments of the Interstate System as well as the identification of acceptable non-Interstate System routes that would substantially reduce mileage and travel time. Judgments of state officials on such matters are most useful when supported with quantitative evidence.

In controversies over routing, states and others repeatedly have criticized DOE for failure to carry out comparative quantitative risk evaluations of alternative routes. DOE could respond to this concern by developing improved risk evaluation tools for comparative route analysis and by giving the results of such analyses appropriate weight in decision making.

In planning routes for the foreign research reactor program, trade-offs were made whose safety implications were not explicitly analyzed. For example, instances were described to the committee in which route selection was influenced by state officials' preference for one proposed route over an alternative with a lower population density that was believed to have a higher accident rate, without a quantitative assessment of the risk implications. Also, states have sometimes expressed preference for routes on which emergency response personnel have already received the necessary training. This could favor the selection of routes through more densely populated areas if emergency responders there have higher levels of training. As a final example, complying with schedule restrictions on shipments might conceivably entail delaying a vehicle en route rather than driving through a city during peak traffic periods (rush hour). DOE does not have a methodology for quantitatively evaluating the risk implications of such trade-offs.

For rail route selection, DOE's practice of negotiating routes with carriers in consultation with states is analogous to its interaction with states on highway routing. There is no indication that enacting regulations governing rail route selection for spent fuel shipments would improve safety compared with the present DOE practice of contractually specifying routes that have been negotiated with carriers, after consultation with the states and tribes.

Procedures for shipments from university reactors differ from DOE's practices for planning foreign research reactor spent fuel shipments. In particular, route planning for university shipments generally has not included procedures similar to DOE's formal consultations with working groups of state and tribal representatives or DOE's publication of transportation plans specifying routes, shipment procedures, and responsibilities and commitments of all parties. University shipments are not a direct DOE responsibility, so DOE's policies regarding preparation of transportation plans do not apply to them. Nonetheless, if similar plans were published for university shipments, there might be some gains in safety (e.g., because they would help to facilitate discussions between shippers, states, and tribes and could thereby lead to more coordinated shipping operations) and in public understanding of these shipments.

4.4 RESPONSES TO THE STUDY CHARGE

The charge for the committee's task regarding routing of spent fuel shipments from research reactors (Sidebar 1.2) was defined by Congress in Section 334 of the Consolidated Appropriations Resolution, 2003. Each of the main provisions of that charge appears in italics below, followed by a summary of the committee's response. The charge refers only to shipments for which DOE is responsible. These include shipments from DOE reactors and shipments of spent fuel from foreign research reactors. The committee also examined practices for shipments from other domestic research reactors (which are not carried out by DOE) because that experience is relevant to the problem of route selection and to the committee's original task of identifying technical and societal issues concerning spent fuel transport.

> *Sec. 334 (b) . . . the National Academy of Sciences shall analyze the manner in which the Department of Energy—*
>
> *(1) selects potential routes for the shipment of spent nuclear fuel from*

research nuclear reactors between or among existing Department facilities currently licensed to accept such spent nuclear fuel.

DOE has selected potential routes for domestic shipments of foreign research reactor spent fuel since 1996 according to the following procedure: First, quantitative risk analyses of representative routes and alternative modes were conducted as part of the NEPA process and published in EISs. The objective of these analyses was to determine whether the transportation activity would meet a threshold standard of acceptable safety. Second, DOE issued RODs that narrowed the range of potential origins, destinations, modes, and routes, citing considerations of practicality, safety, and preferences expressed in public comments. Third, a set of potential routes was specified in detail, based primarily on DOT regulations (for highways) and following a formal consultation process with affected states and tribes, and after discussions with railroads and states (for rail routes). Finally, potential routes were published in transportation plans that were reviewed by the states and tribes. Planning for shipments from DOE research reactors has followed a similar procedure. The committee's analysis of selection of potential routes appears in Section 4.3.1.

(2) selects such a route for a specific shipment of such spent nuclear fuel.

For shipments of spent fuel from foreign research reactors, the route for each shipment is selected from among potential routes defined in the transportation plans. All such shipments that are now planned will be between Charleston Naval Weapons Station and DOE's Savannah River Site in South Carolina or between the Savannah River Site and Idaho National Laboratory. States and tribes are consulted as part of this selection procedure, and state and tribal preferences are taken into account in making final route selections. DOE may also in the past have taken into account the quality of its working relationships with states along the potential routes in making final selections. Once a route has been selected, states and tribes are required to receive advance notification of shipments and are to inform DOE of conditions within their jurisdictions that may affect these shipments. Before a route is selected for a specific shipment, it is submitted to the USNRC for approval of security provisions. USNRC evaluations have influenced final route selections (see, for example, Section 4.1.1).

Procedures for shipments from DOE research reactors are similar, except that routes for these shipments are not submitted to the USNRC for approval. For shipments from university reactors, normally an agent of the licensee who is managing the transportation submits a proposed route complying with DOT regulations for USNRC review and coordinates the movement with jurisdictions along the route. The committee's analysis of the selection of routes for specific shipments appears in Section 4.3.2.

FINDING: The Department of Energy's procedures for selecting routes within the United States for shipments of foreign research reactor spent fuel appear on the whole to be adequate and reasonable. These procedures are risk informed; they make use of standard risk assessment methodologies in identifying a suite of potential routes and then make final route selections by taking into account security, state and tribal preferences, and information from states and tribes on local transport conditions. The Department of Energy's procedures reflect the agency's position (which is consistent with Department of Transportation regulations) that the states are competent and responsible for selecting highway routes. For rail route selection, the Department of Energy's practice of negotiating routes with carriers in consultation with states is analogous to its interaction with states on highway routing.

RECOMMENDATION: The Department of Energy should continue to ensure the systematic, effective involvement of states and tribal governments in its decisions involving routing and scheduling of foreign and DOE research reactor spent fuel shipments.

(3) conducts assessments of the risks associated with shipments of such spent nuclear fuel along such a route.

The EISs included quantitative risk assessments that produced estimates of expected numbers of fatalities for representative routes from transportation accidents, radiation exposure during incident-free transportation, and as a consequence of releases of radioactive material during accidents. The risk assessments did not seek minimum-risk routes or compare risks of alternative routes. DOE stated that more refined assessments were not possible in the EISs because actual routes could not be selected until shipments are actually made, after consultation with states and review by USNRC. DOE presented additional estimates of risks of alternative routes to the states and tribes during development of the foreign research reactor spent fuel transportation plans, but these estimates appear not to have influenced route selections. The committee's analysis of DOE's assessments of risks of shipments appears in Sections 4.3.1, 4.3.2, and 4.3.3.

(c) The analysis under subsection (b) shall include a consideration whether, and to what extent, the procedures analyzed for purposes of that subsection take into account the following:

(1) The proximity of the routes under consideration to major population centers and the risks associated with shipments of spent nuclear fuel from research nuclear reactors through densely populated areas.
(2) Current traffic and accident data with respect to the routes under consideration.
(3) The quality of the roads comprising the routes under consideration.

(4) Emergency response capabilities along the routes under consideration.
(5) The proximity of the routes under consideration to places or venues (including sports stadiums, convention centers, concert halls and theaters, and other venues) where large numbers of people gather.

DOE's practice has been to place responsibility for detailed review of highway routes on the states and tribes. The quantitative risk analyses in the EISs take into account population density, but not traffic conditions, accident rates for specific route segments, road quality, emergency response capabilities, or places of public gatherings. Rather, DOE has considered these factors through its consultations with the states and tribes and in consultations on rail routes involving carriers, states, and the Federal Railroad Administration. DOE has modified routes according to state proposals derived from consideration of these factors. In some instances, states themselves have examined accident data or conducted risk analyses of specific routes, but most state proposals on routing have been based primarily on judgments of state officials. The committee's analysis of DOE's consideration of the listed factors in route selection appears in Section 4.3.

FINDING: Highway routes for shipment of spent nuclear fuel are dictated by DOT regulations (49 CFR Part 397). The regulations specify that shipments normally must travel by the fastest route using highways designated by the states or the federal government. They do not require the carrier or shipper to evaluate risks of portions of routes that meet this criterion. These regulations are a satisfactory means of ensuring safe transportation, provided that the shipper actively and systematically consults with the states and tribes along potential routes and that states follow the route designation procedures prescribed by the DOT.

RECOMMENDATION: DOT should ensure that states that designate routes for shipment of spent nuclear fuel rigorously comply with its regulatory requirement that such designations be supported by sound risk assessments. DOT and DOE should ensure that all potentially affected states are aware of and prepared to fulfill their responsibilities regarding highway route designations.

5

Improving Spent Fuel and High-Level Waste Transportation in the United States

The focus of this chapter is on the last two charges of the original statement of task for this study (see Sidebar 1.1):

- What are likely to be the key principal technical and societal concerns for radioactive waste transportation in the future, especially over the next two decades?
- What options are available to address these concerns, for example, options involving changes to planned transportation routes, modes, procedures, or other limitations/restrictions; or options for improving the communication of transportation risks to decision makers and the public?

The task statement makes a clear distinction between current and future transportation activities. This distinction was made in recognition of the fact that the federal government is planning to initiate a large-scale, multidecade program to transport much of the nation's commercial spent nuclear fuel and Department of Energy (DOE) spent fuel and high-level radioactive waste to a federal repository. To this end, DOE plans to submit an application to the U.S. Nuclear Regulatory Commission (USNRC) for a license to construct and operate a repository at Yucca Mountain, Nevada (see Chapter 1). At the time the task statement was developed, the National Academies anticipated that there were likely to be specific technical and societal concerns associated with a transportation program to a federal repository. This is in fact the case as shown in this chapter.

The committee was not directed by the statement of task to undertake a detailed programmatic review of the federal repository transportation program, nor did it attempt to do so. While many of the concerns raised in this chapter apply specifically to this federal transportation program, they could also apply to other transportation programs designed to move large quantities of spent fuel and(or) high-level waste within the United States to other federal repositories or to interim storage—for example, the Private Fuel Storage, LLC program. A detailed description of the transportation system for this federal repository is provided in Appendix C.

5.1 PRINCIPAL FINDINGS AND RECOMMENDATION

PRINCIPAL FINDING ON TRANSPORTATION SAFETY: The committee could identify no fundamental technical barriers to the safe[1] transport of spent nuclear fuel and high-level radioactive waste in the United States. Transport by highway (for small-quantity shipments) and by rail (for large-quantity shipments) is, from a technical viewpoint, a low-radiological-risk activity with manageable safety, health, and environmental consequences when conducted in strict adherence to existing regulations. However, there are a number of social and institutional challenges to the successful[2] initial implementation of large-quantity shipping programs that will require expeditious resolution as described in this report. Moreover, the challenges of sustained implementation should not be underestimated.

Spent fuel has been transported in the United States and several other countries for several decades; the committee knows of no releases of radioactive materials from package containments above regulatory limits.[3] This safety record can be attributed to the robust design and construction of the packages used for transport and the rigorous regulatory oversight of transportation operations. Studies of package performance have demonstrated the effectiveness of package containment over a wide range of transport conditions, including most severe accident conditions (Chapter 2). Similarly, studies of the health and safety risks of spent fuel transportation (Chapter 3) indicate that such risks are generally well characterized and are

[1]As noted in Chapter 1, *safety* refers to measures taken to protect spent fuel and high-level waste during transport operations from failure, damage, human error, and other inadvertent acts.

[2]The committee defines "success" in terms of the program's ability, under existing statutes, regulations, agreements, and budgets, to transport spent fuel and high-level waste in a safe, secure, timely, and publicly acceptable manner.

[3]As described in Section 3.1, however, there are well-documented instances in which radioactive contamination on the external surfaces of packages have exceeded regulatory limits. The committee is aware of no documented instances in which this contamination has resulted in exposures of workers or the public above regulatory limits.

generally low. However, the social risks and related institutional challenges may impinge on the successful implementation of large-quantity shipping programs. Transportation programs can take proactive steps to identify and manage these risks and challenges as discussed in Chapter 3 and elsewhere in this chapter.

The wording of this finding—"The committee could identify no fundamental technical barriers to the safe transport of spent fuel and high-level radioactive waste in the United States"—is carefully and narrowly constructed. This finding is focused on the technical aspects of transportation programs: package and conveyance design, fabrication, and maintenance and the conduct of transportation operations. It is predicated on the assumption that these technical tasks are being carried out with a high degree of care and in strict adherence to regulations. The finding also is based on an assessment of past and present transportation programs and would apply to future programs only to the extent that they continue to exercise appropriate care and adhere to applicable regulations. Continued vigilance by all parties involved in these transportation programs—planners, implementers, and regulators—will be required to ensure that transportation operations in the United States continue to be conducted in a safe manner, especially as large-quantity shipping programs to interim storage and a federal repository are ramped up over the next one to two decades. Some issues of particular concern are discussed in Section 5.2.

In Chapter 2, the committee notes concerns about the potential impacts of very long duration fires on package containment effectiveness. Specifically, the committee notes that there may be a very small number of credible accident conditions involving long-duration, fully engulfing fires that are potentially capable of damaging the seals on transportation packages if such fires are allowed to burn in an uncontrolled manner for long periods of time (many hours to days). The committee also recommends that additional investigations be carried out to obtain a bounding-level understanding of the risks and consequences of such accidents. In Chapter 3, the committee describes a relatively simple operational step that can be taken to mitigate these risks. Consequently, the committee judges that very long duration fires do not present a technical barrier to transportation safety.

PRINCIPAL FINDING ON TRANSPORTATION SECURITY: Malevolent acts against spent fuel and high-level waste shipments are a major technical and societal concern, especially following the September 11, 2001, terrorist attacks on the United States. The committee judges that some of its recommendations for improving transportation safety might also enhance transportation security. The Nuclear Regulatory Commission is undertaking a series of security studies, but the committee was unable to perform an

in-depth technical examination of transportation security because of information constraints.

RECOMMENDATION: An independent examination of the security of spent fuel and high-level waste transportation should be carried out prior to the commencement of large-quantity shipments to a federal repository or to interim storage. This examination should provide an integrated evaluation of the threat environment, the response of packages to credible malevolent acts, and operational security requirements for protecting spent fuel and high-level waste while in transport. This examination should be carried out by a technically knowledgeable group that is independent of the government and free from institutional and financial conflicts of interest. This group should be given full access to the necessary classified documents and Safeguards Information to carry out this task. The findings and recommendations from this examination should be made available to the public to the fullest extent possible.

Several participants at the committee's information-gathering meetings highlighted security[4] as an important current concern for transportation of spent fuel and high-level waste in the United States. The committee concurs with this view and judges that such concerns are likely to grow in the future, especially once shipments commence to centralized interim storage or a federal repository.

As reported in Section 1.2, the committee was unable to perform an examination of transportation security risks because of information restrictions: much of the information available on this topic is either classified or otherwise restricted. The committee concluded that it would be difficult to provide a substantive assessment of security issues because not all committee members have the necessary clearances to access this information.

Four members of the committee and one staff member with appropriate security clearances were given a classified briefing by USNRC staff on investigations under way within that agency to assess the security of transportation packages. Some of these members also have some knowledge of the extensive classified and unclassified literature on this topic. There appears to be sufficient information available to undertake a substantive review of spent fuel and high-level waste transportation security by a cleared group if it is given unrestricted access to the relevant literature and information. The cooperation of several federal agencies (USNRC, DOE, and the

[4]As noted in Chapter 1, *security* refers to measures taken to protect spent fuel and high-level waste during handling and transport from sabotage, attacks, and theft.

Department of Homeland Security [DHS]) would be required to obtain the information needed to carry out this study.

The committee's recommendation that an examination of spent fuel and high-level waste transportation security be carried out independently of the government and by a group free of financial and institutional conflicts of interest is made in the spirit of improving its objectivity and public credibility. The committee's recommendation that the findings and recommendations of this examination be presented in a format that can be shared with the public is made in the spirit of improving the quality of informed dialogue on this sensitive but important issue. The preparation of findings and recommendations that are suitable for public release will require that the group charged with this examination be given access to appropriate and timely classification guidance.

While the recommendations in this report are focused primarily on improving transportation safety, the committee judges that some of these might also improve transportation security. For example, the recommended operational changes to reduce the number of total shipments to a federal repository (Section 5.2.1), to limit shipment travel times and stops (Section 5.2.3), and to encourage transport of older (and radiologically cooler) spent fuel (Section 5.2.4) would help to reduce the opportunities for some types of malevolent acts or limit their potential consequences.

5.2 TRANSPORTATION OPERATIONS

The committee uses the term *transportation operations* to refer to the spectrum of activities associated with the actual shipments of spent fuel and high-level waste in the United States. The committee provides findings and recommendations on the following six operational issues in this section:

1. Mode (road vs. rail) for transporting spent fuel and high-level waste to a federal repository (Section 5.2.1)
2. Route selections for transport to a federal repository (Section 5.2.2)
3. Use of dedicated trains for transport to a federal repository (Section 5.2.3)
4. Acceptance order for commercial spent fuel transport to a federal repository (Section 5.2.4)
5. Emergency response planning and training (Section 5.2.5)
6. Information sharing and openness (Section 5.2.6)

Although these recommendations are focused on DOE's program for transporting spent fuel and high-level waste, they also apply to any large-quantity shipping programs whether federally or privately operated. *The committee intends that these recommendations would also apply to the*

Private Fuel Storage program for transporting large quantities of commercial spent fuel to centralized interim storage in Utah, if that facility is constructed and opened.

5.2.1 Mode for Transporting Spent Fuel and High-Level Waste to a Federal Repository

FINDING: Transport of spent fuel and high-level waste by rail has clear safety, operational, and policy advantages over highway transport for large-quantity shipping programs. The committee strongly endorses DOE's selection of the "mostly rail" option for the Yucca Mountain transportation program for the following reasons:

- It reduces the total number of shipments to the federal repository by roughly a factor of five, which reduces the potential for routine radiological exposures, conventional traffic accidents, and severe accidents (Table 3.8).
- Rail shipments have a greater physical separation from other vehicular traffic and reduced interactions with people along transportation routes, which also contributes to safety.
- Operational logistics are simpler and more efficient.
- There is a clear public preference for this option.

The committee does not endorse the development of an extended truck transportation program to ship spent fuel cross-country or within Nevada should DOE fail to complete construction of the Nevada rail spur or procure the necessary rail equipment by the time the federal repository is opened.

RECOMMENDATION: DOE should fully implement its mostly rail decision by completing construction of the Nevada rail spur, obtaining the needed rail packages and conveyances, and working with commercial spent fuel owners to ensure that facilities are available at plants to support this option. These steps should be completed before DOE commences the large-quantity shipment of spent fuel and high-level waste to a federal repository to avoid the need to procure infrastructure and construct facilities to support an extended truck transportation program. DOE should also examine the feasibility of further reducing its needs for cross-country truck shipments of spent fuel through the expanded use of intermodal transportation (i.e., combining heavy-haul truck, legal-weight truck, and barge) to allow the shipment of rail packages from plants that do not have direct rail access.

Mode selection is of special concern for the federal repository transportation program given its size and multidecade duration. DOE has decided that the mostly rail alternative defined in the final Yucca Mountain Environmental Impact Statement (EIS; DOE, 2002a) is its preferred alternative for transporting spent fuel and high-level waste to a federal repository (DOE, 2004d). In this EIS, DOE noted that it referred to rail as its preferred mode as early as 1998 in the draft request for proposals for contractor support for waste acceptance and transportation (DOE, 2002a, p. M-9). In identifying mostly rail as the preferred mode, DOE evidently does not mean that it prefers each of the detailed site-specific mode choices assumed in the final EIS (DOE, 2002a), but rather that it will seek to employ rail transportation to the extent practicable.

DOE summarized its evaluation of transportation mode options in the final EIS as follows (DOE, 2002a, p. 2-97–2-98):

> DOE believes that the EIS provides the environmental impact information necessary to make certain broad transportation-related decisions, namely the choice of a national mode of transportation outside Nevada (mostly rail or mostly legal-weight truck), the choice among alternative transportation modes in Nevada (mostly rail, mostly legal-weight truck, or heavy-haul truck with use of an associated intermodal transfer station), and the choice among alternative rail corridors or heavy-haul truck routes with use of an associated intermodal transfer station in Nevada.
>
> DOE has identified mostly rail as its preferred mode of transportation, both nationally and in Nevada. The environmental impacts for mostly rail are expected to be less overall than the impacts for mostly truck. For the mostly rail scenario, 9,600 rail and 1,100 truck shipments[[5]] are expected for shipping 70,000 MTHM [metric tons heavy metal] and, for the mostly truck scenario, 53,000 truck and 300 rail shipments are expected. The reduced number of shipments to move 70,000 MTHM and corresponding expected reduction in environmental impacts are the basis for preferring the mostly rail scenario.

The impacts that weighed most heavily in DOE's mode preference are safety related, primarily involving fatalities from exposure to ionizing radiation and conventional traffic fatalities (see Table 3.8). DOE noted that

---

[5]For commercial spent fuel, one shipment, as the final EIS uses the term, apparently is equal to one package moving to the repository. Although the risk analysis in the EIS assumes one package per train, DOE states that in practice up to five railcars, each carrying one spent fuel package, could move together in one train (DOE, 2002a, p. J-14). DOE also states that its present plans call for three packages per train, or about 3000 trains entering Nevada (DOE, 2004d, p. 18559). A truck would carry only one package. This is based on a rail package capacity of 6 to 12 metric tons (about 7 to 13 short tons) heavy metal (MTHM), compared with 1.8 MTHM (2 short tons) for a legal-weight truck package (DOE, 2002a, Table J-2); the ratio of the number of shipments in the mostly rail scenario, compared with mostly truck, is about the ratio of truck package capacity to rail package capacity.

security considerations also support its rail preference. The analysis in the final EIS indicates that in the event of a terrorist attack on a transport package in transit, the likely consequences for a rail shipment would be less than for a legal-weight truck shipment, mainly because of the rail package's thicker wall.[6] The committee did not examine the risks of terrorist attacks on spent fuel packages and therefore cannot confirm this conclusion.

Table 3.8 shows that the mostly rail option results in about a factor of five decrease in the number of shipments to the federal repository (53,000 mostly truck shipments versus 9600 mostly rail shipments[7]) over a period of 24 years. The mostly rail option also results in almost a factor of four reduction in expected radiation-related fatalities during routine transport, from 15 fatalities to 4. The number of expected fatalities for the maximally reasonably foreseeable accident is higher for the rail option than for the truck option (five fatalities versus less than one fatality), but the likelihood of occurrence of such a rail accident is very small (less than 3 in 10 million chances of occurring per year).

The committee also sees clear operational advantages to the mostly rail option. Railroads in the United States are privately owned and operated, which allows for greater control over other activities on the rail line and a more coordinated regime for carrying out safety inspections. Rail transport can also result in reduced shipment travel times, especially if dedicated trains are used, which allows for the more efficient utilization of transport packages and conveyances.[8]

DOE's mostly rail transportation strategy will require the development of a rail spur within Nevada, because the Yucca Mountain site is currently without direct rail access.[9] The final EIS (DOE, 2002a) examined three alternative, but not mutually exclusive, provisions for transportation within Nevada: (1) a rail alternative that would entail constructing a rail spur from

[6]While it is true that rail packages have thicker walls, they also hold greater inventories of spent fuel. It is not immediately obvious to the committee how these two factors would trade off for various types of terrorist attacks.

[7]It should be noted that these shipment numbers are small compared to other types of hazardous material transport that occur on the nation's highways, railroads, and waterways each year. About 400,000 large trucks are dedicated to hazardous materials service, including most tank trucks. About 115,000 railroad tank cars and more than 3000 tank barges operating on the inland and coastal waterways are in hazardous materials service. See NRC (2005c) for details.

[8]In principle, reducing the travel time allows more round trips to be carried out per unit time or, at a fixed throughput, reduces the required numbers of packages and conveyances.

[9]The EIS for the rail spur was under way when the present report was being finalized (December 2005). The transportation risks on rail spur will not be known publicly until the EIS and spur design are released. The committee presumes that the rail spur will be constructed up to modern-day standards and therefore will have risks similar to or less than other rail lines used to transport spent fuel and high-level waste to the repository.

a point on an existing rail line to the repository; (2) a heavy-haul truck alternative, in which full-size rail packages would be removed from railcars at a terminal constructed near an existing rail line and loaded onto heavy-haul trucks for transport to the repository; and (3) a legal-weight truck alternative involving the transport of spent fuel and high-level waste from commercial and DOE sites to the repository by truck. The final EIS stated that mostly rail is also the preferred mode within the State of Nevada (DOE, 2002a, p. S-2). DOE examined five possible rail corridors in Nevada in its final EIS (Figure 5.1) but it did not express a preference among them.

In December 2003, DOE published a notice[10] announcing that one of the five corridors, the Caliente corridor, is its preferred corridor in which to construct a rail line, and that a second route, the Carlin corridor, is its secondary preference. The notice explains the significance of this designation as follows: "If the Department adopts the mostly rail mode in Nevada, DOE will issue a Record of Decision selecting a rail corridor. . . . If the Department selects a rail corridor, DOE will issue a Notice of Intent in the Federal Register to initiate the preparation of a rail alignment EIS . . . to consider alternative alignments within the selected corridor. . . ." Concerning the basis for the selection, the notice states: "The Department's preference for Caliente takes into account many factors, including its more remote location, the diminished likelihood of land use conflicts, concerns raised by Nevadans, and national security issues raised by the U.S. Air Force on the Caliente-Chalk Mountain corridor [another corridor analyzed in the EIS]." DOE further explained its preferred mode designation and selection of the corridor in a Record of Decision (ROD) published in April 2004 (DOE, 2004d) and at the same time announced (DOE, 2004e) that it was beginning an EIS covering the selection of the alignment within the Caliente corridor and construction and operation of the rail line, with a draft to be issued in early 2005.

The Caliente corridor begins at a point on the Union Pacific rail line near Caliente Nevada, 120 miles northeast of Yucca Mountain; runs west, passing north of the Nevada Test and Training Range, a military facility; and then turns south to Yucca Mountain (Figure 5.1). The final Yucca Mountain EIS estimates the cost of building the rail line in this 319-mile (513-kilometer) corridor to be $880 million in 2001 dollars and the construction time to be 46 months (DOE, 2002a).[11] The three nonpreferred corridors are shorter and have lower estimated construction costs than the

---

[10]DOE, "Notice of Preferred Nevada Rail Corridor," 68 FR 74951–74952, December 29, 2003.

[11]In December 2005, DOE acknowledged that construction costs had increased to $2 billion, presumably in current year (2005) dollars.

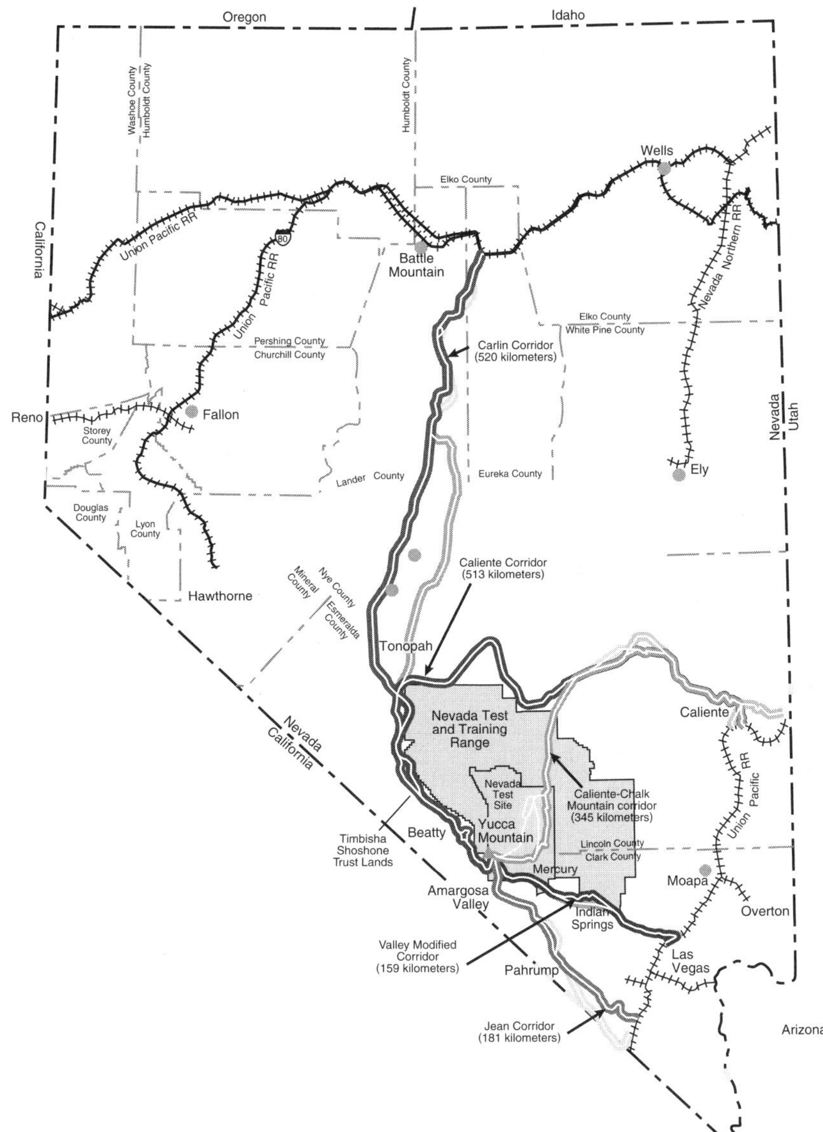

FIGURE 5.1 Potential corridors for the Nevada rail spur. The Caliente corridor starts near the town of Caliente at the lower right and runs north and west of the Nevada Test and Training Range. SOURCE: Modified from DOE (2002a).

Caliente route, but would have routed shipments through or near Las Vegas or through the Nellis Range.

In its notice on beginning the alignment EIS (DOE, 2004e), DOE invited comments on whether it should allow private entities to ship commercial commodities on its rail line. The possibility of commercial use might make construction of the route more acceptable to some local interests. The choice of whether the line should be available for public use has implications for regulatory oversight of the line's construction and operation. According to federal law, the Surface Transportation Board[12] has jurisdiction over rate and service issues for common carrier rail lines (i.e., rail lines available for public use). The board must approve construction, and all state and local environmental or permitting requirements are preempted. Construction of a track for exclusive DOE use would not require board approval. Federal law would not preempt state and local regulation of construction of such a track.

DOE's preferred rail transportation strategy will also require the execution of an ambitious intermodal transportation scheme to move transportation packages to railheads from commercial sites lacking direct rail connections. The final Yucca Mountain EIS notes that 24 commercial spent fuel storage sites have no rail service but do have facilities to load rail packages. Of these 24, 17 have access to waterways, and the other 7 can ship to railheads by heavy-haul truck. There are an additional 6 sites that lack crane capacity or other facilities to load rail packages (DOE, 2002a, p. J-15, Tables J-5 and J-26).

Of the 9600 train shipments in the final EIS mostly rail scenario, on the order of 2000 would be moved from sites without rail access to a railhead by heavy-haul truck or barge over distances of 6 to 256 kilometers (4 to 160 miles) (DOE, 2002a, Tables J-5, J-26, J-27). DOE's illustrative heavy-haul truck (DOE, 2002a, Figure 2-29) is 67 meters (220 feet) long and weighs 90 metric tons (100 short tons) empty and around 180 metric tons (200 short tons) loaded. For comparison, a legal-weight truck carrying a spent fuel package would be about 18 meters (60 feet) in length and have a loaded weight of 36 metric tons (40 short tons).

The possibility of a large volume of barge shipments of spent nuclear fuel has been a point of controversy. State representatives reported to the committee that there is opposition in the upper Midwest to spent fuel barge shipments on the Great Lakes. The final EIS estimates were not based on an evaluation of whether barge would be preferred from the standpoint of cost or risk at any sites of origin. Until up-to-date and detailed local site access

---

[12]The board is an independent agency and the successor agency to the Interstate Commerce Commission.

assessments are available, it is not possible to say whether any significant volume of local barge transport will occur.

The final EIS assessments of spent fuel origin site (i.e., commercial nuclear plant sites and independent spent fuel storage installations) capabilities to ship by rail, either directly or with local truck or barge haul to a nearby railhead, were based on two DOE studies published in 1992: the Facility Interface Capabilities Assessment (FICA) (Viebrock et al., 1992) and Near Site Transportation Interface (NSTI) study (Viebrock and Mote, 1992). FICA was a study of the capability of each nuclear generator site to handle spent fuel packages of four dimensions: legal-weight truck package, an overweight truck package, and two rail or barge packages. The NSTI study examined on-site and near-site rail and barge infrastructure at 76 reactor sites.

It is likely that changes have occurred at some origin sites since these studies were completed. The NSTI warned that rail line abandonment was tending to curtail rail access to reactor sites. Development in areas surrounding origin sites since 1990 may also have restricted transportation options. On the other hand, facilities that have developed on-site dry spent fuel storage facilities (see Sidebar 1.4) may have improved their package-handling capabilities.

The State of Nevada has argued that DOE is underestimating the significance of origin site access and handling capabilities as constraints on transportation options and, as a consequence, DOE's estimate of the total number of shipments to the repository in the mostly rail scenario (9600 rail and 1000 truck, as described above) is overly optimistic. A 1996 analysis for the Nevada Nuclear Waste Project Office concluded that substantial truck shipping would be likely, even after Nevada rail access was constructed.[13]

The Nevada analysis used data from the FICA and NSTI studies to construct scenarios for possible Yucca Mountain shipping campaigns, similar to the transportation scenarios in the final EIS for Yucca Mountain, but with more specificity about modes and routes and with explicit consideration of how each origin site would go about analyzing its mode choice. It concluded that with Nevada rail access to the repository in place, it would be reasonable to expect at least 17 utilities to ship by truck for the entire journey to the repository, rather than the 6 assumed in the final Yucca Mountain EIS analysis. Nevada officials stated to the committee that it is their expectation that, even with a Nevada rail link, at least 25 percent of

[13]Planning Information Corporation, The Transportation of Spent Nuclear Fuel and High-Level Radioactive Waste: A Systematic Basis for Planning and Management at the National, Regional, and Community Levels, September 1996, *http://www.state.nv.us/nucwaste/trans/1pichome.htm*.

all commercial spent fuel would travel by truck,[14] implying 10 times the volume of truck movements in the final EIS mostly rail scenario.

Nuclear power industry participants dispute the state's estimates.[15] They note that shipment by truck would be unattractive to commercial operators: The cost and time required for loading and shipping a truck package would be comparable to that for a rail package that held seven times the quantity of spent fuel. Moreover, loading large amounts of spent fuel for truck transport would overburden origin site facilities. These representatives believe that origin sites will have a strong preference to ship by rail whenever possible.

The degree to which DOE must defer to the spent fuel owners' preferences regarding transport mode is governed by the terms of the standard contracts (see Appendix C and Sidebar 5.1). The owners' delivery commitment schedule, required under the contract, will specify a "proposed shipping mode" (truck, rail, or barge). DOE can disapprove a delivery commitment schedule, in which case "the parties shall promptly seek to negotiate mutually acceptable schedules" (Standard Contract, Article V.B). However, the contract also specifies that DOE must provide transport packages "suitable for use at the Purchaser's [spent fuel owner's] site" (Article IV.B), so DOE apparently could not require a utility to use a rail package if doing so would require site modifications.

DOE is now updating the FICA information regarding on-site capabilities to load and handle shipping packages of various dimensions that will be needed to order shipping packages; DOE plans to update the NSTI data as well.[16] Until these updates are complete, estimates of the likely number of rail and truck shipments must be considered uncertain.

Other modal options that might conceivably play a useful role have not yet been evaluated fully. These include intermodal transportation employing conventional trucks; that is, placing spent fuel at the point of origin into standard truck packages, transporting the containers by truck to a nearby rail terminal, transferring them to railcars for transportation to the repository, and finally transferring them back to truck at a point near the repository if the repository does not have rail access. Trucks could be of the

---

[14]Statement of Robert R. Loux to the committee, July 25, 2003.

[15]Tiffany Wlazlowski. 2002. U.S. seeks ways to safely transport radioactive waste to Yucca Mountain. *Transport Topics*, June 3. p. 1.

[16]"The data on transportation infrastructure in the vicinity of sites, . . . will . . . be updated. . . . Updating this information close to the time of actual shipment ensures that the latest information is used for identifying site-specific transportation needs" (DOE, 2003c, p. 9). Presumably DOE recognizes the necessity of updating this information immediately, because it may be critical to its transportation plan and because interventions to preserve or develop facilities may be needed. It would also be necessary to further update the information as shipping dates approach.

SIDEBAR 5.1 Standard Contract for Disposal of Spent Fuel and High-Level Waste

Title 10, Part 961 of the Code of Federal Regulations, Standard Contract for Disposal of Spent Nuclear Fuel and/or High-Level Radioactive Waste, specifies the responsibilities of DOE and purchasers of its services (i.e., the owners of commercial spent fuel) for transportation of spent fuel to the federal repository:

- DOE is responsible for providing transport packages to the purchaser at the purchaser's site, and also for providing procedures and training for handling and loading such packages.
- The purchaser is responsible for preparing, packaging, and loading spent fuel into the packages in accordance with all applicable laws and regulations. DOE may designate a representative to observe these activities. The purchaser must notify DOE at least 60 days in advance of commencement of such activities. This notification must provide DOE with a description of the material to be loaded.
- DOE must accept title to the loaded packages "freight-on-board" at the purchaser's site. Once title has passed, the purchaser is no longer responsible for the spent fuel. DOE has the right to dispose of this spent fuel as it sees fit and is not obligated to provide compensation to the purchaser for this material.
- DOE must provide the means for transporting the packages from the purchaser's site to the repository.

The standard contract also establishes the priority order for DOE acceptance of commercial spent nuclear fuel as follows:

- DOE acceptance will be based on the date and amount of fuel discharged from the reactor (see Sidebar 5.2 for further details), except as noted in the next two bullets.
- DOE may accord priority for acceptance of spent nuclear fuel from reactors that have been permanently shut down.
- The purchaser also has the right to exchange delivery schedules with parties to other contracts, but DOE has the right to approve or disapprove such exchanges in advance.

DOE and the purchaser are also responsible for exchanging the following information on an annual basis:

- DOE must provide the purchaser with the projected annual receiving capacity for spent nuclear fuel at the repository.
- DOE also must provide the purchaser with other pertinent information on the waste disposal program, including cost projections, project plans, and project reports.
- The purchaser must provide DOE with information on actual discharges of spent fuel to date and projected discharges for the next 10 years.

dimensions that are legal to operate on all main highways (80,000 pounds [approximately 36 metric tons] total weight, limiting the weight of the container and spent fuel to about 50,000 pounds [approximately 22 metric tons]) or that exceeded the legal weight but were within the range of weights that states routinely approve for operation after issuance of special permits (at least 120,000 pounds [approximately 53 metric tons] in most states, allowing packages that weigh 80,000 pounds [approximately 36 metric tons]).

Nevada representatives, among other observers, have expressed concern that in spite of DOE's expressed rail preference, events may be directing it toward commencing shipments to the repository by truck, possibly with the intention of a later conversion to mostly rail transportation.[17] The Western Governors' Association (WGA) also acknowledged the prospect of interim transportation arrangements for initial shipments in a 2005 resolution (WGA, 2005, p. 1): "For many years, Western Governors have consistently urged the federal government to develop a comprehensive transportation plan, including the preparation of contingency plans for events such as the early shipment of waste."

DOE has acknowledged that to deal with the uncertainties and complexities of the Yucca Mountain program, the goal of its transportation planning has been to maintain as much flexibility as possible, including the flexibility to ship by truck as well as by rail.[18] DOE's desire for flexibility conflicts with the states' preference for early and specific decisions on transportation. The ROD on mode and corridor choices states DOE's current position on the use of trucks (DOE, 2004d):

> The Department would use truck transport where necessary, depending on certain factors such as the timing of the completion of the rail line proposed to be constructed in Nevada. This could include building an intermodal capability at a rail line in Nevada to take legal-weight truck casks from rail cars and transport them the rest of the way to the repository via highway, should the rail system be unavailable at the time of the opening of the repository. In addition, since some commercial utilities are not able to accommodate rail casks, they would ship by legal-weight truck to the repository.

The option of shipping in legal-weight truck packages on railcars to Nevada and then on truck to the repository was considered in the final Yucca Mountain EIS, but dismissed as impractical, at least as the primary

[17]Robert M. Halstead, Testimony to the U.S., House of Representatives, Committee on Transportation and Infrastructure, April 25, 2002; Robert L. Loux from the State of Nevada made a similar statement to the committee as its meeting on Las Vegas on July 25, 2003.

[18]Statements of Margaret Chu and Jeff Williams to the Nuclear Waste Technical Review Board, January 28, 2003.

method of operation over the life of the transportation program (DOE, 2002a, pp. J-74–J-75). Nevada has raised questions about the safety of shipping truck packages by rail, especially with respect to the performance of these packages under extreme accident conditions. (Such shipments are allowed under current USNRC regulations.) The risk analyses described in Chapter 3 have not considered this transportation scenario, so additional analyses might be prudent if DOE were to ship truck packages by rail.[19]

The costs of initiating the transport of spent fuel and high-level waste to Yucca Mountain using trucks followed by the initiation of rail could be considerable, especially if an extended, high-throughput initial truck campaign was planned. It would require setting up essentially two independent transportation systems, each requiring funding for planning, scheduling, package and conveyance procurement, and package maintenance. It might also require the modification of package-receipt facilities at the federal repository to handle large numbers of smaller truck packages. If the costs of standing up and operating the truck program were high, DOE might not have the resources to complete the construction and procurement of rail infrastructure, and DOE and the nation could be saddled with a long-term truck transportation program.

Several challenging tasks remain before DOE will be in a position to fully implement the mostly rail option:

- Completion of the Nevada rail EIS and construction of the Nevada rail line
- Completion of the NSTI study on rail and barge infrastructure at commercial sites
- Completion of the FICA surveys on infrastructure to load and handle shipping packages at commercial sites
- Construction of any needed infrastructure improvements at commercial sites and along planned routes
- Acquisition of a transport package and conveyance fleet to support the mode decision

Time and adequate resources will be required to complete these tasks in advance of the opening of a federal repository at Yucca Mountain. Failure to complete these tasks before the repository is opened may create pressure on DOE to initiate its transportation program using legal-weight or overweight trucks. Similarly, an early congressional directive to open an interim storage facility for commercial spent fuel at another government site could also result in large numbers of truck shipments. The extensive use of trucks

[19]As noted in Section 2.2.3, the USNRC is examining the performance of a truck package in a rail tunnel fire.

in the early stages of these programs could divert time and resources away from implementation of the mostly rail option, especially if DOE is forced to procure truck packages and conveyances. It could also affect the receiving facilities at the repository, which presumably will be designed primarily to handle rail packages. Finally, it could increase opposition to the transportation program and reduce trust and confidence in DOE's transportation program and its managers, which could conflict with DOE's stated desire (as noted previously) to maintain flexibility in mode selection.

Even under the mostly rail option envisioned in the final EIS for Yucca Mountain, there will still be 1100 truck shipments (DOE, 2002a). DOE has an opportunity to further reduce the truck shipments, especially cross-country shipments, through an expanded use of intermodal transport for those sites that lack direct rail access.

5.2.2 Route Selection for Transportation to a Federal Repository

FINDING: DOE has not made public a specific plan for selecting rail and highway routes for transporting spent fuel and high-level waste to a federal repository. DOE also has not determined the role of its program management contractors in selecting routes or specific plans for collaborating with affected states, tribes, and other parties.

RECOMMENDATION: DOE should identify and make public its suite of preferred highway and rail routes for transporting spent fuel and high-level waste to a federal repository as soon as practicable to support state, tribal, and local planning, especially for emergency responder preparedness. DOE should follow the practices of its foreign research reactor spent fuel transport program of involving states and tribes in these route selections to obtain access to their familiarity with accident rates, traffic and road conditions, and emergency responder preparedness within their jurisdictions. Involvement by states and tribes may improve the public acceptability of route selections and may reduce conflicts that can lead to program delays.

Implementation of DOE's transportation program for Yucca Mountain will be a daunting task, given its size coupled with its lack of control over budgets and schedules (see Section 5.3), the large number of involved parties, the geographic extent of the transportation system (Figure 1.1), and the long time frames for transportation operations. This argues for simplification, and one of the best ways to simplify the program is to ship by rail whenever practicable. There will be many fewer shipments that need to be routed, and transport will take place over private rights of way. Railroads have established procedures in place for selecting routes for hazardous materials carriage. These procedures, however, do not take into account all of the concerns of the states and tribes (e.g., avoiding large population centers and being sensitive to environmental justice concerns). The DOE program for transporting foreign research reactor spent fuel (Chapter 4) is a good model for involving states and tribes in rail routing decisions.

Once DOE selects the suite of routes[20] it will use for its shipments, it may be required to undertake selected route infrastructure improvements before any shipments can be made, as described in Appendix C. It also must decide on safety and security measures, procedures for notification of jurisdictions through which shipments pass, arrangements for state inspections of shipments, communications and tracking, and handling contingencies en route. Most of these operations would be carried out by contractors. DOE has proposed and withdrawn two proposals for the organization of contractor support for management of transportation operations.

Comments submitted on the final Yucca Mountain EIS revealed frustration over DOE's lack of specificity on route designations. DOE has asserted that it was not required in the EIS to specify routes for each site or the decision process for route selection. The EIS does, however, state DOE's conception of route selection as follows (DOE, 2002a, p. J-23):

> Approximately 4 years before shipments to the proposed repository begin, the Office of Civilian Radioactive Waste Management plans to identify the preliminary routes that DOE anticipates using in state and tribal jurisdictions so it can notify governors and tribal leaders of their eligibility for assistance under the provisions of Section 180(c) of NWPA.

Section 180(c) refers to assistance for emergency responder training. This issue is discussed in more detail in Section 5.2.5 of this chapter.[21]

DOE summarizes its rail routing practices for spent fuel as follows (DOE, 2002a, p. M-6):

> Except for requirements contained in 10 CFR 73.37, there are no Federal regulations pertaining to rail routes for shipment of spent nuclear fuel or high-level radioactive waste. The shipper and railroad companies (carriers) determine rail routes based on best available route and track conditions, schedule efficiency, and cost effectiveness.

The regulation cited, 10 CFR 73.37, provides the USNRC's require-

---

[20]The NWPA requirements for emergency responder training requires that DOE select routes well in advance of shipments (see Section 5.2.5). For security reasons, DOE will not provide advanced public notice of shipments along these routes, only advanced notification to state governors and law enforcement officials. See Appendix C, Section C.1.4, for more details.

[21]Emergency responder preparedness has been used as a route selection criterion in the research reactor transport program as described in Chapter 4. DOE could in principle use emergency responder preparedness as a route selection criterion in the federal repository program, but to the committee's knowledge has not announced any plans to do so.

ments for physical protection of spent nuclear fuel in transit (see Appendix C). Specifically, it gives the Commission the authority to preapprove proposed routes for spent fuel shipments based on security considerations. However, DOE is not subject to these regulations because it accepts ownership of commercial spent fuel at the plant gate.

The final Yucca Mountain EIS describes the rail route evaluation process envisioned as follows (DOE, 2002a, p. M-10):

> The Regional Servicing Contractor would identify rail transportation routes in conjunction with the appropriate rail carriers. Because railroad companies determine the routing of shipments, the Contractor would rely on the rail carrier to provide primary and secondary route recommendations consistent with safe railroad operating practices. Guidelines would include . . . use of key routes as described in [the Association of American Railroad's] *Recommended Railroad Operating Practices for Transportation of Hazardous Materials*. . . .

Although DOE has not proceeded with the Regional Servicing Contractor arrangement, this statement presumably reflects its general plan for rail route selection through consultation with railroads. It is not clear from this statement whether DOE would in any circumstances impose route choices on rail carriers, for example, for security reasons or to avoid large population centers. No specific mention is made of consultation with the states on rail routes. Some states are seeking greater specificity and control of the process for reaching decisions on rail routing or on ensuring that the appropriate notification procedures and emergency response capabilities are in place as noted later in this section.

In 1996, DOE stated that it did not intend to produce general routing criteria for spent fuel and high-level waste shipments and would instead rely on standard railroad practice to determine rail routes (WIEB, 1995, p. 44). Such practices do not normally provide for much state and tribal consultation. On the other hand, DOE has shown a willingness to consult with states and tribes in planning for and carrying out spent fuel shipments. As described in some detail in Chapter 4, DOE consulted with South Carolina in developing rail routes for foreign research reactor spent fuel shipments to Savannah River. DOE even agreed to identify an alternate rail route at the state's request.

DOE has also imposed routing restrictions on carriers, including requirements to minimize time, distance, and number of interchanges; to use the best track; to avoid population centers; and to schedule movements through cities outside peak commuter hours for other spent fuel transportation campaigns. Examples of such special provisions include the rail shipping arrangements for transport of core debris from Three Mile Island in 1986 to 1990 and several series of rail shipments from commercial nuclear plants to temporary storage (WIEB, 1991, pp. 90–92).

For highway routing, the Regional Servicing Contractor would first select a route consistent with U.S. Department of Transportation (DOT) regulations and with consideration of infrastructure adequacy along routes. Then (DOE, 2002a, p. M-9),

> . . . the Contractor would submit the route plan to DOE for approval. DOE would interact with states and Native American governments concerning these selections. . . . With DOE approval, the Contractor would then submit the route plans to the Nuclear Regulatory Commission. . . .

Although the Regional Contractor plan is no longer current, the statement provides an indication of DOE's intentions on how route selections would be made.

In past DOE highway routing debates, the states have tended to urge DOE in the direction of taking an active role in route designation and of reducing the discretion of carriers to choose routes. DOE worked closely with affected states to select highway routes for foreign research reactor shipments from the Savannah River Site to Idaho National Laboratory (see Chapter 4). In preparing for transportation of transuranic waste[22] to the Waste Isolation Pilot Plant (WIPP), an underground repository in New Mexico, DOE specified a set of routes in consultation with the states, and these routes were incorporated in state alternate preferred route designations. DOE then specified routes in its contracts with carriers (see TEC, 2002). The WGA (2005) cites the WIPP transportation program as the model DOE should follow in planning for the Yucca Mountain transportation program.

The WGA has repeatedly expressed its concern that DOE is not on track to produce a transportation plan that is sound, timely, and built on adequate consultation with the states. It recommended that DOE undertake a series of actions that include the following[23] (see also WGA, 2005):

- Develop criteria and a methodology for evaluating and selecting routes and modes;
- Propose a set of shipping routes to the affected states and tribes for review and comment;
- Through this consultation with states and tribes, identify a set of primary and secondary routes for each site of origin to each destination; and
- Require the use of these routes through contract provisions with the private parties engaged in transportation.

---

[22]Primarily materials such as clothing, containers, and debris generated during nuclear weapons development and production.

[23]WGA, Policy Resolution 02-05: Transportation of Spent Nuclear Fuel and High-level Radioactive Waste, June 25, 2002.

While the WIPP transportation program in some ways offers useful lessons for how DOE could collaborate with states, there are some important differences between transuranic waste and spent fuel or high-level waste that limit the relevance of this model. The cooperative effort to develop the WIPP transportation program was led by western governors who had a strong interest in moving waste out of their states to WIPP. Moreover, the development of the WIPP repository was supported by its host state (New Mexico). In contrast, for the Yucca Mountain program, some important transit states and the repository state (Nevada) have no spent fuel or high-level waste to be shipped. Additionally, the requirements for transport to the Nevada repository are more demanding, given the greater hazards of the materials being shipped. Consequently, security arrangements and emergency responder preparation will require more resources and pose more difficult logistical problems.

The WIPP transportation program is not a useful model for the transportation operations addressed by this study in other ways. Transportation to WIPP has so far been entirely by truck, whereas transportation to Yucca Mountain (and to Private Fuel Storage, LLC) is planned to be mostly by rail. The schedule for opening a WIPP repository was aggressive and did not allow much time for transportation planning. There were unanticipated delays in opening the repository, however, which allowed enough time for the transportation program to develop. While the outcome was good, the process for getting there was not. Recent delays in the aggressive schedule for opening the federal repository at Yucca Mountain program might also provide additional time for transportation planning and route selection.

Past DOE experience in routing spent fuel and transuranic waste shipments has demonstrated the benefits of state and tribal consultations. Such consultations provide DOE with information on accident rates for specific routes, road conditions that could affect shipments, and emergency responder preparedness—information that DOE could not easily obtain on its own. The selection of routes will ultimately require that DOE, in consultation with states, balance these factors and its own programmatic and security considerations in a transparent and supportable fashion.

5.2.3 Use of Dedicated Trains for Transport to a Federal Repository

FINDING: Studies carried out to date on transporting spent fuel by dedicated versus general trains have failed to show a clear radiological risk-based advantage for either option. However, the committee finds that there are clear operational, safety, security, communications, planning, programmatic, and public preference advantages that favor dedicated trains. The committee strongly endorses DOE's decision to transport spent fuel and high-level waste to a federal repository using dedicated trains.

RECOMMENDATION: DOE should fully implement its dedicated train decision before commencing the large-quantity shipment of spent fuel and high-level waste to a federal repository to avoid the need for a stopgap shipping program using general trains.

In July 2005, DOE's Office of Civilian Radioactive Waste Management (OCRWM) announced that DOE was adopting the following policy regarding rail operating practices for shipments to Yucca Mountain: "The Department of Energy will use dedicated train service . . . for its usual rail transport of spent nuclear fuel . . . and high-level radioactive waste . . . to the Yucca Mountain Repository site in Nevada when the repository is operational" (DOE 2005b).

This declaration follows a long-running controversy in the United States about whether rail shipments of spent fuel and high-level waste should be carried out using

- *Dedicated trains*, which would carry only spent fuel and high-level waste; or
- *General trains*, which could carry other freight in addition to spent fuel and high-level waste, just as other routine hazardous materials shipments are handled. Certain additional practices and precautions would still be applied to such trains.[24]

During the 1960s, some railroads in the eastern United States announced that they would refuse to handle spent nuclear fuel and some other radioactive waste shipments under the rates and terms of common carriage (i.e., by general train service). The railroads announced that they would require special contracts for these shipments. Such contracts typically include a hold-harmless clause covering the railroad, an obligation on the part of the shipper to guarantee connecting line service, and a stipulation that the service would be made by dedicated trains (Klassen, 1982, pp. 1–3).

The Energy Research and Development Administration (a DOE predecessor agency), the USNRC, and commercial nuclear utilities challenged the legality of the railroads' tariff actions in 1975. These plaintiffs argued that dedicated trains could not be shown to significantly improve safety and that their use added to costs. The Interstate Commerce Commission ruled against

[24]For example, the Association of American Railroads has developed and issued "Performance Specification for Trains Used To Carry High-Level Radioactive Material" (Standard S-2043). This standard establishes requirements for coupling systems, brakes, and dynamic load tests for railcars used to transport spent fuel and high-level waste. These cars could be used on both dedicated and general trains.

the railroads, and the railroads appealed the case to the Supreme Court, which upheld the ruling in 1980. The railroads were thus required to publish tariffs for transporting spent fuel and to cancel mandatory special train tariffs (Klassen, 1982, p. 2-3).

Although some spent fuel shipments have moved in general service trains in recent years,[25] dedicated trains appear to be standard practice in the United States.[26] DOE and the railroads have agreements in place on a set of safety and security protocols and special operating restrictions for spent fuel movements. These cover the make-up of trains as well as security provisions and operating restrictions and controls. There appears to be strong industry support for the use of dedicated trains.[27] The Association of American Railroads has adopted a policy recommending shipment of spent fuel by dedicated train.[28] Private Fuel Storage, LLC, a consortium of commercial nuclear utilities that plans to build an interim storage facility for spent fuel in Utah (see Chapter 1), plans to ship spent fuel primarily using dedicated trains. A representative from the Nuclear Energy Institute, the nuclear energy industry's policy and lobbying arm, told the committee at one of its information-gathering meetings that the industry also prefers dedicated trains for shipping its spent fuel to Yucca Mountain.

DOE's final EIS for Yucca Mountain (DOE, 2002a) did not provide a detailed analysis of the benefits of dedicated trains to support a decision on the issue. The EIS noted (p. J-76) that "DOE has not determined the commercial arrangements it would request from railroads for shipment of spent nuclear fuel and high-level radioactive waste." It acknowledged the policy of the railroad industry favoring dedicated trains but also cited a 1998 study by the Research and Special Programs Administration[29] (RSPA) of the U.S. Department of Transportation (DOT, 1998), which concluded that shipments by dedicated train would result in a greater number of nonradiological accident fatalities than shipments in general trains because there would be more trains on the rails.

---

[25]Specifically, some naval spent fuel shipments have been transported to the Idaho National Laboratory in general freight.

[26]Allan Rutter, testimony before the Subcommittees on Railroads and on Highways and Transit, Committee on Transportation and Infrastructure, U.S. House of Representatives, April 25, 2002.

[27]Private Fuel Storage, LLC and the Association of American Railroads are cooperating on the development of an advanced railcar for transporting spent fuel. A subgroup of the committee had the opportunity to see this railcar when it visited the Federal Railroad Administration's Transportation Technology Center in Pueblo, Colorado, in October 2003.

[28]Edward R. Hamberger, "Transportation of Spent Rods to the Proposed Yucca Mountain Storage Facility," testimony before the Subcommittees on Highways and Transit and on Railroads, Committee on Transportation and Infrastructure, U.S. House of Representatives, April 25, 2002.

[29]Now the Research and Innovative Technology Administration.

The Government Accountability Office (GAO) examined the question of the relative safety of dedicated and general trains for the transport of spent fuel in a July 2003 report (GAO, 2003). The report identifies three arguments that are commonly made by proponents of dedicated trains that the use of such trains lowers transportation risks. First, dedicated trains have a shorter travel time from origin to destination than a railcar in general service, thus reducing exposure to terrorist attacks. GAO cites a rail industry estimate that a spent fuel package would travel from the East Coast to Nevada in 3 to 4 days by dedicated train versus 8 to 10 days by general train. In the industry estimate, the extra time for the general rail movements would be spent primarily in rail yards. Second, the use of dedicated trains would ensure that cars carrying spent fuel packages are not mixed with cars carrying flammable materials. This reduces the consequences of potential accidents involving fires and explosions. Third, because cars carrying spent fuel packages are much heavier than typical freight cars (470,000 pounds [210,000 kilograms] versus about 200,000 pounds [90,000 kilograms], according to GAO), mixing spent fuel cars with ordinary cars in a train creates stability problems and increases the likelihood of train derailments. GAO concluded (p. 23) that "it is not clear that the advantages of dedicated trains outweigh the additional costs"; however, it did not cite cost data.

The July 2005 dedicated train policy statement identifies the following grounds for DOE's decision (DOE, 2005b):

- "[R]adiological risk resulting from transport [via dedicated train] without incident may be lower due to decreased time in transit."
- Dedicated trains have "potential advantages" for security, although "DOE shipments have been and will continue to be made securely using both [dedicated train service] and general freight service."
- The "primary benefit" of dedicated trains would be "significant cost savings over the lifetime of the Yucca Mountain project" because any higher costs of dedicated train operations would be offset by savings from shorter transit and turnaround times, which allow operations with fewer packages and railcars.

Although not cited in DOE's policy statement, a study by the Federal Railroad Administration (FRA) on use of dedicated trains provides support for DOE's arguments. The study was prepared in response to a congressional directive and transmitted by the Secretary of Transportation to Congress in September 2005 (FRA, 2005).[30] The study concluded that

[30]This study was released after the committee's last meeting, so the committee did not have an opportunity to review and analyze it in detail.

dedicated train service would result in lower risk to the general public than general train service from radiological exposure in incident-free operations; the probability that a package carried in a dedicated train will be involved in an accident is lower than for a regular service train; a package on a dedicated train has lower risk of being exposed to an engulfing fire in the event of an accident; and radiological exposure in the event of a severe accident involving a dedicated train would be less, compared with a comparable accident involving a regular train, because wreckage could be cleared more quickly. The study also concluded that "regardless of the type of train, the potential exposures are essentially benign when compared to a lifetime of normal background radiation exposure . . ." (FRA, 2005, p. 3).

FRA acknowledges that its study did not compare the risks, for dedicated trains and regular service, of injuries and fatalities that result from train accidents and are unrelated to radiation exposure. As noted above, the 1998 RSPA study (DOT, 1998) concluded that the use of dedicated trains would lead to greater losses of this kind, compared with regular train service, because the use of dedicated trains would increase the total nationwide annual train-miles of traffic. However, the FRA report argues that the disadvantage of dedicated trains in this respect would be less than proportional to the increase in train-miles because superior equipment and operational requirements would be placed on dedicated trains and because the frequency of some operations that carry higher risks, such as switching at the origin point of the shipment, would be the same whether dedicated trains or regular service were employed.

Although DOE's policy statement on dedicated trains, along with the railroads' professed unwillingness to handle spent nuclear fuel in regular service, would appear to settle the question, critics of DOE's Yucca mountain repository plan have noted that the wording of the policy statement is vague on some points. In particular, the meaning of the term "usual rail transport" and the significance of the statement that "DOE shipments have been and will continue to be made securely using [dedicated trains] and general freight service" have been questioned.[31]

In summary, the committee's recommendation that DOE implement its dedicated train decision before commencing the large-quantity shipment of spent fuel and high-level waste to Yucca Mountain is based on the following operational, safety, security, communications, planning, and programmatic advantages:

- *Increased efficiency of operations.* The use of dedicated trains would allow greater control over schedules and routes, reduce travel times

[31]Senator Harry Reid and Senator John Ensign, letter to Hon. Samuel W. Bodman, Secretary, Department of Energy, August 17, 2005, *http://reid.senate.gov/record2.cfm?id=244114*.

to the repository, and simplify tracking. This could increase throughputs and reduce costs for equipment (conveyances and packages) and escort personnel, and allow more flexibility in the planning and deployment of those escorts.

- *Increased safety of operations.* Reduced travel times would help limit worker[32] and public radiation exposures during incident-free transport (see Chapter 3). If desired and feasible, such trains also could be routed to avoid major population centers. The separation of spent fuel and high-level waste shipments from other freight would reduce the potential for accidents involving very long duration fires (see Chapters 2 and 3). Dedicated trains can be designed and operated to higher safety standards than are currently required for general trains, further reducing the potential for accidents, especially train derailments.
- *Increased security of operations.* Security concerns have become more central to this issue since the September 11, 2001, attacks on the United States. These concerns, especially with respect to layovers in rail yards and routing through large population centers, have not been fully addressed by previous studies. The use of dedicated trains would help reduce transit times and therefore reduce opportunities for malevolent acts, especially in rail yards. Additional security escorts also could be added more easily to dedicated trains when needed.
- *Reduced program risks.* An accident involving spent fuel or high-level waste transport packages could cause substantial program delays, particularly if it involves fatalities and/or results in damage to the transportation package, even in the absence of any radiation releases. The use of dedicated trains would allow DOE greater control over avoidance of situations (e.g., avoid tunnels, switching yards, and peak traffic conditions on heavily traveled corridors; minimize stops in yards and on sidings) that could contribute to such accidents.

5.2.4 Acceptance Order for Commercial Spent Fuel Transport to a Federal Repository

FINDING: The order for accepting commercial spent fuel that is mandated by the Nuclear Waste Policy Act (NWPA) was not designed with the transportation program in mind. In fact, the acceptance order prescribed by the NWPA could require DOE to initiate its transportation program with long cross-country movements of younger (i.e., radiologically and thermally hotter) spent fuel from multiple commercial sites. There are clear transportation operations and safety advantages to be gained from shipping older

[32]As shown in Table 3.8, some estimated worker exposures for the Yucca Mountain transportation program are at DOE administrative limits; however, public doses are much lower.

(i.e., radiologically and thermally cooler) spent fuel first and for initiating the transportation program with relatively short, logistically simple movements to gain experience and build operator and public confidence.

RECOMMENDATION: DOE should negotiate with commercial spent fuel owners to ship older fuel first to a federal repository or federal interim storage, except in cases (if any) where spent fuel storage risks at specific plants dictate the need for more immediate shipments of younger fuel. Should these negotiations prove to be ineffective, Congress should consider legislative remedies. Within the context of its current contracts with commercial spent fuel owners, DOE should initiate transport through a pilot program involving relatively short, logistically simple movements of older fuel from closed reactors to demonstrate the ability to carry out its responsibilities in a safe and operationally effective manner. DOE should use the lessons learned from this pilot activity to initiate its full-scale transportation program from operating reactors.

The NWPA, as amended, establishes the order in which DOE must accept spent fuel from commercial reactors for transport to Yucca Mountain (Appendix C and Sidebar 5.2). The NWPA specifies that DOE must accept spent fuel based on the amount and order in which it was discharged from the owner's reactors. Each time a nuclear plant discharges fuel from its reactor, its owner receives an allocation in the "acceptance queue" to ship an equivalent amount of spent fuel to the federal repository. DOE will accept commercial spent fuel for shipment to the federal repository starting at the beginning of the queue and will work its way through the queue during the planned 24-year life of the transportation program. The NWPA allows owners to ship any spent fuel from any of their sites for each of their allocations in the acceptance queue.

There are two exceptions to this requirement (see Sidebar 5.2):

1. DOE may accord priority for acceptance of spent nuclear fuel from reactors that have been permanently shut down.
2. The owners of spent fuel can exchange positions in the acceptance queue, but only with the approval of DOE.

The latest report on acceptance priority ranking and annual capacity, which DOE is required to issue annually under its standard contract with commercial spent fuel owners (see Sidebar 5.1), was issued in July 2004 (DOE, 2004f). This report provides DOE's current plans for accepting commercial spent fuel for transport to Yucca Mountain. This ranking shows that DOE will nominally be required to accept 400 MTHM of commercial spent fuel during the first year of repository operations (estimated in the

SIDEBAR 5.2 Order for Acceptance of Commercial Spent Fuel for Shipment to a Federal Repository

The NWPA establishes the order in which DOE must accept commercial spent fuel from its owners for disposal at a federal repository. The NWPA specifies that DOE must accept spent fuel based on the amount and order in which it was discharged from the owner's reactors. Each time a nuclear plant discharges fuel from its reactor, its owner receives an allocation in the "acceptance queue" to ship an equivalent amount of spent fuel to the federal repository. DOE will accept commercial spent fuel for shipment to the federal repository starting at the beginning of the queue and will work its way through the queue during the planned 24-year life of the transportation program. There are two exceptions to this rule, which are discussed in Sidebar 5.1.

The main advantage of this acceptance system is that it provides an equitable order for acceptance of spent fuel. Owners who have allocations near the beginning of the acceptance queue made earlier payments into the Nuclear Waste Fund and have had to bear the costs for on-site spent fuel storage for a longer period of time. These owners are able to ship their spent fuel to the federal repository first, relieving them of further on-site storage costs. Owners who have more recent allocations in the acceptance queue will be required to maintain on-site spent fuel storage until they move to the front of the queue.

This acceptance system has two significant disadvantages, however. First, it will require that DOE make multiple shipments from multiple reactor sites during its planned 24-year transportation program. Second, the NWPA allows owners to ship any spent fuel from any of their sites for each of their allocations in the acceptance queue. Owners are likely to have a strong preference to ship spent fuel that is stored in spent fuel pools to free up space and reduce the need for (and expense of) future movements of spent fuel into dry casks. Some of the fuel offered for shipment could be recently discharged and radiologically active. Both of these factors could greatly complicate planning and execution of the transportation program and increase overall transportation risks.

report to be 2010) in 16 allocations from spent fuel owners (Table 5.1). The acceptance rate ramps up to 3000 MTHM in 39 allocations by 2014.

Some of the allocations (see column 4 of Table 5.1) are not large enough to fill even a single transportation package. Unless multiple allocations are combined,[33] there could be a large number of partially filled packages transported to Yucca Mountain. This could greatly increase the number of total shipments to the repository. For rail shipments, DOE may have to marshal packages from several reactors in rail yards to reduce the total number of trips to Yucca Mountain.

[33]In some cases, a single owner may hold several allocations in a given year. In other cases, allocations can be traded or purchased, subject to DOE review and approval. See Sidebar 5.1.

TABLE 5.1 Spent Fuel Annual Capacity and Acceptance Priority Ranking

Year of Repository Operation[a]	Quantity of Spent Fuel to Be Shipped to Repository (MTHM)	Number of Allocations in Acceptance Queue[b]	Range of Allocation Quantities[c] (MTHM)
1	400	16	0.1–145
2	600	23	0.1–72
3	1200	26	0.1–251
4	2000	37	0.1–467
5	3000	39	0.1–488
6	3000	39	0.1–452
7	3000	43	0.1–420
8	3000	47	0.1–375
9	3000	50	0.1–524
10	3000	46	0.1–444

[a]After Yucca Mountain is opened.

[b]Number of allocations made to owners of commercial spent fuel. In some cases, owners may have multiple allocations.

[c]Minimum and maximum allocated quantities of spent fuel for the number of allocations listed in column 3.

SOURCE: DOE (2004f).

The term "nominal" is used to describe these acceptances because, as described previously, the owners of the spent fuel will ultimately decide which fuel DOE will be required to transport. Some owners may have multiple allocations in a given year and may own multiple reactors. Under the standard contract, they have the right to designate fuel from any of their reactors for transport up to the combined quantity limits of their annual allocations. Owners are not required to inform DOE of which sites will ship spent fuel until about 5 years prior to the transport date and also are not required to finalize these plans until 12 months in advance of the shipping date. This makes DOE's planning assumptions especially tenuous.

The order for acceptance of spent fuel from commercial owners prescribed by the standard contract could require DOE to initiate its transportation program with movements of spent fuel from multiple, geographically dispersed sites. It gives DOE limited control over the age and radiological content of the fuel that is offered for transport.[34] It provides little opportu-

[34]The lack of DOE control over the age and radiological content of spent fuel offered for transport also has cost and operational implications for the repository receiving facility. DOE is planning to build above-ground spent fuel storage facilities and purchase dry casks so that spent fuel can be aged before being packaged for disposal. Aging is necessary because DOE has limited control over the radiological content of the spent fuel that is shipped for disposal.

nity for optimizing the transportation program to reduce the total number of shipments or the need for maintaining large numbers of transportation routes and for emergency responder training.

The order in which fuel is shipped to Yucca Mountain has important implications for safety, operational efficiency, and possibly the security of a Yucca Mountain transportation program. For reasons described in more detail below, it would be preferable from a safety standpoint to ship older spent fuel first to the federal repository. Under the current standard contract system, owners of commercial spent fuel are not required to ship older fuel first, however. The oldest fuel at many operating commercial reactors now resides in dry casks.[35] These casks were purchased and loaded by the owners, and some of these casks are licensed only for storage. They represent significant sunk costs,[36] which continue to grow as more plants resort to dry storage. Owners are likely to have a strong preference to give DOE their more recently discharged spent fuel, which is stored in pools at nuclear plant sites. This would free up pool space and reduce the need for (and the expense of) future movements of spent fuel into dry casks.[37]

Under current USNRC regulations, owners are required to store spent fuel in their pools (see Sidebar 1.4) for one year[38] before it can be transported. Owners could in principle designate one-year-old fuel for transport to Yucca Mountain. This could necessitate the purchase by DOE of transport packages with heavy shielding and active cooling systems[39] to maintain adequate cooling and reduce external radiation doses to workers and

[35]Current industry practice is to store only spent fuel that is older than five years in dry casks. See Sidebar 1.4.

[36]Each dry cask can cost $1 million or more to purchase and load, and the storage facilities can cost several tens of millions of dollars to license, construct, and protect. The federal government may be legally liable for those costs that occurred as the result of the failure of DOE to begin accepting spent fuel from commercial nuclear power plants in 1998.

[37]In December 2005, legislation was introduced into Congress that would require plant operators to move spent fuel from pools to dry casks at their sites within six years of its discharge from the reactor. The legislation would also require DOE to take title to the spent fuel once it is moved to dry casks and full responsibility for maintaining dry storage. The intent of this legislation is to eliminate the short-term need for a federal repository. However, if the legislation were to become law and a federal repository were to be opened, DOE could presumably ship spent fuel in whatever order desired because it would already have title to that fuel.

[38]This requirement dates to the 1970s when industry planned to ship spent fuel for reprocessing within 90–120 days of its discharge from the reactor. This would have required heavily shielded packages with active cooling systems. The requirement was established to provide enough time for iodine-131 (which has an 8-day half-life) in the fuel to decay.

[39]These packages have pumps and heat exchangers to remove decay heat from the spent fuel. Active cooling is generally required for spent fuel that is less than about three years old. Currently, there are no certified spent fuel transportation packages with active cooling systems in the United States.

the public. Designation of any fuel younger than five years old might require DOE to seek additional transport package approvals from the USNRC and to ship only partially filled packages to meet external dose requirements.

A recent Government Accountability Office study for Congress explores some of these issues (GAO, 2003). The GAO concluded that transport risks would be reduced by moving larger quantities of spent fuel per shipment and by deliberate selection of the order for picking up spent fuel from commercial nuclear plant sites. The report argues that reducing the number of shipments reduces risks because moving spent fuel in fewer, larger shipments would present fewer chances for accidents and fewer targets for terrorists. It would also simplify the problems of tracking and protecting shipments, and it would reduce routine exposure of workers to radiation. GAO found that the standard contracts between DOE and industry constrain DOE's ability to minimize the number of shipments or to control the order in which stocks of spent fuel are picked up.

GAO analyzed DOE's current plans for acceptance of spent fuel from commercial nuclear plant sites. It estimated that the 12 utilities with the largest quantities of spent fuel would make 576 shipments, with each shipment consisting of up to three rail packages, based on this current plan. GAO determined that if each owner consolidated its fuel into shipments of five full rail packages, total shipments would be reduced by roughly half to 287 (GAO, 2003, pp. 18–19).

The GAO study also found that properly selecting the order of shipment of spent fuel could reduce transportation and storage risks in three ways:

1. Early shipments of fuel from shut-down reactors would reduce the number of spent fuel storage sites. GAO found that nine spent fuel storage sites are not accumulating any additional fuel and could be cleared of their stocks, eliminating them as potential terrorist targets.
2. Early shipment of fuel from storage pools would reduce the likelihood of a pool fire, which could result from sustained loss of coolant from a spent fuel storage pool (see NRC, 2005d).
3. Shipping the older fuel first would reduce radiological transportation risks.

Selecting the shipment schedule that minimized risk would require an analysis to find the right balance between the advantages of moving older fuel first and of removing fuel from pool storage. However, GAO observed that under the terms of the standard contract, DOE cannot choose the order and locations of shipments according to age or form of present storage of the spent fuel. Contracts guarantee utilities positions in the queue for shipping specified quantities of spent fuel, and each utility can decide which fuel it wishes to ship within each of its allocated slots.

GAO did not recommend that DOE necessarily change the standard contracts. Rather, it noted that the absolute magnitude of risk reduction that could be obtained is unknown; therefore, it is unknown whether the benefits would justify the costs entailed in renegotiating the contracts and changing spent fuel owners' disposal plans. GAO recommended that DOE evaluate the potential benefits of selecting the order of shipments to minimize risk (GAO, 2003, p. 24).

In the committee's judgment, there are substantial transportation safety advantages (and possibly security advantages[40]) to be gained from moving the older fuel first.[41] This fuel has relatively low burn-ups (typically 25,000–30,000 megawatt-days per metric ton) relative to more recently discharged spent fuel (with burn-ups approaching 60,000 megawatt-days per metric ton). The low burn-up fuel emits less decay heat and radiation. Some of the oldest spent fuel has been in storage for several decades, enough time for the shortest-lived radionuclides to decay to background levels (Figure 5.2). Shipping this fuel first would provide an additional margin of safety, especially in reducing the potential hazards to workers and the public during both normal[42] and accident conditions.

The wording of the committee's recommendation was carefully constructed in recognition that there could conceivably be at-plant storage risks that might dictate earlier-than-desirable (from a transportation standpoint) movements of younger spent fuel to a federal repository or federal interim storage. The committee has not examined at-plant storage risks and

---

[40]As noted in Chapter 1, the committee was unable to perform an assessment of transportation security. If security threats do turn out to be a serious concern for spent fuel transport, then shipping older fuel first could help to reduce those threats by reducing inventories of radioactive materials in transportation packages.

[41]While not the subject of this report, shipping younger spent fuel first is also not optimal from a repository operations standpoint because that fuel may have to be stored for aging in surface facilities before being emplaced underground. This could require the construction of additional above-ground storage pads, the acquisition of additional storage packages, and additional handling steps. See also Footnote 34.

[42]While public exposures from routine transportation are already very low, this is not necessarily the case for transport workers. As shown in Table 3.8, for example, some transport workers involved in the program to ship spent fuel and high-level waste to the federal repository are estimated to receive the maximum doses allowed under DOE administrative guidelines. However, these guidelines also specify that exposures should be as low as reasonably achievable (ALARA). Shipping older fuel could help reduce these exposures.

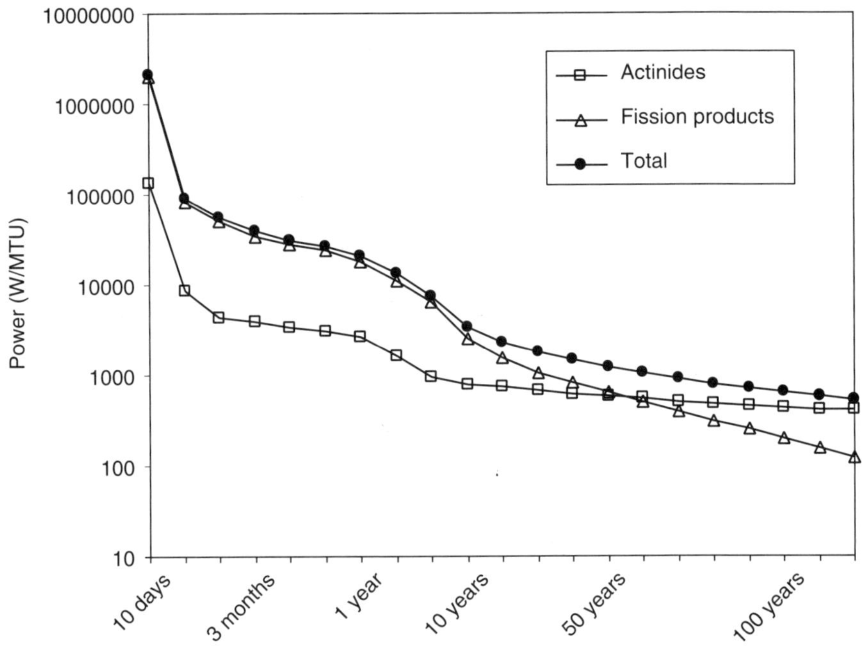

FIGURE 5.2 Plot of decay heat power (watts per metric ton of uranium) as a function of time (on a logarithmic scale) for commercial spent fuel after discharge from a reactor. Decay heat drops by about a factor of 100 during the first year after its removal from a reactor. SOURCE: NRC (2005d).

does not know if they are significant.[43] Nevertheless, the committee sees several practical difficulties in carrying out analyses to identify such risks as part of DOE's acceptance order negotiations with spent fuel owners:

- USNRC regulations require that spent fuel be stored at plant sites in a safe and secure manner, and the Commission and industry have repeatedly asserted that at-plant storage is safe and secure (see NRC, 2005d).
- Consequently, a plant owner could place itself in legal and regula-

[43]A recent National Academies report (NRC, 2005d) examined the safety and security risks of spent fuel storage at commercial power plant sites. That report concluded that immediate steps should be taken to improve the security of pool storage and that additional analyses should be undertaken.

tory jeopardy if it told DOE that it needed to ship younger spent fuel to reduce on-site storage risks.

- DOE lacks the data or authority to carry out at-plant storage risk analyses.

Moving old fuel first has a collateral benefit: Much of this fuel is in pool or dry storage at 15 of 23 permanently shut-down reactors (Table 5.2). These reactors are located throughout the United States. There would be several benefits of shipping spent fuel from these shut-down sites as early as possible in the transportation program:

- As noted previously, DOE has the authority under its standard contracts with spent fuel owners to choose the time and order in which it will ship from these shut-down sites. This provides operational flexibility during the early years of the transportation program, when DOE needs it most.
- DOE could initiate its transportation program by shipping spent fuel from one or two sites that were located close to main line rail routes, thereby reducing burdens on federal agencies and/or states for route security inspections, emergency responder training, and en route inspections.
- Some of these shut-down sites have enough stored spent fuel to support an extended shipping campaign, which would further reduce burdens for inspections and emergency responder training.
- Some of these shut-down sites have enough stored spent fuel to support greater than three-package rail shipments, which is currently the nominal DOE shipping configuration. Such shipments could help reduce program schedules and costs, reduce the total number of shipments to the federal repository, increase the cost-effectiveness of dedicated trains, and decrease any long-term safety or security concerns from leaving this material at these shut-down sites.
- Some of the spent fuel at these shut-down sites has already been placed into transport-ready form, and most of these sites have rail access.
- Removal of spent fuel from these sites would allow earlier license termination and site closure.

In short, shipping the older fuel first from shut-down plants would give DOE a better ability to optimize routing, scheduling, and emergency responder planning and training, especially during the early phases of its transportation program.[44] As experience is gained, longer movements could

---

[44]Of course, it might also subject DOE to increased political pressures to accept spent fuel first from certain sites or states, which could result in a suboptimal transportation program.

TABLE 5.2 Permanently Shut-down Commercial Nuclear Reactors in the United States

Reactor Unit Name	Location	Reactor Type (Thermal Output, MW)
Vallecitos	California	BWR (50)
CVTR	South Carolina	Heavy water (65)
Pathfinder	South Dakota	Superheat BWR (190)
Fermi 1	Michigan	Breeder (200)
Saxton	Pennsylvania	PWR (28)
Indian Point 1	New York	PWR (615)
Peach Bottom 1	Pennsylvania	HTGR (115)
Humbolt Bay 3	California	BWR (200)
Dresden 1	Illinois	BWR (700)
Three Mile Island 2	Pennsylvania	PWR (2772)
Shippingport[b]	Pennsylvania	PWR (~200)
LaCrosse	Wisconsin	BWR (165)
Fort St. Vrain	Colorado	HTGR (842)
Rancho Seco	California	PWR (2772)
Shoreham	New York	BWR (2436)
Yankee Rowe	Massachusetts	PWR (600)
San Onofre 1	California	PWR (1347)
Trojan	Oregon	PWR (3411)
Millstone 1	Connecticut	BWR (2011)
Haddam Neck	Connecticut	PWR (1825)
Maine Yankee	Maine	PWR (2772)
Big Rock Point	Michigan	BWR (67)
Zion 1, 2	Illinois	PWR (3250 each)

NOTE: BWR = boiling water reactor; HTGR = high-temperature gas reactor; PWR = pressurized water reactor.

[a]As of December 31, 2002.

[b]DOE was responsible for decommissioning this reactor and moving spent fuel into off-site storage.

be added. Such an operation could be part of a pilot program by DOE to gain experience and build public confidence[45] by demonstrating an ability to transport spent fuel to Yucca Mountain in a safe, secure, and operationally effective manner. These advantages are further elaborated in another National Research Council report (NRC, 2003).

---

[45]A University of New Mexico survey suggests that public concerns about the program to transport transuranic waste to the WIPP repository in New Mexico were reduced once the shipping program was under way and waste was being shipped on a routine basis. See Section 3.2.1.

Shutdown Year	Location of Spent Fuel Storage	Quantity in On-site Storage (MTHM)[a]
1963	Off-site	NA
1967	Off-site	NA
1967	Off-site	NA
1972	Off-site	NA
1972	Off-site	NA
1974	On-site	31
1974	Off-site	NA
1976	On-site	29
1978	On-site	91
1979	Off-site	NA
1982	Off-site	NA
1987	On-site	39
1989	On-site	15[c]
1989	On-site	229
1989	Off-site	NA
1991	On-site	128
1992	On-site	245
1992	On-site	360
1995	On-site	526
1996	On-site	448
1996	On-site	543
1997	On-site	71
1998	On-site	1021

[c]The spent fuel is stored in an on-site ISFSI licensed by the USNRC. DOE took title to the fuel in 1999.

SOURCES: USNRC. 2004. Fact Sheet on Decommissioning Nuclear Power Plants. *http://www.nrc.gov/reading-rm/doc-collections/fact-sheets/decommissioning.html*; DOE (2004f, Appendix A); information on Shippingport and Ft. St. Vrain from various sources.

5.2.5 Emergency Response Planning and Training

FINDING: Emergency responder preparedness is an essential element of safe and effective programs for transporting spent fuel and high-level waste. Emergency responder preparedness has so far received limited attention from DOE, states, and tribes for the planned transportation program to the federal repository. DOE has the opportunity to be innovative in carrying out its responsibilities for emergency responder preparedness. Emergency responders are among the most trusted members of their communities. Well-trained responders can become important emissaries for DOE's trans-

portation program in local communities and can enhance community preparedness to respond to other kinds of emergencies.

RECOMMENDATION: DOE should begin immediately to execute its emergency responder preparedness responsibilities defined in Section 180(c) of the Nuclear Waste Policy Act. In carrying out these responsibilities, DOE should proceed to (1) establish a cadre of professionals from the emergency responder community who have training and comprehension of emergency response to spent fuel and high-level waste transportation accidents and incidents; (2) work with the Department of Homeland Security to provide consolidated "all hazards" training materials and programs for first responders that build on the existing national emergency response platform; (3) include trained emergency responders on the escort teams that accompany spent fuel and high-level waste shipments; and (4) use emergency responder preparedness programs as an outreach mechanism to communicate broadly about plans and programs for transporting spent fuel and high-level waste to a federal repository with communities along planned shipping routes.

The transportation of spent nuclear fuel to a federal repository would utilize the same state, tribal, and local emergency response capabilities that are in place to deal with existing hazardous materials transport accidents and incidents (see Appendix C). DOE has special responsibilities under the NWPA to ensure that emergency response capabilities and training are adequate to support its repository transportation program. It is responsible under the NWPA for providing technical assistance and funding to states and tribal nations for training on both routine transportation procedures and emergency response. These responsibilities are enumerated in Section 180(c):

> The Secretary shall provide technical assistance and funds to States for training for public safety officials of appropriate units of local government and Indian tribes through whose jurisdiction the Secretary plans to transport spent nuclear fuel or high-level radioactive waste. . . . Training shall cover procedures required for safe routine transportation of these materials, as well as procedures for dealing with emergency response situations. The [Nuclear] Waste Fund shall be the source of funds for work carried out under this subsection.

DOE issued a statement of policies and procedures for providing such assistance in an April 1998 *Federal Register* notice (DOE, 1998b). This statement was the product of more than three years of work by DOE staff and involved several rounds of *Federal Register* notices and public comments. DOE has determined that this latest notice will remain in draft form "until program progress or legislation provides definitive guidance as to when shipments will commence. At that time, OCRWM may finalize these policy and procedures or will consider promulgating regulations on Section 180(c) implementation" (DOE, 1998b, p. 27353).

This notice addresses training for both safe routine transportation and emergency response situations. The draft notice states (DOE, 1998b, p. 23754):

> It is OCRWM's policy that, for NWPA shipments, each responsible jurisdiction will have the training necessary for safe routine transportation of spent nuclear fuel or high-level waste and to respond to NWPA transportation incidents or accidents. OCRWM will provide training and technical assistance, subject to annual appropriations, to assist states and tribes to obtain access to the increment of training necessary to prepare for NWPA shipments. . . . If Congress does not fully appropriate the funds requested, the funding to eligible jurisdictions will be decreased proportionately.
>
> For safe routine transportation of spent nuclear fuel and high-level waste, it is OCRWM's policy to provide each eligible state and tribe the funding and technical assistance to prepare for safety and enforcement inspections of NWPA highway shipments, for rail measures that complement FRA inspection procedures, and for access to satellite tracking equipment and training on that equipment in cases where the capability does not already exist.

To carry out this policy, DOE intends to make two types of grants available to eligible states and tribal nations:

1. One-time planning grants of $150,000[46] to help eligible jurisdictions determine their needs for training funds and technical assistance. The amount of the grant is based on DOE's experience with planning for shipments to WIPP in New Mexico.
2. Base grants for safety and enforcement planning and training activities. Such grants will be available on an annual basis for up to five years and will consist of two parts: The first part will be for safety and enforcement inspection training and awareness-level training, awareness-level refresher training, and emergency responder trainer training (see Sidebar 5.3). The second part will be for enhanced emergency responder training such as operations- or technician-level training. The amount of funding available to a particular jurisdiction is based on need as determined by DOE.

States and tribal nations will be eligible to apply for grants approxi-

[46]The committee observes that this is a very small amount of funding to cover an assessment of statewide needs.

SIDEBAR 5.3 Emergency Responder Training

Emergency responder training requirements are established by federal regulation. Occupational Safety and Health Administration regulations (29 CFR 1910.120) require that emergency responder training for "hazardous substances" (which includes radioactive materials) be based on the duties and function of each person in the emergency response organization. These regulations describe five levels of training proficiency and require that individuals receive sufficient annual refresher training to maintain their competence or else demonstrate competence on at least an annual basis.

1. *First-responder awareness level.* This training is for individuals who are likely to discover the hazardous release and who have been trained to initiate the emergency response by notifying the proper authorities. These individuals would include most public safety personnel, including police officers and firefighters, and possibly highway maintenance workers. Such individuals are required to have training or sufficient experience to demonstrate competence in recognizing the presence of hazardous substances in an emergency; the risks and potential outcomes associated with these substances; and the ability to recognize the need for additional resources and to make appropriate notifications.
2. *First-responder operations level.* This training is for individuals who respond to actual or suspected releases of hazardous substances and are responsible for defensive actions to protect people, property, and the environment. They are responsible for keeping the release from spreading and preventing exposures. These individuals would likely be part of a hazmat team associated, for example, with a fire company. Operations-level responders are required to have at least eight hours of training or sufficient experience to demonstrate competence beyond a basic awareness level in the following: knowledge of basic hazard and risk assessment techniques; selection and use of operational-level protective personnel equipment; basic hazardous material terms; basic containment, confinement, or control operations; basic decontamination techniques; and relevant operating procedures.
3. *Hazardous materials technician.* This training is for individuals who are responsible for stopping hazardous substance releases to the environment. They

mately four years before shipments are scheduled to commence through or along the borders of their jurisdictions. DOE estimates that it will take as long as one year for the application process to be completed, which will leave approximately three years for the assistance program to be implemented. DOE will notify the governor or tribal leader in writing when the jurisdiction becomes eligible to apply for these grants. Jurisdictions will be asked to select an agency or representative to apply for and administer the grants.

Only state and tribal organizations are eligible to apply for these grants. However, states and tribes are required to coordinate their planning with local jurisdictions and indicate in the grant application how the needs of take aggressive actions in approaching the points of release to plug or patch them. These individuals are required to have at least 24 hours of training at the operations level and, in addition, to demonstrate competence in the following: ability to implement the emergency response plan; knowledge of the classification and verification of known and unknown materials by using survey instruments; ability to function within an assigned role in the incident command system; ability to select and use chemical personnel protection equipment; understanding of hazard and risk assessment techniques; ability to perform advance control, containment, or confinement operations in the capabilities of available protective equipment and resources; understanding and implementing decontamination procedures; understanding basic chemical toxicological terminology and behavior.

4. *Hazardous materials specialist.* This training is for individuals who have similar responsibilities to a hazardous materials technician but with more directed knowledge of the hazardous substances that they are called upon to contain. They are required to have at least 24 hours of training at the technician level and be able to demonstrate competence in the following: ability to implement the local emergency response plan and knowledge of the state response plan; understanding classification, identification, and verification of known and unknown substances using advanced survey instruments; ability to select and use specialized personnel protective equipment; an in-depth understanding of hazard and risk assessment techniques; ability to perform specialized control, containment or confinement operations; ability to determine and implement decontamination procedures; ability to develop a site safety and control plan; understanding of chemical, radiological, and toxicological terminology and behavior.

5. *On-scene incident commander.* This training is for individuals who will assume control of the incident scene. These individuals must receive at least 24 hours of training at the operations level and demonstrate competence in the following: ability to implement the incident command system and emergency response plan; knowledge of the hazards and risks associated with employees working in chemical protective clothing; knowledge of the state emergency response plan and Federal Regional Response Team; knowledge and understanding of the importance of decontamination procedures.

local safety officials have been considered and how incremental training will be provided. This would include a description of where training would be obtained, what drills and exercises would be included in the training, what equipment and supplies would be purchased, and what other technical assistance would be needed from DOE. DOE anticipates that awareness-level training (see Sidebar 5.3) will be made available to local public safety officials and that enforcement training will be made available to state-level and tribal employees. However, it is largely up to the state or tribal nation to determine, in consultation with local governments and first responders, who receives training and at what frequency such training is provided.

DOE received 19 sets of comments from states and organizations on its proposed policies and procedures. These are enumerated in a *Federal Register* notice.[47] According to the notice (p. 23757), "the large majority of commenters emphasized that they believe that additional change is still needed in key areas, primarily more cooperative route selection and a more cooperative transportation planning process." Several organizations identified the WIPP transportation planning process as a good example of cooperative planning. Some specific comments addressed the following unmet needs, specifically for:

- Improved communications, including early and substantive public outreach, with DOE taking the responsibility for interacting with states and tribes rather than relegating these responsibilities to private contractors.
- Early selection of routes to allow sufficient time for states to designate alternate routes and assess their planning and training needs.
- Policies on early and cooperative selection of routes, especially to avoid the possibility that the selection of too many routes would dilute planning and training resources.
- Training of emergency responders equivalent to Occupational Safety and Health Administration (OSHA) operations-level training (see Sidebar 5.3).

Concerns about transportation planning and emergency responder training continue to be voiced at DOE-sponsored forums such as the Transportation External Coordination (TEC) Working Group meetings (see Section 3.4). A consistent theme expressed at these meetings is the need for DOE to get on with the process of route selection so that states and tribal nations can begin planning and training activities.

DOE's Office of Environmental Management, which is responsible for cleanup of the nation's nuclear weapons sites, has developed the Transportation Emergency Preparedness Program (TEPP) to support training of federal, state, tribal, and local authorities in emergency preparedness and response to transportation incidents involving DOE radioactive materials shipments. The program has designated coordinators in each of DOE's eight regional offices and has been used extensively for WIPP-related emergency preparedness and response. The TEPP provides a number of planning and training tools for state, tribal, and local governments. The planning tools include models for developing needs assessments, preparedness plan-

[47]Office of Civilian Radioactive Waste Management; Safe Routine Transportation and Emergency Response Training; Technical Assistance and Funding. 63 FR 23757–23766, April 30, 1998.

ning, and response procedures. They also include a "drill-in-a-box" that provides scenarios and materials for preparing and conducting tabletop exercises and drills for transportation incidents. The program provides technical assistance to state and tribal governments for developing, updating, and testing emergency response plans.

TEPP also provides training materials (Modular Emergency Response Radiological Transportation Training [MERRTT]) for emergency managers and responders, firefighters, law enforcement, and related personnel. These materials provide information on fundamental concepts and procedures for responding to radioactive materials transport incidents. The training is organized into a modular format and includes manuals, instructor guides, and overheads. The program offers several train-the-trainer sessions at various locations across the United States for qualified instructors.

The WIPP emergency responder training program is often cited as a good model for transportation planning and emergency response. WIPP's success was somewhat fortuitous, however. Transportation routes to WIPP were not identified early in the program. When DOE made an initial and ultimately unsuccessful effort to open WIPP in 1988, little planning for transportation had been carried out. It took DOE another 11 years (until 1999) to get the repository open, which gave the agency more time to identify routes and provide the necessary training to emergency responders.

Transportation to WIPP ramped up over a period of five years, which provided DOE with time for planning and training. Initially, only one site shipped waste to WIPP: Rocky Flats, near Denver, Colorado. WIPP has now received waste from eight sites. A relatively small number of routes are used for WIPP shipments (Figure 5.3). This has allowed DOE to keep the demand for emergency responder training and coordination to a manageable level.

The WIPP program was willing to go beyond its legally mandated requirements in planning and implementing its transportation program. For example, DOE executed a memorandum of understanding with the WGA and Southern States Energy Board that included emergency response operations. The WGA, in cooperation with DOE, developed the WIPP *Transportation Safety Program Implementation Guide* (WGA, 2003), which governs the conduct of transuranic waste shipments through the western states, including the conduct of emergency response operations. The Southern States Energy Board (SSEB, 1994) developed the *Transuranic Waste Transportation Handbook*, which serves as a primer for waste transportation through its member states.[48] The Council of State Governments' Midwest-

[48]DOE also provided funding to New Mexico to help construct a highway bypass around Sante Fe for WIPP shipments.

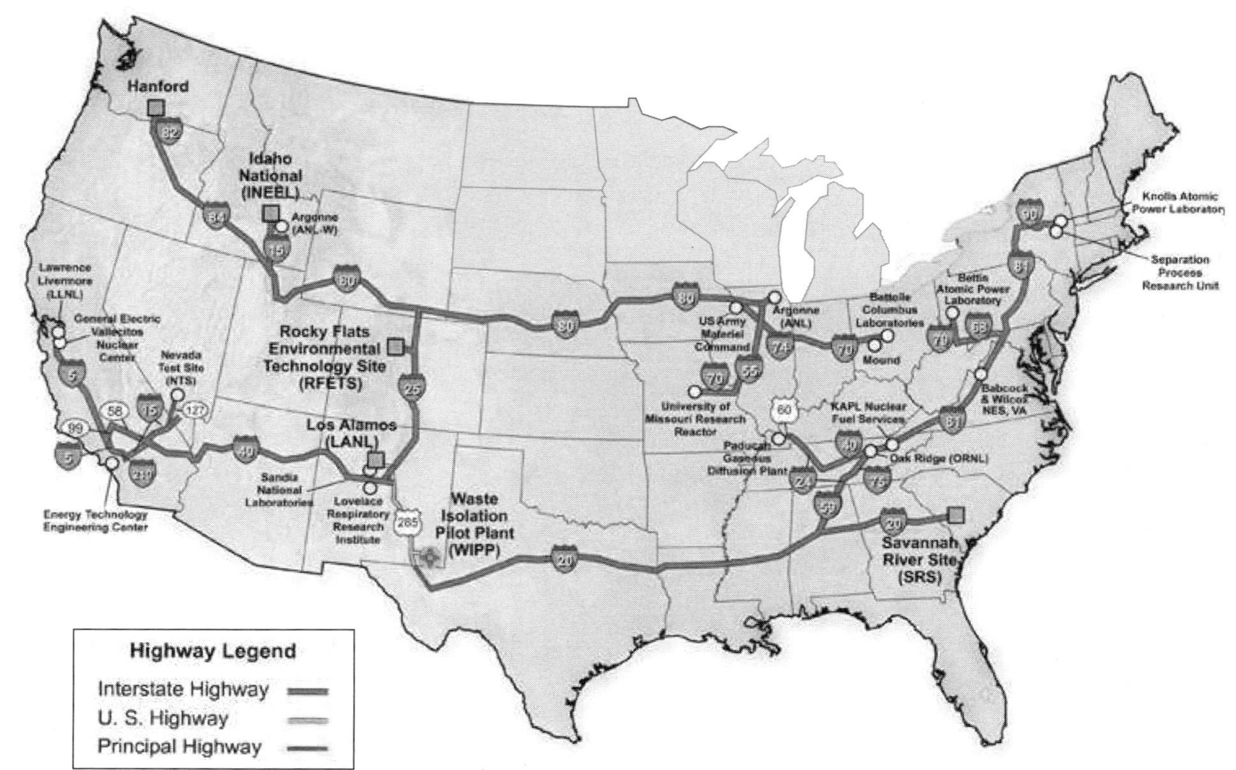

FIGURE 5.3 Routes used for shipping transuranic waste to the Waste Isolation Pilot Plant near Carlsbad, New Mexico. SOURCE: Modified from *http://www.wipp.ws/routes.htm*.

ern Office has also issued guidance for radioactive waste shipments through its member states (CSG, 2004).

WIPP is providing direct training in emergency preparedness and response to first responders along WIPP routes through TEPP. While Section 180(c) of the NWPA requires DOE to provide funds for emergency response training for Yucca Mountain, direct training is not required.

Many lessons from the WIPP transportation program could potentially be applied to Yucca Mountain. DOE's Yucca Mountain Program issued its safe transportation strategic plan in November 2003 (DOE, 2003c). In it, DOE pledged to (p. 2) "approach its transportation planning cooperatively, using a collaborative process that incorporates the successful elements from transportation systems developed for other DOE programs." WIPP is mentioned as such a program.

While the WIPP experience offers some useful lessons, the experience may not be scalable to the Yucca Mountain transportation program. DOE will have to provide training assistance to emergency responders along planned shipping routes in up to 45 states. Even during the early phases of the program, DOE is likely to face demands for training assistance from a dozen or more states along identified transportation routes.

DOE will not begin providing Section 180(c) technical assistance and funding to states until it identifies the routes for shipping spent fuel and high-level waste to Yucca Mountain. The committee sees a clear strategic advantage to DOE in making these decisions and providing at least a base level of assistance at the earliest possible date.

Volunteer and paid emergency responders have been the institutional foundations in many communities since Benjamin Franklin's era and are among their communities' most trusted members. If adequately involved, they can become important emissaries for DOE's program with the local community. When elected officials or members of the public ask "Are these shipments safe?" and "Can our community handle an emergency?" they often look to local emergency response officials for answers. It is in DOE's interests to provide the training and technical assistance necessary for these officials to feel confident that they have the equipment and training needed to respond to any accident or incident involving spent fuel and high-level waste. This training and technical assistance might also help to mitigate some of the social risks described in Chapter 3 (see Section 3.2).

A DOE representative told the committee that training may be ineffective if started too early, given the expected turnover of emergency responder personnel, especially volunteer firefighters, who may experience a 20 percent or higher annual turnover.[49] Although it is true that turnover in some

[49]About 75 percent of the United States is protected by volunteer or only part-paid fire departments. The membership rules for training and proficiency in most volunteer fire depart-

segments of the emergency responder community is high, especially among volunteers, the committee nevertheless judges that it is never too early to start training. There is a cadre of professionals who will stay in their positions over the long term; they would benefit from early training and they could help DOE with planning for such training. Such early training will be essential for developing appropriate organizational memory and culture and for setting expectations. DOE could focus its early assistance on training-the-trainer and other long-term activities such as planning for equipment procurements, calibrations, upgrades, and replacements of radiation detection instruments used by emergency responders.[50]

There are even benefits, albeit more indirect, to be gained from training volunteer emergency responders who may leave their posts before the transportation program begins operating: these people live in the communities through which these shipments will pass and some are community leaders. They can serve as informal but important sources of information to their communities about DOE's transportation program.

Since DOE's Section 180(c) assistance will reach the majority of states and many tribal nations, the committee sees clear opportunities for innovation. DOE could work with the Department of Homeland Security to provide consolidated all hazards training materials and programs for emergency responders that would provide at least awareness-level training (see Sidebar 5.3). Such programs could reach large numbers of emergency responders well in advance of any route determinations and would help to leverage DOE's limited 180(c) funding. Such training could also help counteract the possible perception among emergency responders, local officials, and members of the public that if special training is needed for spent fuel and high-level waste shipments, transporting such materials must be an especially risky activity.

Another opportunity for innovation involves the deployment of trained emergency responders on the escort teams that accompany spent fuel and high-level waste shipments to Yucca Mountain. These individuals could be given the responsibility for establishing liaisons with tribal and local government emergency responder organizations along transportation routes. Such individuals would provide the first line of emergency response in an

---

ments are as rigorous as those for full-time paid fire service, and most volunteer firefighters take enormous pride in their service and competence. While it is true that turnover among volunteers is high, many volunteer departments have a cadre of long-serving members who retain the corporate memory and help to train newcomers.

[50]Some of the nontraining activities could take place under the "technical assistance" clause of the NWPA if sufficient funding is made available by Congress for such purposes. These activities could also be carried out in cooperation with the Department of Homeland Security as part of its activities to upgrade emergency responder preparedness for terrorist attacks.

accident or incident and would function as a known and trusted resource for local incident commanders. This innovation could be most readily implemented for rail shipments, because these will be fewer in number and will have more space to accommodate escorts.

DOE could also use this state and local interest to communicate more broadly with the public about its transportation program by

- Opening its emergency responder training sessions to selected individuals from local communities, especially opinion leaders,
- Providing emergency responder information through a Web site designed for interested non-experts, and
- Working with local communities and their schools to develop programs to monitor environmental conditions (e.g., radiation levels) along transportation routes.

Although the discussion in this section has focused on the federal repository transportation program operated by DOE, the committee's suggested innovations are also potentially applicable to the transportation program operated by Private Fuel Storage, LLC. This private transportation program has no legal responsibility under the NWPA to support emergency responder preparedness along its shipping routes. To the committee's knowledge, Private Fuel Storage, LLC, has no plans to provide financial aid to communities along its planned transportation routes to support emergency responder training. However, state, tribal, and local communities are likely to take a keen interest in emergency response preparedness in this program. The committee judges that there would be significant benefits to the program in terms of capacity and public confidence building through early and innovative actions to support emergency responder preparedness.

5.2.6 Information Sharing and Openness

FINDING: There is a conflict between the open sharing of information on spent fuel and high-level waste shipments and the security of transportation programs. This conflict is impeding effective risk communication and may reduce public acceptance and confidence. Post–September 11, 2001, efforts by transportation planners, managers, and regulators to further restrict information about spent fuel shipments make it difficult for the public to assess the safety and security of transportation operations.

RECOMMENDATION: The Department of Energy, Department of Homeland Security, Department of Transportation, and Nuclear Regulatory Commission should promptly complete the job of developing, applying, and disclosing consistent, reasonable, and understandable criteria for protecting

sensitive information about spent fuel and high-level waste transportation. They should also commit to the open sharing of information that does not require such protection and should facilitate timely access to such information: for example, by posting it on readily accessible Web sites.

This finding and recommendation are motivated by several factors. In a representative democracy, citizens have a general right, subject to legitimate privacy and national security restrictions, to obtain information about government programs that affect their communities. Such sharing of information might also help to build community trust and confidence in transportation programs and help implementers to identify and manage the social risks described in Chapter 3 (Section 3.2).

The current study was conceptualized before the September 11, 2001, terrorist attacks on the United States, and its original focus was intended to be on the safety of spent fuel and high-level waste shipments. During the committee's information-gathering meetings, however, several participants expressed concern about the security of spent fuel and high-level waste transport, especially the potential consequences of terrorist attacks on transport packages in highly populated areas. Other participants expressed concerns about the efforts of federal agencies to use the September 11 attacks as a pretext for withholding information that could help the public to evaluate the safety and security of spent fuel and high-level waste shipments. These presenters expressed concern that such withholding could allow the government to operate such programs with little public scrutiny.

The committee itself encountered information restrictions during this study. In compiling historical data on spent fuel shipments in the United States, the committee discovered that the USNRC had removed some of the needed information (USNRC, 1978) from its Web site and document reading room because it was deemed to be too sensitive for public release. The committee saw no reason for withholding this information given that it involved only past shipping campaigns. However, at the committee's request, Commission staff provided the summary data on historical spent fuel shipments for use in this report (Chapter 3). The staff also told the committee that it is now updating and reviewing this historical information to determine what is appropriate for public release.

The committee discussed the possibility of expanding its report to include information on the security of spent fuel and high-level waste transportation. This effort was supported by the federal study sponsors. As noted previously, four members of the committee and one staff member with appropriate security clearances were given a classified briefing by USNRC staff on investigations under way within that agency to assess the security of transportation packages. However, the committee was not able to receive timely written guidance from the study sponsors on what information from these studies and other documents[51] could be released to the public in the committee's final report.

The USNRC has provided guidance to its staff for communicating security risks since the September 11, 2001, terrorist attacks (USNRC, 2004e, f). The guidance notes that "[p]eople who live near power plants and other nuclear facilities have a different sense of the risks they are asked to bear on behalf of the rest of the country" (USNRC, 2004e, p. 54) and recommends that agency officials "clearly establish what information can be shared and what can't" (USNRC, 2004e, p. 52). The guidance notes that information can be shared about additional security personnel, new equipment, and upgraded procedures put in place, but without specifics being disclosed. The guidance also recommends that agency officials clearly state why they cannot share detailed security information. The agency asserts, "Because the public is keenly aware of security concerns most will understand and respect the need to keep certain information classified" (USNRC, 2004e, p. 52).

DOE's transportation program for Yucca Mountain apparently has not yet confronted this issue to the same extent, largely because it is still in the planning stages. However, DOE's Environmental Management program, which is responsible for cleaning up the U.S. nuclear weapons complex, has long grappled with the issue of openness (e.g., Ashford and Rest, 1999; Bradbury et al., 1996a,b, 1997a,b, 2003; Bradbury and Branch, 1999; Drew et al., forthcoming). The TEC Working Group, which advises the Yucca Mountain transportation program, has provided suggestions on best practices to assist DOE program managers "in their efforts to communicate about radioactive materials transportation in a manner that is responsive to the needs and concerns of stakeholders" (TEC, 2002, p. 1). TEC recommends sharing information such as the number of shipments, mode(s), possible route(s), time frame, quantity, type of material being shipped, and reason for making the shipments. Web sites are also considered "good tools for making information available" (TEC, 2002, p. 3).

Some of the information recommended for sharing by TEC is considered by other parts of DOE to be too sensitive to share with the public. For example, the program for transporting transuranic waste to WIPP has restricted information sharing about its shipments. At its Albuquerque meeting, the committee received comments from the representative of a stakeholder group concerning public restrictions on information sharing about these shipments after the terrorist attacks of September 11, 2001 (Hancock, 2004). This representative noted that State of New Mexico officials were no longer allowed to provide shipment schedule information to members of

---

[51]Some of this information exists in published reports that were removed from public circulation by the government after the September 11 attacks.

the public, and DOE no longer responded to requests for information about incidents involving these shipments.

Clearly, the public has a right to receive some timely information about the shipments of spent fuel and high-level waste that pass through their communities. The TEC guidance on information sharing described above strikes the committee as a reasonably complete and appropriate guidance for sharing at different times in the life cycle of a shipment or shipping campaign. Some general information is appropriate to share before shipments commence: This includes reasons for making the shipments; general information about the materials to be shipped; possible shipping modes and routes; and general shipping time frames. Appropriate post-shipment information includes more details on the shipments, such as quantities of materials shipped; specific modes and routes used for the shipments; timing of shipments; accidents and incidents during shipments; and any resulting response actions.

Federal agencies need to develop and then abide by clear and consistent guidelines for protecting information about spent fuel and high-level waste transportation activities. The class of information to be protected should be defined clearly and should be small, encompassing only that information that is truly in need of protection. The remaining information should be made freely available to the public through agencies' Web postings and other dissemination channels.

5.3 ORGANIZATIONAL STRUCTURE OF THE FEDERAL TRANSPORTATION PROGRAM

FINDING: Successful execution of DOE's program to transport spent fuel and high-level waste to a federal repository will be difficult given the organizational structure in which it is embedded, despite the high quality of many current program staff. As currently structured, the program has limited flexibility over commercial spent fuel acceptance order (Section 5.2.4); it also has limited control over its budget and is subject to the annual federal appropriations process, both of which affect the program's ability to plan for, procure, and construct the needed transportation infrastructure. Moreover, the current program may have difficulty supporting what appears to be an expanding future mission to transport commercial spent nuclear fuel for interim storage or reprocessing. In the committee's judgment, changing the organizational structure of this program will improve its chances for success.

RECOMMENDATION: The Secretary of Energy and the U.S. Congress should examine options for changing the organizational structure of the Department of Energy's program for transporting spent fuel and high-level

waste to a federal repository. The following three alternative organizational structures, which are representative of progressively greater organizational change, should be specifically examined: (1) a quasi-independent DOE office reporting directly to upper-level DOE management; (2) a quasi-government corporation; or (3) a fully private organization operated by the commercial nuclear industry. The latter two options would require changes to the Nuclear Waste Policy Act. The primary objectives in modifying the structure should be to give the transportation program greater planning authority; greater budgetary flexibility to make the multiyear commitments needed to plan for, procure, and construct the necessary transportation infrastructure; and greater flexibility to support an expanding future mission to transport spent fuel and high-level waste for interim storage or reprocessing. Whatever structure is selected, the organization should place a strong emphasis on operational safety and reliability and should be responsive to social concerns.

The Yucca Mountain transportation program operates within the larger milieu of the repository development effort (Chapter 1). Consequently, its success will depend to a large degree on the decisions made within DOE, other agencies, and Congress on whether and when to license, construct, and open a federal repository. If completed, the federal repository program will be the most expensive waste disposal effort in U.S. history. Current projected life-cycle costs are $58 billion in 2000 dollars (DOE, 2001d). The transportation program would also be the most expensive effort to ship spent fuel and high-level waste in the nation's history. Its share of the life cycle costs is about $6 billion in 2000 dollars.

Certain characteristics of the Yucca Mountain transportation program will make it exceptionally challenging to carry out successfully: The transportation program

- Will last for more than two decades;
- Is decentralized, encompassing more than 70 sites in 31 states;
- Involves a large number of parties (i.e., industry, regulators, and state, tribal, and local governments) over which it has limited control;
- Must operate with a high degree of consistency and reliability;
- Has limited flexibility over schedules because of the standard contract requirements for acceptance of spent fuel (see Section 5.2.4); and
- Has limited budgetary control within DOE and is subject to the annual congressional appropriations process.

The transportation system is not only physically and logistically complex, but also has a "nested complexity" that derives from the institutional architecture in which it is embedded. The transportation program is embedded in DOE's OCRWM, which is responsible for constructing and operating a federal repository. The transportation program must compete with the much larger repository development program for personnel, funding, and management attention.

OCRWM has embarked on what some have described (and the committee agrees) is an ambitious schedule to open a Yucca Mountain repository. Its last announced schedule for opening the repository included the following milestones:

- Submission of a repository construction license application to the USNRC in December 2004.
- Approval by the Commission in 2008 to begin construction of the repository and ancillary surface facilities.
- Submission of a license amendment to the USNRC in 2009 to begin receiving waste.
- Approval of this amendment by the USNRC and start-up of repository operations by the end of 2010.

DOE missed the first milestone because of problems with the Environmental Protection Agency Agency's standard for Yucca Mountain and completion of DOE's licensing support network (see Section 1.3.2). A new schedule had not been formally announced by December 2005 when work on this report was being completed, although DOE has stated that Yucca Mountain will not open until 2012 at the earliest.

The transportation program's schedule is linked to the schedule for opening the repository. The last announced schedule for the transportation program included the following milestones:

- Complete work on the Nevada rail EIS and issue a ROD on the specific alignment in early 2006.
- Award the design contract for the rail spur in early 2005 and begin construction of the rail spur in early 2006. According to the final Yucca Mountain EIS (DOE, 2002a), construction is planned to last no more than 46 months at an estimated cost of about $880 million (see Section 5.2.1).[52]
- Issue a request for proposals on rolling stock (i.e., railcars) in early 2005 and begin receiving equipment deliveries in early 2007.
- Undertake an assessment of industry's ability to provide transportation packages of the sizes and quantities needed for the program. Package

[52]As noted previously, DOE acknowledged that the estimated cost for constructing the complete 319-mile (513-kilometer) rail spur had increased to about $2 billion.

design, certification, and fabrication activities would be carried out in fiscal years 2005 and 2006.

- Complete work on an operational plan and issue a final concept of operations in early fiscal year 2005. This would presumably detail how DOE plans to conduct its transportation operations, including the role of contractors.
- Issue final routing selection criteria in late fiscal year 2005 and its transportation operations plan in late fiscal year 2006.
- Identify a suite of transportation routes in early fiscal year 2006, which will then trigger the awarding of Section 180(c) planning grants (see Section 5.2.5) to states later that fiscal year. Awards of Section 180(c) base grants would be made later in fiscal year 2007.

In October 2005, OCRWM announced that it intended to adopt a standardized package design to transport, store, and dispose of commercial spent fuel. This will require certification of the new package design and modification of the design for the fuel receipt and handling facilities at the federal repository. DOE has not indicated what additional delays may be encountered in the repository program and its associated transportation program to implement these changes.

Funding for the Yucca Mountain Program comes from a combination of direct federal appropriations to cover the costs of disposing of defense spent fuel and high-level waste and the Nuclear Waste Fund (Appendix C) to cover the costs of commercial spent fuel disposal. Both funding sources are controlled by Congress through the annual appropriations process.[53] The transportation program's annual budget request to Congress is submitted as part of the OCRWM budget request based on a target set by the Office of Management and Budget in consultation with DOE management. As shown by Table 5.3, the transportation program's budget has not received high priority within OCRWM, possibly because the overall OCRWM budget has itself been underfunded relative to requested levels during several of the past years. The transportation program will require substantially higher future budgets to construct a Nevada rail line, procure transportation equipment, and make other necessary infrastructure improvements.

While the apparent delay in opening a federal repository would potentially provide more time for the transportation program to attain operational readiness, it introduces other complications. For example, the com-

---

[53]The Bush administration made an unsuccessful attempt in fiscal year 2005 to exempt the Nuclear Waste Fund from the annual appropriations process. This attempt, which was supported by the nuclear industry, would have provided DOE with a great deal more budgetary discretion.

TABLE 5.3 Congressional Appropriations for OCRWM and Its Transportation Program

Federal Fiscal Year	OCRWM Request ($ millions)	OCRWM Appropriation ($ millions)	Transportation Program Request ($ millions)	Transportation Program Appropriation ($ millions)
1999	380	353	2	2
2000	409	351	2	2
2001	437	401	3.8	2.7
2002	445	375	5.9	4.6
2003	591	457	30.2	9.4
2004	591	577	73.1	63.6
2005	131[a]	572	186	30.7
2006	651	450[b]	85.4	19.9

[a]DOE requested $880 million, but the Bush administration requested only $131 million for defense waste disposal. The administration intended to obtain the remaining funding from the Nuclear Waste Fund and unsuccessfully attempted to have that fund taken "off budget," which would have freed it from the annual congressional appropriations process.

[b]Does not include the $50 million appropriated by Congress in fiscal year 2006 for initiation of a site selection process for an integrated spent fuel recycling facility.

SOURCE: DOE budget documents and written communications.

mercial nuclear industry could begin shipments of spent fuel to Private Fuel Storage, LLC, in Utah once that facility is constructed and opened.[54] Under current plans, this fuel will be placed in packages (see Sidebar 1.4) for transport and storage. However, unless the industry adopts the new DOE standardized package, which has not yet been designed, licensed, or manufactured, it may be required to repackage the fuel before it can be transported to the federal repository. In the meantime, additional nuclear plants are expected to establish dry-cask storage to relieve growing storage pressures in their pools, and additional plants may be closed and decommissioned. These changes could further affect the transportation program's pickup schedules and require more repackaging of spent fuel for transport to the federal repository.

The recent decision by Congress to promote the development of one or more federal interim storage sites for commercial spent fuel potentially

[54]A license for this facility was submitted to the USNRC in 1997. As noted in Chapter 1, the Commission authorized its staff to issue a license to construct and operate this facility under the conditions in 10 CFR 72.40.

further complicates DOE's transportation mission. The Energy and Water Development Fiscal Year 2006 report[55] provides the following direction to DOE (italics added):

> Integrated spent fuel recycling.—Given the uncertainties surrounding the Yucca Mountain license application process, the conferees provide $50,000,000, not derived from the Nuclear Waste Fund, for the Department to develop a spent nuclear fuel recycling plan. Under the Nuclear Energy account, the conferees provide additional research funds to select one or more advanced recycling technologies and to complete conceptual design and initiate pre-engineering design of an Engineering Scale Demonstration of advanced recycling technology. Coupled with this technology research and development effort, funds are provided under the Nuclear Waste Disposal account to prepare the overall program plan *and to initiate a competition to select one or more sites suitable for development of integrated recycling facilities (i.e., separation of spent fuel, fabrication of mixed oxide fuel, vitrification of waste products, and process storage) and initiate work on an Environmental Impact Statement. The site competition should not be limited to DOE sites, but should be open to a wide range of other possible federal and non-federal sites on a strictly voluntary basis. The conferees remind the Department that the Nuclear Waste Policy Act prohibits interim storage of nuclear waste in the State of Nevada.* To support the development of detailed site proposals for this competition, the conferees make a total of $20,000,000 available to the site offerors, with a maximum of $5,000,000 available per site. To be eligible to receive these funds, each applicant site must be able to identify all state, regulatory, and environmental permits required for permitting this facility, including identifying any legislative or regulatory prohibitions that might prevent siting such a facility. The conferees direct the Secretary to submit a detailed program plan to the House and Senate Committees on Appropriations not later than March 31, 2006, and to initiate the site selection competition not later than June 30, 2006. The target for site selection is fiscal year 2007, and the target for initiation of construction of one or more integrated spent fuel recycling facilities is fiscal year 2010.

The development of an integrated spent fuel recycling facility will likely require interim storage at the recycling facility as well as additional transportation capacity, possibly involving a different mix of transportation packages, conveyances, and routes than for the federal repository.

Even if Yucca Mountain fails to receive a license and an integrated spent fuel recycling facility is never constructed, the federal government may still require a transportation capability to move commercial spent fuel

---

[55]House Report 109-275 Making Appropriations for Energy and Water Development for the Fiscal Year Ending September 30, 2006, and for Other Purposes.

to one or more centralized storage sites to meet its commitments under the NWPA (see Chapter 1 and Appendix C).[56] This fact alone argues for the establishment of a generic federal transportation capability that could service a repository and possibly other government transportation needs. Under the current organizational structure for the transportation program, all of the federal government's transportation "eggs" have been placed in the federal repository basket.

The current transportation program is unusual in another sense: The committee knows of no other federal government-run program that has a mission to take ownership of private-sector waste for the purposes of transport and disposal.[57] Such programs are usually private-sector responsibilities. The government's usual role is to control and regulate such activities.

Several members of the committee have extensive experience with the design and operation of large transportation programs. Their experience suggests that such programs are more likely to be successful when they have the following:

- An appropriately focused mission;
- A systems-driven focus on the mission;
- Authority to carry out the mission and accountability for failure;
- Independent and strong regulatory oversight;
- Continuity and predictability of funding; and
- Alert, flexible, and responsive management.

OCRWM's transportation program lacks some of these attributes. The current program is focused on transport to Yucca Mountain; it is not organized to provide a generic transportation capacity that could serve the government's other transportation needs. The transportation program director's authority for carrying out the mission is limited because priorities

---

[56]The fiscal year 2006 House Energy and Water Development Appropriations Bill directed DOE to begin accepting commercial spent fuel for interim storage at a government site within 12 months, but this language was not included in the final conference report.

[57]Amtrak, a government-chartered corporation, has carried private-sector freight. The federal government has programs for the shipment of government-owned materials and wastes. For example, the military has transportation programs for moving military materials and wastes. DOE has transportation programs for shipping nuclear weapons and naval spent fuel. It also has transportation programs for shipping the wastes from its environmental cleanup programs at defense sites. Transport of non-DOE domestic and some foreign research reactor spent fuel is the responsibility of reactor operators (see Chapter 4). DOE is responsible for the transport of some foreign research reactor fuel, but this fuel is of U.S. origin and, because it contains HEU (see Chapter 4), is a proliferation concern.

are set at higher levels in the department. In principle, accountability follows authority. The program is partially self-regulating (see Table 1.3). There is little continuity, predictability, or rationality of funding in the annual appropriations process, as noted previously, and the transportation program has historically received a low priority for funding within OCRWM (see Table 5.3).

The committee judges that there are several options for changing the current organizational structure to improve its chances for success; the principal attributes and advantages of these structures are summarized in Sidebar 5.4. One of these options could be implemented within the current structure of the NWPA, whereas two others would likely require fundamental changes to the NWPA. All of the options involve transferring the transportation program out of OCRWM with the explicit goal of increasing management authority and accountability for executing the program's mission(s). Such a transfer could also be advantageous to OCRWM because it would allow that program to focus its staff and resources on its primary near-term mission, which is to license and construct a federal repository.

There are of course some advantages to the current organizational structure: The transportation and repository development programs are closely coupled. In principle, this promotes coordination, cooperation, and systems-driven integration, helping to ensure a match-up between transportation supply, demand, and schedules. However, the committee has not seen much evidence that these advantages are being realized in the current transportation program. Programmatic decisions appear to be based more on funding availability (Table 5.3) than on technical or schedule considerations.

The committee did not perform an exhaustive analysis of alternative organizational structures for the transportation program. However, based on the expertise and experience of its members, at least three different organizational structures seem feasible. The following paragraphs describe these options and their potential advantages and disadvantages.

First, within the current structure of the NWPA, the transportation program could be organized as the Nuclear Waste Transportation Administration, a quasi-independent DOE office program reporting directly to upper-level DOE management (i.e., the DOE secretary, deputy secretary, or under secretary). The main advantages of this structure are that it would free the transportation program from the budget, personnel, and schedule constraints imposed by OCRWM management and give program staff greater authority to execute its mission. The effectiveness of this structure would be enhanced by giving the program more predictability and continuity of funding so that it could make long-term commitments to

SIDEBAR 5.4 Principal Attributes of Potential Organizational Structures for the Federal Transportation Program

Nuclear Waste Transportation Administration

- Independent DOE program reporting directly to upper-level DOE management
- Could serve all of the federal government's commercial spent fuel transportation needs
- Permissible under the current NWPA
- Effectiveness of this organizational structure could be enhanced by giving it authority to tap the Nuclear Waste Fund without annual congressional authorizations
- Organizational model: Federal Highway Administration (FHWA) within DOT

Quasi-Government Corporation

- Private-sector organization with partial government ownership
- Would have exclusive authority to take title to commercial spent fuel for the purposes of transport to a federal repository or federal interim storage
- Would be subject to the full regulatory authority of government for corporate finances and governance, worker and public health and safety, and transportation safety and security
- Could be chartered to be responsive to public participation and the social risk concerns (Chapter 3)
- Would require changes to the current NWPA to implement
- Organizational models: British Nuclear Fuels Limited and AREVA

Private Company

- Similar to quasi-government corporation but with full private ownership
- Could be the most effective option for solving the spent fuel acceptance order problem
- Would require changes to the current NWPA to implement
- Organizational model: Private Fuel Storage, LLC

construct the Nevada rail spur[58] and purchase transportation packages and conveyances. This could be accomplished by giving the program more

[58]Under this and the other two organizational options described in this section, the responsibility for constructing the Nevada rail spur and making other transportation-related infrastructure improvements within Nevada could continue to reside within OCRWM following the recommended organizational restructuring. The transportation program could be given the responsibility for purchasing transportation packages and conveyances and making the needed infrastructure improvements at commercial power plant sites (see Appendix C) and on transportation routes outside of Nevada. This could provide a better separation between repository-specific responsibilities and other responsibilities for a generic federal transportation capability.

direct access to the Nuclear Waste Fund and authority to tap that fund without prior congressional authorization.

A possible model for this arrangement is the Federal Highway Administration (FHWA) within DOT, which reports directly to the Secretary of Transportation. The FHWA is funded mainly from congressional appropriations from the Highway Users Trust Fund, a government account replenished by highway user fees. While FHWA's main mission is to provide programmatic grants to states for highway construction, it also sets selected standards for federally funded roads and provides technical support to state departments of transportation. Additionally, it funds and conducts research to develop and improve relevant technology and enhance the effectiveness of its grant programs.

FHWA oversees a highly decentralized set of activities in which oversight is provided by the federal government but roads are owned by states and local governments, and built and maintained largely by hundreds of private contractors. Some of these activities are also subject to review and approval by other agencies—for example, the Environmental Protection Agency and the Department of Defense. Many of these activities also involve considerable public outreach and participation. The FHWA must operate in a collaborative fashion to execute its missions and has had a long and successful partnerships with state, contractor, material supplier, and academic institutions. The administration has developed a skilled staff with the core competencies required to carry out its missions.

A successful Nuclear Waste Transportation Administration would be similar in many respects to the FHWA. It would be largely dependent on congressional appropriations for its funding and, like FHWA, would rely on a permanent fund (the Nuclear Waste Fund) for most of its budget. It would be involved heavily in cooperative relationships with other federal agencies, states, tribes, local governments, nuclear utilities, contractors, and other nongovernmental organizations. Even though it would likely contract out many of its functions, it would need to have a strong staff with competencies in transportation planning, design, operations, materials handling, and public outreach. However, it would also be different from FHWA in some respects: It would have operational responsibilities and would be much smaller in dollar terms.

The other two organizational models would require changes to the NWPA. The second option would be to reorganize the transportation program as a quasi-government corporation. Such a corporation would operate like a private-sector organization and would be subject to the full regulatory authority of the government for corporate finances and governance, worker and public health and safety, and transportation safety and security. The charter for this corporation would give it exclusive authority to take title to commercial spent fuel for transport to the federal repository or federal interim storage.[59] DOE would take title to shipped materials at the gate of the federal repository or interim storage facility.

This quasi-corporate model has two primary advantages. First, it would bring some private-sector efficiencies to the transportation program, which could help the program operate in a more timely and cost-effective manner. The corporation could freely draw upon existing worldwide transportation capabilities, thereby reducing development costs and schedules. Second, this arrangement would get the government out of the business of regulating itself, which creates both real and perceived conflicts of interest. It would also reduce the potential for political pressures on program plans and operations—for example, on the acceptance order for commercial spent fuel. If desired, DOE could also contract with this corporation to transport its defense spent fuel and high-level waste to the repository.

The main potential disadvantages of this model are that a corporation could be perceived to have less accountability to the public and would be freed from many of the public participation processes in which the government is required to engage.[60] If chartered correctly, however, a quasi-government corporation might actually be more accountable than a government agency for meeting legal and regulatory requirements: the corporation could be fined and its staff subject to civil and criminal penalties for violating statutes and regulations. Moreover, in chartering this corporation, Congress could establish requirements for outside consultation and public participation and make it responsive to the social risk concerns described in Chapter 3 (Section 3.2).

There is a precedent for this organizational model in the United Kingdom and France. British Nuclear Fuels Limited (BNFL) and the French company AREVA are private companies with a high level of government ownership.[61] These companies provide a wide range of nuclear services, including spent fuel and high-level waste transportation services. They are subject to the full regulatory authorities in all of the countries in which they operate, including their home countries.

---

[59]This take-title provision would be workable only if the corporation were covered by the Price Anderson Act.

[60]For example, Flynn et al. (1998) reported the results of a national survey that asked respondents whether they preferred having the federal government manage nuclear waste transportation directly or contract with private companies to manage it. A majority of respondents (about 52 percent) preferred to have the federal government manage transportation directly. The WGA has also expressed a clear preference that the federal government not delegate key transportation responsibilities to contractors (WGA, 2005).

[61]The committee cites these companies as examples of government-owned organizations that transport commercial spent fuel. The citation is not an endorsement of these companies, their business models, or their performance records.

A third option for reorganizing the transportation program is to make it a fully private company operated by the commercial nuclear industry. There is already an industry consortium in existence (Private Fuel Storage, LLC; see Chapter 1) that is developing a transportation capability that could serve as a model for such a company. This arrangement would have some of the same advantages and disadvantages of the quasi-government corporation option, although requirements for outside consultations and public participation might be harder to enforce. However, this might be the most effective option for addressing the commercial spent fuel acceptance order issue (Section 5.2.4), especially if owners had collective economic incentives to maximize the efficiency of the transportation program,[62] and individual owner interests did not trump these collective incentives. Under this option, DOE would be responsible for transporting its own spent fuel and high-level waste but could contract this activity out to the private entity if desired.

The selection of a specific organizational model for the federal transportation program is a policy decision that goes well beyond the task for this study. In making this decision, the federal government will have to consider factors beyond Yucca Mountain. The government is encouraging the construction of new nuclear plants in the United States. If such construction occurs on a large scale, the federal repository and the transportation program that supports it will have to be expanded. In this case, it might make sense for the government to turn over the transportation program to a quasi-private or private entity if it does not wish to be in the permanent business of transporting the industry's spent fuel. As noted previously, a generic transportation capability could also be useful if the federal government decides to transport commercial fuel to one or more centralized sites for storage or reprocessing to meet its commitments under the NWPA (see Chapter 1 and Appendix C). This argues for the establishment of a generic transportation capability that could service a range of government transportation needs.

If the federal government decides to maintain the transportation program in its current organizational form, the committee judges that it will at the very least need a greater commitment to continuity in funding and programmatic direction from the Secretary of Energy and Congress to successfully execute its mission. This may not be possible in the current fiscal and political climate, which is why the committee is recommending that other organizational structures be examined.

[62]For example, Congress could make adjustments to the Nuclear Waste Fund (either by changing the fee structure or by providing negotiated refunds to cover transportation costs) to provide such an incentive.

Regardless of which option is selected, continuing attention must be paid to ensuring that the transportation program develops and maintains an integrated systems focus. The program is complex from both physical and institutional perspectives: it will involve the movement of large quantities of hazardous materials from multiple locations over long distances for sustained periods of time. It will also involve major construction, equipment acquisition, and training. It must coordinate its activities with a large number of constituencies: Congress; spent fuel owners; state, tribal, and local governments; and other nongovernmental organizations. The successful operation of various components and functions of the transportation system is a necessary but not a sufficient condition for overall system effectiveness. The interconnections among the components must also be explicitly thought through and managed. An integrated systems approach is a proven technique for achieving this goal.

The committee did not review the current OCRWM transportation program to determine if it has an integrated systems focus. The committee did see evidence of integrated systems thinking in one presentation it received from transportation program staff (Lanthrum, 2004). However, the committee also saw clear evidence that the current organizational structure for the transportation program is impeding such an integrated approach because, as noted previously, the program does not have the autonomy and funding necessary to execute its mission.

The industry has developed best practices that could be applied to this program (Meredith et al., 1985; Blanchard and Fabrycky, 2005). One element of such best practices is the development of a continuing review and correction process to ensure that a systems focus is maintained from program conception through operations. The committee strongly encourages the program to seek expert advice (e.g., using consultants and expert advisory groups) to learn about and incorporate best industry practices for designing and operating this transportation system using an integrated systems approach. This encouragement is in addition to the recommendations in Section 3.4 for an expert committee to advise transportation implementers on social risk.

Finally, the committee's comments in this section should not be interpreted to reflect on the quality of the federal staff in OCRWM's Office of National Transportation. The committee has had the opportunity to interact with several of these staff during the course of this study and judges that they are capable and dedicated individuals. However, they are working within a difficult organizational structure and in a political environment that could make success close to impossible.

References

Adeola, F. 1994. Environmental hazards, health and racial inequity in hazardous waste distribution. Environ and Behav 26(1):99–126.

Adkins, H. E., Jr., J. M. Cuta, and B. J. Koeppel. 2005. Spent Fuel Transportation Package Response to the Baltimore Tunnel Fire Scenario (draft for public comment). NUREG/CR-6886. Prepared for Pacific Northwest National Laboratory. Richland, WA.

AEC (Atomic Energy Commission). 1972. Environmental Survey of Transportation of Radioactive Materials to and from Nuclear Power Plants. WASH-1238. Washington, DC: Atomic Energy Commission.

American Bar Association. 2004. Environmental Justice for All: A Fifty State Survey of Legislation, Policies and Initiatives. Chicago, IL: American Bar Association and Hastings College of the Law.

Ammerman, D. J., and K. W. Ginn. 2004. Collapse of the cypress street viaduct—effect on a postulated spent fuel truck cask. Proceedings of the INMM 45th Annual Meeting, Orlando, FL, July 18–22.

Ammerman, D. J., J. A. Koski, C. L. Lopez, A. Kapoor, K. Sorenson, J. Sprung, and R. Luna. 2002. Comparison of Selected Highway and Railway Accidents to the 10CFR71 Hypothetical Accident Sequence and NRC Risk Assessments. Prepared for the U.S. Department of Energy Transportation Program. Albuquerque, NM.

Ammerman, D. J., C. L. Lopez, J. A. Koski, and A. Kapoor. 2003. Comparison of selected highway and railway accidents to the 10CFR71 hypothetical accident sequence and NRC risk assessments. Paper presented at 44th Institute of Nuclear Materials Management Annual Meeting, Phoenix, AZ, July 3–17.

Ashford, N., and K. Rest. 1999. Public Participation in Contaminated Communities. Prepared for Center for Technology, Policy, and Industrial Development, Massachusetts Institute of Technology. Cambridge, MA.

Bajwa, C. S. 2002. An Analysis of a Spent Fuel Transportation Cask Under Severe Fire Accident Conditions. Proceedings, Transportation, Storage, and Disposal of Radioactive Materials, PVP-Vol. 469, ASME Pressure Vessels and Piping Conference, Vancouver, British Columbia, Canada, August.

Barnfield, J. H., and P. J. Donelan. 1985. Demonstration Drop Test. Pp. 79–88 in The Resistance to Impact of Spent Magnox Fuel Transport Flasks. London: Mechanical Engineering Publications.

Bazant, Z. P. 2004. Scaling theory for quasibrittle structural failure. Proc Natl Acad Sci USA 101:13400–13407.

Bell, B., Jr. 2001. Holden says radioactive shipment was bungled: Governor charges that federal agency broke promises on moving waste. St. Louis Post-Dispatch, November 1.

Berrens, R., A. Bohara, H. Jenkins-Smith, and C. Silva. 2002. Information disclosure requirements and the effect of soil contamination on property values. J Environ Plan Manage 45:323–339.

Blanchard, B., and W. Fabrycky. 2005. Systems Engineering and Analysis. Englewood Cliffs, NJ: Prentice Hall.

Blythe, R. A., J. D. Hart, J. C. Miles, M. Shears, and E. P. S. Tufton. 1984. The Central Electricity Generating Board Flask Test Project. IAEA-SM-286/78P. Vienna, Austria: International Atomic Energy Agency.

Bradbury, J., and K. Branch. 1999. An Evaluation of the Effectiveness of Local Site-Specific Advisory Boards for U.S. Department of Energy Environmental Restoration Programs. Report PNNL-12139. Washington, DC: Pacific Northwest National Laboratory.

Bradbury, J., K. Branch, and M. Zalesny. 1996a. Site Specific Advisory Board Initiative 1996 Evaluation Survey Results: Volume I, Summary Report. Prepared for the U.S. Department of Energy, Office of Environmental Management, Office of Intergovernmental and Public Accountability.

Bradbury, J., K. Branch, and M. Zalesny. 1996b. Site Specific Advisory Board Initiative 1996 Evaluation Survey Results: Volume II, Individual Site Results. Prepared for the U.S. Department of Energy, Office of Environmental Management, Office of Intergovernmental and Public Accountability.

Bradbury, J., K. Branch, and M. Zalesny. 1997a. Site Specific Advisory Board Initiative 1997 Evaluation Survey Results: Volume I, Summary Report. Pacific Northwest National Laboratory. Prepared for the U.S. Department of Energy, Office of Environmental Management, Office of Intergovernmental and Public Accountability.

Bradbury, J., K. Branch, and M. Zalesny. 1997b. Site Specific Advisory Board Initiative 1997 Evaluation Survey Results: Volume II, Individual Site Results. Prepared for the U.S. Department of Energy, Office of Environmental Management, Office of Intergovernmental and Public Accountability.

Bradbury, J., K. Brach, and E. Malone. 2003. An Evaluation of DOE-EM Public Participation Programs. PNNL-14200. Prepared for Pacific Northwest National Laboratory. Washington, DC.

Brenner, D. J., and C. D. Elliston. 2004. Estimated radiation risks potentially associated with full-body CT screening. Radiology 232(3):735–738.

Bryant, B., and P. Mohai, eds. 1992. Race and the Incidence of Environmental Hazards. Boulder, CO: Westview Press.

Bullard, R. D. 1990a. Dumping in Dixie: Race, Class, and Environmental Quality. Boulder, CO: Westview Press.

Bullard, R. D. 1990b. Ecological inequalities and the new south: Black communities under siege. J Ethnic Studies 17(4):101–115.

Buren, M. A. 1998. Transportation of radioactive material: General scope of federal and state authority. Presentation at the NEI Lawyers Committee Meeting, Washington, DC, February 26.

California Energy Commission. 1998. Fact Sheet on the U.S. Department of Energy Shipments of Foreign Research Reactor Spent Nuclear Fuel via the Concord Naval Weapons Station. December 29.

Cashwell, J. W., K. S. Neuhauser, P. C. Reardon, and G. W. McNair. 1986. Transportation Impacts of the Commercial Radioactive Waste Management Program. SAND85-2715. Prepared for Sandia National Laboratories. Albuquerque, NM.

CEQ (Council on Environmental Quality). 1997. Environmental Justice: Guidance Under the National Environmental Policy Act. Washington, DC: Executive Office of the President.

Clark County Nuclear Waste Division. 2004. Radioactive Waste Transportation State Laws. Clark County, NV: Department of Comprehensive Planning, Nuclear Waste Division.

Collins, A. H., G. T. Duncan, and K. Poole. 1985. Execution of the crash demonstration by British Rail. Pp. 205–211 in The Resistance to Impact of Spent Magnox Fuel Transport Flasks. London: Mechanical Engineering Publications.

Covello, V. T., D. B. McCallum, and M. T. Pavlova, eds. 1989. Effective Risk Communication: The Role and Responsibility of Government and Nongovernment Organizations. New York: Plenum Press.

Crouch, E., and R. Wilson. 1982. Risk/Benefit Analyses. Cambridge, MA: Ballinger Publishing Company.

CSG (Council of State Governments' Midwestern Office). 2004. Planning Guide for Shipments of Radioactive Materials through the Midwestern States. Lombard, IL: Council of State Governments.

Dake, K. 1991. Orienting dispositions in the perception of risk: An analysis of contemporary worldviews and cultural biases. J Cross-Cult Psychol 22:61–82.

Dake, K. 1992. Myths of nature: Culture and the social construction of risk. J Soc Issues 48:21–37.

Dallard, P. R. B. 1985. Flask analytical studies. Pp. 47–64 in The Resistance to Impact of Spent Magnox Fuel Transport Flasks. London: Mechanical Engineering Publications.

DOE (U.S. Department of Energy). 1980. Final Environmental Impact Statement for the Storage of U.S. Spent Nuclear Power Reactor Fuel. DOE/EIS-0015. Washington, DC.

DOE. 1995a. U.S. Department of Energy Environmental Justice Strategy Executive Order 12989. On-line. Office of Environmental Management. Available at *http://web.em.doe.gov/stake/envjus.html*. Accessed January 2, 2004.

DOE. 1995b. Programmatic Spent Nuclear Fuel Management and Idaho National Engineering Laboratory Environmental Restoration and Waste Management Programs Final Environmental Impact Statement. Federal Register 60:20979. April 28.

DOE. 1995c. Record of Decision on the Programmatic Spent Nuclear Fuel Management and Idaho National Engineering Laboratory Environmental Restoration and Waste Management Programs Final Environmental Impact Statement. Federal Register 60:28680. June 1.

DOE. 1996a. Final Environmental Impact Statement on a Proposed Nuclear Weapons Nonproliferation Policy Concerning Foreign Research Reactor Spent Nuclear Fuel. DOE-EIS-0218-F.

DOE. 1996b. Record of Decision for the Final Environmental Impact Statement on a Nuclear Weapons Nonproliferation Policy Concerning Foreign Research Reactor Spent Nuclear Fuel. Federal Register 61:25091–25103. May 17.

DOE. 1996c. Charleston: Potential Highway and Rail Routes. May 20. On-line. Available at *http://web.em.gov/rod/charles.htm*.

DOE. 1998a. Viability Assessment of a Repository at Yucca Mountain Overview. DOE/RW-0508. Washington, DC: Office of Civilian Radioactive Waste Management.

DOE. 1998b. Office of Civilian Radioactive Waste Management; Safe Routine Transportation and Emergency Response Training; Technical Assistance and Funding. Federal Register 63:23753–23766.

DOE. 1999. DOE Standard—Radiological Control. DOE-STD-1098-99. Washington, DC: U.S. Department of Energy.

DOE. 2001a. Nuclear Waste Fund Fee Adequacy: An Assessment. DOE/RW-0534. Washington, DC: Office of Civilian Radioactive Waste Management.

DOE. 2001b. DOE and State of Missouri Reach Agreement for Spent Fuel Transportation on Interstate 70. Press release No. R-01-080. May 25.

DOE. 2001c. DOE Prepared for Cross-Country Shipment of Spent Nuclear Fuel: Missouri State Officials Complete Safety Training. Press release No. R-01-95. June 15.

DOE. 2001d. Analysis of the Total System Life Cycle Cost of the Civilian Radioactive Waste Management Program. DOE/RW-0533. Washington, DC: Office of Civilian Radioactive Waste Management.

DOE. 2001e. Transportation Packaging Quick Facts: Model 2000 Transport Package. Washington, DC: Office of Environmental Management.

DOE. 2002a. Final Environmental Impact Statement for a Geologic Repository for the Disposal of Spent Nuclear Fuel and High-Level Radioactive Waste at Yucca Mountain, Nye County, Nevada. DOE/EIS-0250F. Washington, DC: Office of Civilian Radioactive Waste Management.

DOE. 2002b. A Resource Handbook on DOE's Transportation Risk Assessment. DOE/EM/NTP/HB-01. Washington, DC: Office of Environmental Management.

DOE. 2002c. Radioactive Material Transportation Practices Manual. DOE M 460.2-1. September 23.

DOE. 2003a. U.S. Department of Energy Foreign Research Reactor Nuclear Spent Fuel Shipments Transportation Plan for Motor Carrier Transport: Savannah River Site to Idaho National Engineering and Environmental Laboratory: Revision 1. U.S. Department of Energy, Idaho Operations Office.

DOE. 2003b. U.S. Department of Energy Foreign Research Reactor Nuclear Spent Fuel Shipments Transportation Plan for Rail or Motor Carrier Transport: Charleston, SC to Savannah River Site: Revision 1. U.S. Department of Energy: Savannah River Operations Office.

DOE. 2003c. Strategic Plan for the Safe Transportation of Spent Nuclear Fuel and High-Level Radioactive Waste to Yucca Mountain: A Guide to Stakeholder Interactions. Washington, DC: Office of Civilian Radioactive Waste Management.

DOE. 2004a. Revision of the Record of Decision for a Nuclear Weapons Nonproliferation Policy Concerning Foreign Research Reactor Spent Nuclear Fuel. Federal Register 69: 69901–69903.

DOE. 2004b. Table 1—Spent Nuclear Fuel Transportation Casks Available for Use at the INEEL. On-line. U.S. Department of Energy. Available at *http://www.id.doe.gov/DOEID/RFPSharedLibrary/PDF/CaskSpecifications4.pdf*.

DOE. 2004c. Global Threat Reduction Initiative: Foreign Research Reactor Spent Nuclear Fuel Shipments. December 7.

DOE. 2004d. Record of Decision on Mode of Transportation and Nevada Rail Corridor for the Disposal of Spent Nuclear Fuel and High-Level Radioactive Waste at Yucca Mountain, Nye County, Nevada. Federal Register 69:18557–18565. April 8.

DOE. 2004e. Notice of Intent to Prepare an Environmental Impact Statement for the Alignment, Construction, and Operation of a Rail Line to a Geologic Repository at Yucca Mountain, Nye County, Nevada. Federal Register 69:18565–18569. April 8.

DOE. 2004f. Acceptance Priority Ranking and Annual Capacity Report. Washington, DC: Office of Civilian Radioactive Waste Management.

DOE. 2005a. FRR MTR, TRIGA, and Target Receipts Quick Reference. Unpublished table. March 1.

DOE. 2005b. Department of Energy Policy Statement for Use of Dedicated Trains for Waste Shipments to Yucca Mountain. Washington, DC: Office of Civilian Radioactive Waste Management.

Donelan, P. J., and A. R. Dowling. 1985. The use of scale models in impact testing. Pp. 23–46 in The Resistance to Impact of Spent Magnox Fuel Transport Flasks. London: Mechanical Engineering Publications.

DOT (Department of Transportation). 1980. Highway Routing of Radioactive Materials. U.S. Department of Transportation: Research and Special Programs Administration. Federal Register 45:7140–7153.

DOT. 1998. Identification of Factors for Selecting Modes and Routes for Shipping High-Level Radioactive Waste and Spent Nuclear Fuel. Washington, DC: Research and Special Programs Administration.

DOT. 2002. Transportation and Environmental Justice: Effective Practices. FHWA-EP-02-016. Washington, DC: Federal Highway Administration and Federal Transit Administration.

DOT. 2004. Emergency Response Guidebook. ERG 2004.

Drew, C., M. Kern, T. Martin, M. L. Blazek, M. Power, and E. Faustman. Forthcoming. The Hanford Openness Workshops: Fostering open and transparent decision making at the Department of Energy Research in Social Problems and Public Policy.

DuCharme, A. R., Jr., R. E. Akins, S. L. Daniel, D. M. Ericson, Jr., B. H. Finley, N. N. Finley, P. C. Kaestner, D. D. Sheldon, J. M. Taylor, and M. S. Tierney. 1978. Transport of Radionuclides in Urban Environs: Working Draft Assessment. SAND77-1927. Prepared for Sandia National Laboratories. Albuquerque, NM.

Eady, V. 2003. Environmental justice in state policy decisions. In Just Sustainabilities: Development in an Unequal World, J. Agyeman, R. D. Bullard, and B. Evans, eds. Cambridge, MA: MIT Press.

Easterling, D. 1997. The vulnerability of the Nevada visitor economy to a repository at Yucca Mountain. Risk Analysis 17(5):635–647.

Easterling, D., and H. Kunreuther. 1993. The vulnerability of the convention industry to the siting of a high level nuclear waste repository. Pp. 209–238 in Public Reactions to Nuclear Waste: Citizens' Views of Repository Siting, R. E. Dunlap, M. E. Kraft, and E. A. Rosa, eds. Durham, NC: Duke University Press.

Emit, R., R. Riggs, W. Milstead, J. Pittman, and H. Vendermolen. 2003. A prioritization of generic safety issues. NUREG-0933. Washington, DC: Office of Nuclear Regulatory Research.

EO [Executive Order]. 1994. Federal Actions to Address Environmental Justice in Minority Populations and Low-Income Populations. 12898. Federal Register 59:7629. February 11.

EPA (Environmental Protection Agency). 1998a. Final Guidance for Incorporating Environmental Justice Concerns in EPA's NEPA Compliance Analyses. Washington, DC: U.S. Environmental Protection Agency.

EPA. 1998b. Cancer Risk Coefficients for Environmental Exposure to Radionuclides. Federal Guidance Report 13. EPA 402-R-99-001. Washington, DC: Office of Radiation and Indoor Air.

EPA. 1999. Cancer Risk Coefficient Estimates for Environmental Exposure to Radionuclides. Federal Guidance Report 13. Washington, DC. On-line. Available at *http://www.epa.gov/radiation/federal/docs/fgr13.pdf*. Accessed on April 11, 2004.

Finley, N. C., D. C. Aldrich, S. L. Daniel, D. M. Ericson, C. Henning-Sachs, P. C. Kaestner, N. R. Ortiz, D. D. Sheldon, and J. M. Taylor. 1980. Transportation of Radionuclides in Urban Environs: Draft Environmental Assessment. NUREG/CR-0743 (SAND79-0369). Prepared for Sandia National Laboratories. Albuquerque, NM.

Finucane, M. L., P. Slovic, C. K. Mertz, J. Flynn, and T. A. Satterfield. 2000. Gender, race, and perceived risk: The "white male" effect. Health Risk Soc 2(2):159–172.

Fischer, F. 2001. Citizens, Experts and the Environment. The Politics of Local Knowledge. Durham, NC: Duke University Press.

Fischer, L. E., C. K. Chou, M. A. Gerhard, C. Y. Kimura, R. W. Martin, R. W. Mensing, M. E. Mount, and M. C. White. 1987. Shipping Container Response to Severe Highway and Railway Accident Conditions. NUREG/CR-4829 (UCID-20733). Livermore, CA: Lawrence Livermore National Laboratory.

Flynn J., P. Slovic, and C. K. Mertz. 1994. Gender, race, and perception of environmental health risks. Risk Anal 14(6):1101–1108.

Flynn, J. H., C. K. Mertz, and P. Slovic. 1998. Results of a 1997 National Nuclear Waste Transportation Survey. On-line. Eugene, OR: Decision Research. Available at *http://www.state.nv.us/nucwaste/trans/1dr01a.htm*. Accessed July 27, 2004.

FRA (Federal Railroad Administration). 1998. U.S. Department of Transportation, Safety Compliance Oversight Plan for Rail Transportation of High-Level Radioactive Waste and Spent Nuclear Fuel: Ensuring the Safe, Routine Rail Transportation of Foreign Research Reactor Spent Nuclear Fuel. Washington, DC: Federal Railroad Administration.

FRA. 2005. Use of Dedicated Trains for Transportation of High-Level Radioactive Waste and Spent Nuclear Fuel: Report to Congress. Washington, DC: Federal Railroad Administration.

Freudenberg, W. 1993. Risk and recreancy: Weber, the Division of Labor, and the rationality of risk perceptions. Soc Forces 71:909–932.

Freudenburg, W. 2003. Institutional failure and the organizational amplification of risks: The need for a closer look. In The Social Amplification of Risk, N. Pidgeon, R. Kasperson, and P. Slovic, eds. Cambridge, United Kingdom: Cambridge University Press.

Frewer, L. 2003. Trust, transparency, and social context: Implications for the social amplification of risk. In The Social Amplification of Risk, N. Pidgeon, R. Kasperson, and P. Slovic, eds. Cambridge, United Kingdom: Cambridge University Press.

Friedberg, W., and K. Copeland. 2003. What Aircrews Should Know about Their Occupational Exposure to Ionizing Radiation. DOT/FAA/AM-03/16. Prepared for FAA Office of Aerospace Medicine and Civil Aerospace Medical Institute. Washington, DC.

GAO (Government Accountability Office). 1983. Siting of Hazardous Waste Landfills and their Correlation with Racial and Economic Status of Surrounding Communities. GAO-RCED-83-168, B-211461. Washington, DC: GPO.

GAO. 1994. Nuclear Nonproliferation: Concerns with U.S. Delays in Accepting Foreign Research Reactors' Spent Fuel. Washington, DC: Resources Community and Economic Development Division.

GAO. 2003. Spent Nuclear Fuel: Options Exist to Further Enhance Security. GAO-03-426. Washington, DC: U.S. Government Accountability Office.

GAO. 2004a. Nuclear Proliferation: DOE Needs to Take Action to Further Reduce the Use of Weapons-Usable Uranium in Civilian Research Reactors. GAO-04-807. Washington, DC: Natural Resources and Environment.

GAO. 2004b. Nuclear Proliferation: DOE Needs to Consider Options to Accelerate the Return of Weapons-Usable Uranium from Other Countries to the United States and Russia. GAO-05-57. Washington, DC: Natural Resources and Environment.

Garabedian, A. S., D. S. Dunn, and A. H. Chowdhury. 2002. Analysis of Rail Car Components Exposed to a Tunnel Fire Environment. CNWRA 2003–2004. Prepared for Center for Nuclear Waste Regulatory Analyses. San Antonio, TX.

Gawande, K., and H. Jenkins-Smith. 2001. Nuclear waste transport and residential property values: Estimating the effects of perceived risks. J Environ Econ Manag 42:207–233.

Goldman, B. A., and L. Fitton. 1994. Toxic Wastes and Race Revisited: An Update of the 1987 Report on the Racial and Socioeconomic Characteristics of Communities with Hazardous Waste Sites. Prepared for the Center for Policy Alternatives, National Association for the Advancement of Colored People, and United Church of Christ. Washington, DC.

Gonzales, A., J. D. Pierce, and D. R. Stenberg. 1986. Target hardness comparisons with the IAEA unyielding target. IAEA-SM-286/114P. Pp. 545–551 in International Symposium on the Packaging and Transport of Radioactive Materials, Davos, Switzerland, June 16–20. Vienna, Austria: International Atomic Energy Agency.

Gregory, R., J. Flynn, and P. Slovic. 1995. Technological stigma. Am Sci 83(3):220–223.

Greiner, M., N. R. Chalasani, and A. Suo-Anttila. 2005. Thermal protection provided by impact limiters to containment seal within a truck package. Proceedings of 2005 ASME Pressure Vessels and Piping Division Conference, Denver, CO, July 17–21.

Gusterson, H. 2000. How not to construct a radioactive waste incinerator. Sci Technol Hum Val 25(3):332–351.

Hall, J. 2003. Remarks of Jim Hall on behalf of the State of Nevada. Remarks at Transportation of Radioactive Waste, Meeting 2, Las Vegas, NV, July 25.

Hancock, D. 2004. WIPP transportation information issues. Presentation at National Academies Committee on Transportation of Radioactive Waste, Meeting 6, Albuquerque, NM, July 22.

Hart, J. D., R. A. Blythe, I. Milne, and M. Shears. 1985a. Rail crash demonstration scenarios. Pp. 115–124 in The Resistance to Impact of Spent Magnox Fuel Transport Flasks. London: Mechanical Engineering Publications.

Hart, J. D., R. A. Blythe, I. Milne, and M. Shears. 1985b. A summary of the CEGB's flask accident impact studies. Pp. 233–249 in The Resistance to Impact of Spent Magnox Fuel Transport Flasks. London: Mechanical Engineering Publications.

Holden, B. 2001. Letter from the Honorable Bob Holden, Governor, State of Missouri, to the Honorable Spencer Abraham, Secretary of Energy, October 23, 2001. *http://www.ewg.org/reports/nuclearwaste/faq/faq_setroutes.php*.

HSK (Swiss Federal Nuclear Safety Inspectorate). 1998. Surface Contamination of Nuclear Spent Fuel Transports: Common Report of the Competent Authorities of France, Germany, Switzerland and the United Kingdom. Switzerland: Swiss Federal Nuclear Safety Inspectorate.

Huerta, M. 1977. Analysis, Scale Modeling, and Full Scale Tests of a Truck Spent-Nuclear-Fuel Shipping System in High Velocity Impacts Against a Rigid Barrier. SAND77-0270. Prepared for Sandia National Laboratories. Albuquerque, NM.

Huerta, M. 1981. Analysis, Scale Modeling, and Full Scale Test of a Railcar and Spent-Nuclear-Fuel Shipping Package in a High-Velocity Impact Against a Rigid Barrier. SAND78-0458. Prepared for Sandia National Laboratories. Albuquerque, NM.

Huerta, M., and H. R. Yoshimura. 1983. A Study and Full-Scale Test of a High-Velocity Grade-Crossing Simulated Accident of a Locomotive and a Spent-Nuclear-Fuel Shipping Package. SAND79-2291. Prepared for Sandia National Laboratories. Albuquerque, NM.

Huizenga, D. G., T. P. Mustin, E. C. Saris, and J. E. Reilly. 1999. Moving into the 21st century—The United States' Research Reactor Spent Nuclear Fuel Acceptance Program. Presentation at the 22nd International Meeting on Reduced Enrichment for Research and Test Reactors, Budapest, October 3–8.

Hunsperger, W. 2001. Effects of the Rocky Flats Nuclear Weapons Plant on neighboring property values. Pp. 157–171 in Risk, Media, and Stigma: Understanding Public Challenges to Modern Science and Technology, J. Flynn, P. Slovic, and H. Kunreuther, eds. Sterling, VA: Earthscan Publications, Ltd.

IAEA (International Atomic Energy Agency). 2000. Regulations for the Safe Transport of Radioactive Materials. Safety Standards Series No. TS-R-1 (ST-1, Revised). Vienna, Austria: International Atomic Energy Agency.

IAEA. 2002. Advisory Material for the IAEA Regulations for the Safe Transport of Radioactive Material. Safety Standards Series No. TS-G-1.1 (ST-2). Vienna, Austria: International Atomic Energy Agency.

IAEA. 2005a. Regulations for the Safe Transport of Radioactive Material, 2005 Edition. Safety Standards Series No. TS-R-1 (ST-1, Revised). Vienna, Austria: International Atomic Energy Agency.

IAEA. 2005b. Radiological Aspects of Non-Fixed Contamination of Packages and Conveyances. IAEA-TECDOC-1449. Vienna, Austria: International Atomic Energy Agency.

ICRP (International Commission on Radiological Protection). 1991. Recommendations of the International Commission on Radiological Protection. ICRP Publication 60. Oxford, UK: Pergamon Press.

IME (The Institution of Mechanical Engineers). 1985. The Resistance to Impact of Spent Magnox Fuel Transport Flasks. London: Mechanical Engineering Publications.

Intertech Services Corporation. 2001. In Search of Equity: A Preliminary Assessment of the Impacts of Developing and Operating the Yucca Mountain Repository on Lincoln County and the City of Caliente, Nevada. Pioche, NV: Intertech Services Corporation.

IOM (Institute of Medicine). 1999. Toward Environmental Justice: Research, Education, and Health Policy. Washington, DC: National Academy Press.

Jacobson, S. K. 1999. Communication Skills for Conservation Professionals. Chicago, IL: Island Press.

JAI Corporation. 2005. Shipping and Storage Cask Data for Commercial Spent Nuclear Fuel. Fairfax, VA: JAI Corporation.

Jefferson, R. M., and H. R. Yoshimura. 1977. Crash Testing of Nuclear Fuel Shipping Containers. SAND77-1462. Prepared for Sandia National Laboratories. Albuquerque, NM.

Jenkins-Smith, H., and W. Smith. 1994. Ideology, culture, and risk perception. In Politics, Policy and Culture, D. Coyle and R. Ellis, eds. Boulder, CO: Westview Press.

Johnson, B. 2003. Are some risks comparisons more effective under conflict? A replication and extension of Roth et al. Risk Anal 23:767–780.

Johnson, B. 2004. Risk comparisons, conflict, and risk acceptability claims. Risk Anal 24(1): 131–145.

Johnson, B., and C. Chess. 2003. How reassuring are risk comparisons to pollution standards and emission limits? Risk Anal 23(5):999–1007.

Kaplan, S., and B. J. Garrick. 1981. On the quantitative definition of risk. Risk Anal 1(1): 11–27.

Kasperson, R., N. Jhaveri, and J. X. Kasperson. 2001. Stigma and the social amplification of risk: Toward a framework analysis. Pp. 9–27 in Risk, Media, and Stigma: Understanding Public Challenges to Modern Science and Technology, J. Flynn, P. Slovic, and H. Kunreuther, eds. Sterling, VA: Earthscan Publications, Ltd.

Kasperson, R., O. Renn, P. Slovic, H. Brown, J. Emel, R. Goble, J. Kasperson, and S. Ratick. 1988. The social amplification of risk: A conceptual framework. Risk Anal 8:177–187.

Kasperson, R. E., ed. 1983. Equity Issues in Radioactive Waste Management. Cambridge, MA: Oelgeschlager, Gunn and Hain.

Ketkar, K. 1992. Hazardous waste sites and property values in the State of New Jersey. Appl Econ 24:647–659.

Kiel, K. 1995. Measuring the impact of the discovery and cleaning of identified hazardous waste sites on house values. Land Econ 71:428–435.

Klassen, D. A. 1982. Transportation of Radioactive Material by Rail: The Special Train Issue. SAND81-1447-TTC-0226. Prepared for Sandia National Laboratories. Albuquerque, NM.

Kollmuss, A., and J. Agyeman. 2002. Mind the gap: Why do people act environmentally and what are the barriers to pro-environmental behavior? Environ Educ Res 8:239–260.

Lanthrum, G. 2004. Office of National Transportation update. Presentation at National Academies Committee on Transportation of Radioactive Waste, July 21.

MacGregor, D., and R. Fleming. 1996 Risk perception and symptom reporting. Risk Anal 16(6):773–783.

MacGregor, D. G., J. Flynn, P. Slovic, and C. K. Mertz. 2002a. Perception of Radiation Exposure: Part I. Perception of Risk and Judgment of Harm. Eugene, OR: Decision Research.

MacGregor, D. G., J. Flynn, C. K. Mertz, and P. Slovic. 2002b. Perception of Radiation Exposure: Part II. Communicating About Radiation Exposure and Health Effects. Eugene, OR: Decision Research.

Mauro, J., and N. M. Briggs. 2005. Assessment of Variations in Radiation Exposure in the United States. EP-D-05-002. Prepared for the Environmental Protection Agency. McLean, VA: S. Cohen and Associates.

McGrattan, K. B., and A. Hamins. 2003. Numerical Simulation of the Howard Street Tunnel Fire, Baltimore, Maryland, July 2001. NUREG/CR-6793. Washington, DC: National Institute of Standards and Technology.

Mendelsohn, R., D. Hellerstein, M. Huguenin, R. Unsworth, and R. Brazee. 1992. Measuring hazardous waste damages with panel models. J Environ Econ Manag 22:259–271.

Meredith, D., K. Wong, and R. Woodhead. 1985. Design and Planning of Engineering Systems. Englewood Cliffs, NJ: Prentice Hall.

Mettler, F. A., Jr., P. W. Wiest, J. A. Locken, and C. A. Kelsey. 2000. CT scanning: Patterns of use and dose. J Radiol Prot 20:353–359.

NCHRP (National Cooperative Highway Research Program). 2004. Effective Methods for Environmental Justice Assessment. NCHRP Report 532. Washington, DC: Transportation Research Board.

NCRP (National Council on Radiation Protection and Measurements). 1987. Ionizing Radiation Exposure of the Population of the United States. NCRP Report 93. Bethesda, MD: National Council on Radiation Protection and Measurements.

NCRP. 1995. Principles and Application of Collective Dose in Radiation Protection. NCRP Report 121. Bethesda, MD: National Council on Radiation Protection and Measurements.

NCRP. 2001. Management of Terrorist Events Involving Radioactive Material. Scientific Committee 46-14, NCRP Report 138. Bethesda, MD: National Council on Radiation Protection and Measurements.

Nevada (Nevada Agency for Nuclear Projects). 1997. Fact Sheet: Foreign Research Reactor Spent Fuel Shipments. Carson City, NV: Office of Emergency Support.

Nevada. 2000. State of Nevada Comments on the U.S. Department of Energy's Draft Environmental Impact Statement for a Geologic Repository for the Disposal of Spent Nuclear Fuel and High-Level Radioactive Waste at Yucca Mountain, Nye County, Nevada. 2 Volumes. Carson City, NV: Nevada Agency for Nuclear Projects.

Ngutter, L. K., J. M. Kofler, C. H. McCollough, and R. J. Vetter. 2001. Update on patient radiation doses at a large tertiary care medical center. Health Phys 81(5):530–535.

NRC (National Research Council). 1951. Physical, Biological, and Administrative Problems Associated with the Transportation of Radioactive Substances. Report prepared by R. D. Evans for the Subcommittee on Shipment of Radioactive Substances, Committee on Nuclear Sciences. Washington, DC: National Academy of Sciences—National Research Council.

NRC. 1957. The Disposal of Radioactive Waste on Land. Washington, DC: National Academy Press.

NRC. 1984. Social and Economic Aspects of Radioactive Waste Disposal. Washington, DC: National Academy Press.

NRC. 1989. Improving Risk Communication. Washington, DC: National Academy Press.

NRC. 1996. Understanding Risk: Informing Decisions in a Democratic Society. Washington, DC: National Academy Press.

NRC. 2003. One Step at a Time: The Staged Development of Geologic Repositories for High-Level Radioactive Waste. Washington, DC: The National Academies Press.

NRC. 2005a. Health Risks from Exposure to Low Levels of Ionizing Radiation: BEIR VII—Phase 2: Washington, DC: The National Academies Press.
NRC. 2005b. Assessment of the Scientific Information for the Radiation Exposure Screening and Education Program. Washington, DC: The National Academies Press.
NRC. 2005c. Cooperative Research for Hazardous Materials Transportation: Defining the Need, Converging on Solutions. Special Report 283. Washington, DC: The National Academies Press.
NRC. 2005d. Safety and Security of Commercial Spent Nuclear Fuel Storage: Public Report. Washington, DC: The National Academies Press.
O'Connor, R. E. 2001. Are Fear and Stigmatization Likely, and How Do They Matter? Lessons from Research on the Likelihood of Adverse Socioeconomic Impacts form Public Perceptions of the Proposed Yucca Mountain Repository. Prepared for Jason Technologies Corporation, Las Vegas, NV. September 8. (This report is reproduced in Appendix N of the final Yucca Mountain EIS [DOE, 2002a]).
Perrow, C. 1999. Normal Accidents: Living with High-Risk Technologies. New York: Basic Books.
Peters, E., and P. Slovic. 1996. The role of affect and worldviews as orienting dispositions in the perception and acceptance of nuclear power. J Appl Soc Psychol 26(16):1427–1453.
Pope, R. B. 2004. Historical background of the development of various requirements in the international regulations for the safe packaging and transport of radioactive material. Paper presented at the 14th International Symposium on Packaging and Transportation of Radioactive Materials, Berlin, Germany, September 20–24.
Pope, R. B., Z. Bernard-Bruls, and M. T. M. Brittinger. 2001. A worldwide assessment of the transport of irradiated nuclear fuel and high-level waste. Paper presented at 13th International Symposium on Packaging and Transportation of Radioactive Materials, Chicago, IL, September 7.
Pope, R. B., M. W. Wankerl, S. Armstrong, C. Hamberger, and S. Schmid. 1991. Historical overview of domestic spent fuel shipments—update. Pp. 1501–1508 in High Level Radioactive Waste Management: Proceedings of the Second Annual International Conference on High-Level Radioactive Waste Management, Volume 2. LaGrange Park, IL: American Nuclear Society.
Renn, O. 1997. Health impacts of large releases of radionuclides: Mental health, stress, and risk perception: Insights from psychological research. Ciba F Symp 203:205–226.
Resnikoff, M. 1994. Assessment of Transportation Impacts: Use of RADTRAN to Assess Risks of Severe Accidents. Unpublished manuscript.
Rosa, E. A., and D. L. Clark, Jr. 1999. Historical routes to technological gridlock: Nuclear technology as prototypical vehicle. Pp. 21–57 in Research in Social Problems and Public Policy, Volume 7, W. R. Freudenburg and T. I. K. Youn, eds. Stamford, CT: JAI Press.
Roth, E., M. G. Morgan, B. Fischhoff, L. Lave, and A. Bostrom. 1990. What do we know about making risk comparisons? Risk Anal 10(3):375–387.
Sandoval, R. P. 1987. An Assessment of the Safety of Spent Fuel Transportation in Urban Environs. Volume II, Appendix A: High-Energy Device Evaluation Tests: Test Data. SAND82-2365. Prepared for Sandia National Laboratories. Albuquerque, NM.
Sandoval, R. P., J. P. Weber, H. S. Levine, A. D. Romig, J. D. Johnson, R. E. Luna, G. J. Newton, B. A. Wong, R. W. Marshall, Jr., J. L. Alvarez, and F. Gelbard. 1983. An Assessment of the Safety of Spent Fuel Transportation in Urban Environs. SAND82-2365. Prepared for Sandia National Laboratories. Albuquerque, NM.
Satterfield, T., and J. Levin. 2002. Risk Communication, Fugitive Values, and the Problem of Tradeoffs at Rocky Flats. Eugene, OR: Decision Research.
Schill, K. 1996. Fuel rod decision expected. Augusta Chronicle, December 16.
Schill, K. 1997. No more fuel rod appeals, Beasley says. Augusta Chronicle, January 1.
Schill, K. 1998. Less foreign waste likely to go to SRS. Augusta Chronicle, February 25.

Schrader-Frechette, K. 1995. Evaluating the expertise of experts. Risk: Health Safe Environ 6:115–126.

Shepard, P., M. Northridge, S. Prakas, and G. Stover. 2002. Advancing environmental justice through community based participatory research. Environ Health Persp Suppl 110(2): 139–140.

Shields, P. J. 2001. Fuel rift called "Outrageous": Interest groups call for end to impasse. Columbia Daily Tribune, May 3.

Slovic, P. 1987. Perception of risk. Science 236:280–285.

Slovic, P. 1993. Perceived risk, trust and democracy. Risk Anal 13:675–682.

Slovic, P. 2001a. Perception of risk from radiation. Pp. 264–274 in The Perception of Risk. London: Earthscan Publications, Ltd.

Slovic, P. 2001b. Perception of risk. Pp. 220–231 in The Perception of Risk. London: Earthscan Publications, Ltd.

Slovic, P., J. Flynn, and R. Gregory. 1994. Stigma happens: Social problems in the siting of nuclear waste facilities. Risk Anal 14(4):773–777.

Slovic, P., J. Flynn, and M. Layman. 2001. Perceived risk, trust and the politics of nuclear waste. Pp. 275–284 in The Perception of Risk. London: Earthscan Publications, Ltd.

Slovic, P., M. Layman, N. Krause, J. Flynn, J. Chalmers, and G. Gesell. 1991. Perceived risk, stigma, and the potential economic impacts of a high-level nuclear waste repository in Nevada. Risk Anal 11:683–696.

Sprung, J. L., D. J. Ammermann, N. L. Breivik, R. J. Dukart, F. L. Kanipe, J. A. Koski, G. S. Mills, K. S. Neuhauser, H. D. Radloff, R. F. Weiner, H. R. Yoshimura. 2000. Reexamination of Spent Fuel Shipment Risk Estimates. NUREG/CR-6672 (SAND2000-0234). Prepared for Sandia National Laboratories. Albuquerque, NM.

Sprung, J. L., D. J. Ammerman, J. A. Koski, and R. F. Weiner. 2001. Spent Nuclear Fuel Transportation Package Performance Study Issues Report. NUREG/CR-6768 (SAND2001-0821P). Prepared for Sandia National Laboratories. Albuquerque, NM.

SSEB (Southern States Energy Board). 1994. Transuranic Waste Transportation Handbook. On-line. Norcross, GA: Southern States Energy Board. Available at *http://www.sseb.org/publications/truwaste94.pdf*.

Steinman, R. L., and K. J. Kearfoot. 2000. A comparison of the RADTRAN 5 and RISKIND 1. 11 incident-free dose models. Proceedings of Waste Management 2000 Symposium, Tucson, February 27–March 2.

Steinman, R. L., R. F. Weiner, and K. J. Kearfoot. 2002. A comparison of transient dose model predictions and experimental measurements. Health Phys 83(4):504–511.

Supko, E. 2000. Estimated U.S. Used Nuclear Fuel Shipments Regulated by NRC, 1964–2000. Washington, DC : Energy Resources International, Inc.

TEC (Transportation External Coordination Working Group). 2002. Best Practices for DOE's Radioactive Materials Transportation Public Information Programs. On-line. Transportation External Coordination Working Group. Available at *http://www.ntp.doe gov/tec/comm/BestPractices.pdf*. Accessed February 2, 2002.

Tuler, S. 2002. Radiation Risk Perception and Communication: A Case Study of the Fernald Environmental Management Project. SERI Report 5. Greenfield, MA: Social and Environmental Research Institute.

UER (Urban Environmental Research, LLC). 2000. Property Value Impacts from the Shipment of High-Level Nuclear Waste Through Clark County, Nevada. Scottsdale, AZ: Urban Environmental Research, LLC.

UER. 2001. Moapa Band of Paiute Indians: Governmental and Fiscal Impact Report Related to the Shipments of High-Level Nuclear Wastes. Scottsdale, AZ: Urban Environmental Research, LLC.

UER. 2002. Tribal Concerns About the Yucca Mountain Repository: An Ethnographic Investigation of the Moapa Band of Paiutes and the Las Vegas Paiute Colony. Scottsdale, AZ: Urban Environmental Research, LLC.

UK Department of Transport, Railway Inspectorate. 1986. Report on the Derailment and Fire That Occurred on 20th December 1984 at Summit Tunnel. London: HMSO. Available at http://www.railwaysarchive.co.uk/documents/DoT_Summit1984.pdf?PHPSESSID=b050a9414e139d0ac06638a0fd717575

United Church of Christ Commission for Racial Justice. 1987. Toxic Wastes and Race in the United States: A National Report on the Racial and Socio-Economic Characteristics of Communities with Hazardous Waste Sites. New York: United Church of Christ Commission for Racial Justice.

UNMIPP (University of New Mexico Institute for Public Policy). 2004. Policy Implications of Public Survey Work on Radioactive Waste Transportation to the WIPP Site. Washington, DC: National Academies, Office of News and Public Information.

UNSCEAR (United Nations Scientific Committee on the Effects of Atomic Radiation). 2000. Sources and Effects of Ionizing Radiation. UNSCEAR 2000 Report to the General Assembly.

USNRC (U.S. Nuclear Regulatory Commission). 1977. Final Environmental Statement on the Transportation of Radioactive Material by Air and Other Modes. NUREG-0170. Washington, DC: U.S. Nuclear Regulatory Commission.

USNRC. 1978. Design Criteria for the Structural Analysis of Shipping Package Containment Vessels, Regulatory Guide 7.6. Washington, DC: U.S. Nuclear Regulatory Commission.

USNRC. 1998. Public Information Circular for Shipments of Irradiated Reactor Fuel. NUREG-0725, Revision 13. Washington, DC: U.S. Nuclear Regulatory Commission.

USNRC. 2003a. Evaluation of the Effects of the Baltimore Tunnel Fire on Rail Transportation of Spent Nuclear Fuel. SECY-03-0002. Washington, DC: U.S. Nuclear Regulatory Commission.

USNRC. 2003b. Cladding Considerations for the Transportation and Storage of Spent Fuel. Interim Staff Guidance 11, Revision 3. Washington, DC: U.S. Nuclear Regulatory Commission Spent Fuel Project Office.

USNRC. 2003c. United States Nuclear Regulatory Commission Package Performance Study Test Protocols: Draft for Public Comment. NUREG-1768. Washington, DC: U.S. Nuclear Regulatory Commission.

USNRC. 2004a. Compatibility with IAEA transportation safety standards (TS-R-1) and other transportation safety amendments; final rule. 10 CFR 71. Federal Register 69(16): 3697–3814.

USNRC. 2004b. Demonstration Test Plan for Full-Scale Spent Nuclear Fuel Rail Transportation Cask Testing Under the Package Performance Study. SECY-04-0135. Washington, DC: U.S. Nuclear Regulatory Commission.

USNRC. 2004c. Options for Full-Scale Spent Nuclear Fuel Transportation Cask Testing Under the Package Performance Study. SECY-04-0029. Washington, DC: U.S. Nuclear Regulatory Commission.

USNRC. 2004d. Staff Requirements—SECY-04-0135—Demonstration Test Plan for Full-Scale Spent Nuclear Fuel Rail Transportation Cask Testing Under the Package Performance Study. Washington, DC: U.S. Nuclear Regulatory Commission.

USNRC. 2004e. Effective Risk Communication: The Nuclear Regulatory Commission's Guidelines for External Risk Communication. NUREG/BR-0308. Washington, DC: U.S. Nuclear Regulatory Commission.

USNRC. 2004f. The Technical Basis for the NRC's Guidelines for External Risk Communication. NUREG/CR-6840. Washington, DC: U.S. Nuclear Regulatory Commission.

USNRC. 2005a. Details and Projected Cost of a Demonstration Test of a Full-Scale Spent Nuclear Fuel Rail Transportation Cask Under the Package Performance Study. SECY-05-0051. Washington, DC: U.S. Nuclear Regulatory Commission.

USNRC. 2005b. Staff Requirements—SECY-05-0051—Details and Projected Cost of a Demonstration Test of a Full-Scale Spent Nuclear Fuel Rail Transportation Cask Under the Package Performance Study. Washington, DC: U.S. Nuclear Regulatory Commission.

USNRC. 2005c. Backgrounder: Research and Test Reactors. On-line. Available at *http://www.nrc.gov/reading-rm/doc-collections/fact-sheets/research-reactors-bg.pdf.*

Vaughn, E. 1995. The significance of socioeconomic and ethnic diversity for the risk communication process. Risk Anal 15(2):169–180.

Vaughan, E., and B. Nordenstam. 1991. The perception of environmental risks among ethnically diverse groups in the United States. J Cross-Cult Psychol 22:29–60.

Vaughan, E., and M. Seifert. 1992. Variability in the framing of risk issues. J Soc Issues 48(4):119–135.

Viebrock, J. M., and N. Mote. 1992. Near-Site Transportation Infrastructure: Final Report. Chicago Operations Office: U.S. Department of Energy.

Viebrock, J. M., N. Mote, and R. B. Pope. 1992. Facility Interface Capability Assessment (FICA) Summary Report. ORNL/SUB/86-97393/7. Oak Ridge, TN: Oak Ridge National Laboratory.

Waddoups. I. G. 1975. Air Drop of Shielded Radioactive Material Containers. SAND75-0276. Prepared for Sandia National Laboratories. Albuquerque, NM.

Watson, S., and A. Jones. 2004. Radiological Consequences Resulting from Accidents and Incidents Involving the Transport of Radioactive Materials in the UK—2003 Review. On-line. National Radiological Protection Board Report NRPB-W64. Available at *http://www.hpa.org.uk/radiation/publications/w_series_reports/2004/nrpb_w64.htm.*

Webler, T. 2002. Low dose risk perception and communication: A case study of the Tritium controversy at Brookhaven National Laboratory. On-line. Greenfield, MA: Social and Environmental Research Institute.

Weiner, R. F., and G. S. Mills. 1999. Relative Risks of Transporting Spent TRIGA Fuels by Four Different Routes. Prepared for Sandia National Laboratories. Albuquerque, NM.

Weiner, R. F., and H. F. Tenn. 1999. Transportation Accidents and Incidents Involving Radioactive Materials (1971–1998). Milestone Report to Department of Energy National Transportation Program. Prepared for Sandia National Laboratories. Albuquerque, NM.

WGA (Western Governors' Association). 2003. WIPP Transportation Safety Program Implementation Guide. On-line. Denver, CO: Western Governors' Association. Available at *http://www.westgov. org/wga/initiatives/wipp/wipp-pig03.pdf.*

WGA. 2005. Policy Resolution 02-05: Transportation of Spent Nuclear Fuel and High-Level Radioactive Waste, June 25, 2002. On-line. Available at *http://www.westgov.org/wga/policy/05/spent-nuke.pdf.*

WIEB (Western Interstate Energy Board). 1991. Rail Primer: Legal, Technical and Business Aspects of Rail Transportation (Revised). Denver, CO: Western Interstate Energy Board.

Williams, B. L., S. Brown, M. Greenberg, and M. A. Kahn. 1999. Risk perception in context: The Savannah River Site stakeholder study. Risk Anal 19(6):1019–1033.

Zimmermann, R. 1993. Social equity and environmental risk. Risk Anal 13(6):649–666.

APPENDIX

A

Biographical Sketches of Committee Members

Neal Lane, *Chair*, is a nationally recognized leader in science and technology policy development and application. He is now a university professor in the Department of Physics and Astronomy and fellow of the James A. Baker III Institute for Public Policy at Rice University, where he previously served as university provost. He also served as assistant to the president for science and technology, director of the White House Office of Science and Technology Policy, director of the National Science Foundation, and chancellor of the University of Colorado at Colorado Springs. Dr. Lane earned his B.S., M.S., and Ph.D. degrees in physics from the University of Oklahoma and is a fellow of the American Physical Society, the American Academy of Arts and Sciences, the American Association for the Advancement of Science, and the Association for Women in Science.

Thomas B. Deen, *Vice-chair*, is a transportation consultant and former executive director of the Transportation Research Board, a position he held from 1980 to 1994. He is former chairman and president of PRC-Voorhees, a transportation engineering and planning consulting firm. During this period, he was in charge of his firm's activities in major urban highway and rail transit projects, both in this country and abroad. Later he was chairman of the national interagency committee that prepared the strategic plan for America's development of intelligent transportation systems. In recent years, the governor of Maryland has appointed him as chair of several blue ribbon committees investigating significant rail and road projects in the state. He also serves on an advisory board for the federal Department of Transportation's Bureau of

Transportation Statistics. His research interests include intermodal planning of urban transportation systems, integration of transportation and land use in urban areas, national transportation policy, and intelligent transportation systems. He holds a B.S. degree from the University of Kentucky and a certificate from Yale University's Bureau of Highway Traffic; he is a member of the National Academy of Engineering.

Julian Agyeman is currently an assistant professor in the Department of Urban and Environmental Policy and Planning at Tufts University. He also served on the Commonwealth of Massachusetts' Environmental Justice Advisory Committee (MEJAC). He earned a B.Sc. degree in geography and botany from the University of Durham, UK; a postgraduate certificate in geography and environmental studies from the University of Newcastle-on-Tyne, Enfield, UK; an M.A. degree in conservation policy from the Middlesex University, Enfield, UK; and a Ph.D. degree in environmental education from the University of London. He is the editor of *Local Environment*, an international peer-reviewed journal and the author and editor of many books and articles. He is a practitioner and researcher in sustainable development and environmental justice at local, national, and international levels. His practical experience was gained through work in the United Kingdom as a consultant on environmental and sustainable development; environmental education adviser and head of curriculum support for the Education Department in the London Borough of Islington; senior environmental education officer for the Directorate of Environmental Health in the London Borough of Lambeth; and chair and founder of the Black Environment Network.

Lisa M. Bendixen is an expert in hazmat risk and safety and has worked on risk assessment and management problems in numerous industries covering both fixed facilities and transportation systems. She was the project manager and primary author of the *Guidelines for Chemical Transportation Risk Analysis*, published by the American Institute of Chemical Engineers' Center for Chemical Process Safety, and has served on the center's technical steering committee. She is currently a vice president with ICF Consulting, where she provides consulting services to private companies, trade associations, and government organizations on a variety of hazardous materials and transportation safety and security issues. She previously spent 22 years in Arthur D. Little, Inc.'s, environment and risk practice. Ms. Bendixen holds a B.S. degree in applied mathematics and an M.S. degree in operations research, both from the Massachusetts Institute of Technology.

Dennis C. Bley has more than 30 years of experience in applying quantitative risk analysis to engineered facilities. His current research involves the application of risk analysis to diverse technological systems, uncertainty modeling, technical risk communication, and human reliability analysis. He is president of Buttonwood Consulting, Inc., and a principal of the WreathWood Group, a joint venture company that supports multidisciplinary research in human performance. He also serves on the board of directors of the International Association for Probabilistic Safety Assessment and Management. His clients include the Department of Energy's (DOE's) Office of Nuclear Energy (for development of risk assessment methods for Generation IV reactors); the Nuclear Regulatory Commission's Office of Nuclear Regulatory Research (also for the development of risk assessment methods); and the Department of Transportation's Volpe Center (for work on human reliability analysis for railroad accidents). Dr. Bley has a Ph.D. degree in nuclear reactor engineering from the Massachusetts Institute of Technology and is a registered professional engineer in the State of California.

Hank Jenkins-Smith is a nationally recognized expert on public perception of environmental and technical risks. His research involves measurement of public and elite risk perceptions of transportation of hazardous and radioactive materials, nuclear waste, and national security issues. He is currently doing research on public perceptions of nuclear security, terrorism, and weapons of mass destruction that is funded through Sandia National Laboratories and the National Science Foundation. His previous research includes public reactions to hazardous facilities, public perceptions of the risks of shipping foreign research reactor spent fuel, and the effects of spent fuel shipments on property values. He is a professor of public policy and holds the Joe R. and Teresa Lozano Long Chair of Business and Government at the George H.W. Bush School of Government and Public Service at Texas A&M University. Previously, he was professor of political science and director of the Institute for Public Policy at the University of New Mexico. Dr. Jenkins-Smith received a B.S. degree in political science and economics from Linfield College and an M.S. and Ph.D. degrees in political science, both from the University of Rochester. Dr. Jenkins-Smith has published several recent peer-reviewed papers of relevance to this committee's work, including "Mitigation and Benefits Measures as Policy Tools for Siting Potentially Hazardous Facilities: Determinants of Effectiveness and Appropriateness" (with Howard Kunreuther. 2001. *Risk Analysis*, 21: 71–382), and "Nuclear Waste Transport and Residential Property Values: Estimating the Effects of Perceived Risks" (with Kishore Gawande. 2001. *Journal of Environmental Economics and Management*, 42:207–233).

Melvin F. Kanninen is internationally recognized for his expertise in fracture mechanics and its applications to structural integrity and durability.

During his 40-year R&D career he has developed and applied this expertise to a wide range of engineering applications, including aging aircraft, rotorcraft, and spacecraft; nuclear power plant pressure vessel and piping systems; and natural gas transmission and distribution pipelines. Dr. Kanninen is currently the principal of MFK Consulting Services. Previously, he held positions with General Electric at the Hanford Atomic Products Operation, Battelle's Columbus Laboratories, and most recently, the Southwest Research Institute where he served as vice president and director of the structural engineering division. He has published more than 180 technical papers, given more than 100 seminar lectures, and coedited six technical books. He is the coauthor of the well-regarded textbook *Advanced Fracture Mechanics*, published by Oxford Press. Dr. Kanninen, who received his Ph.D. degree in engineering mechanics from Stanford University, is a fellow of the American Society of Mechanical Engineers and a member of the National Academy of Engineering.

Ernest J. Moniz is widely recognized for his work in theoretical nuclear physics and, more recently, science and technology policy formulation. He joined the Massachusetts Institute of Technology (MIT) faculty in 1973 and is currently the Cecil and Ida Green Professor of Physics and codirector of the Laboratory for Energy and the Environment. He previously served as head of the MIT Physics Department; as undersecretary of the U.S. Department of Energy; and as associate director for science in the Office of Science and Technology Policy. His current research-related activities include a foundation-sponsored project on the future of coal, work for Los Alamos National Laboratory on security issues related to weapons of mass destruction, and service on a technical advisory board for EPRI. Dr. Moniz received a B.S. degree in physics from Boston College and a Ph.D. degree in theoretical physics from Stanford University. He received honorary doctorate degrees from the University of Athens, the University of Erlangen-Nurenburg, and Michigan State University. He is a fellow of the American Association for the Advancement of Science, the Humboldt Foundation, and the American Physical Society.

John W. Poston, Sr., is a nationally recognized expert in health physics, occupational dosimetry, and health effects of radiation releases from accidents and terrorist events. He is professor and past chair of the Department of Nuclear Engineering and a consultant at the Veterinary Teaching Hospital at Texas A&M University, where he teaches health physics and conducts research on dosimetry. His dosimetry research is supported by the Department of Energy's Office of Nuclear Energy, Science and Technology, and he consults with Sandia National Laboratories and a Texas nuclear utility on operational safety issues. He chaired the National Council on

Radiation Protection and Measurements committee that produced the 2001 report *Management of Terrorist Events Involving Radioactive Material*, and he served as a peer reviewer for the American Association of Railroads on a risk assessment for rail transport of spent nuclear fuel. Dr. Poston is president emeritus of the Health Physics Society and received the 2003 Loevinger-Berman Award from the Society of Nuclear Medicine.

Lacy E. Suiter has more than three decades of experience in emergency planning and response at both state and federal levels. He spent 30 years as a career employee of the Tennessee Emergency Management Agency, the last 12 years as that agency's director. He also served as executive associate director for Response and Recovery for the Federal Emergency Management Agency until his retirement in 2002. In that capacity he was responsible for planning and executing the federal government's response to major disasters and emergencies and managing that agency's multibillion-dollar individual and public assistance grant programs. Mr. Suiter earned his B.S. degree in business from Middle Tennessee State University and is the recent recipient of the United States Army Meritorious Civilian Service Award and the United States Army Corps of Engineers D. De Fleury Medal.

Joseph M. Sussman is an internationally recognized transportation operations expert whose research has covered a wide range of transportation issues, including transportation systems and institutions; regional strategic transportation planning; intercity freight and passenger rail; intelligent transportation systems; simulation and risk assessment methods; and complex systems analysis. He is currently the Japan Rail East Professor in the Department of Civil and Environmental Engineering and Engineering Systems Division at the Massachusetts Institute of Technology (MIT) and receives some of his research support through the Department of Transportation's Volpe Center and from the Association of American Railroads. He previously served as director of MIT's Center for Transportation Studies and head of the Department of Civil and Environmental Engineering. Dr. Sussman received a B.C.E. degree from City College of New York; an M.S.C.E. degree from the University of New Hampshire; and a Ph.D. degree in civil engineering systems from MIT.

Elizabeth Q. Ten Eyck is an expert in domestic and international nuclear safeguards and security for government-owned and licensed commercial nuclear facilities. She has more than 30 years of career federal service—first as a security engineer for the U.S. Secret Service; then as director of the Office of Safeguards and Security for the U.S. Department of Energy; and, until she retired in 2000, as director of the Division of Fuel Cycle Safety and Safeguards of the U.S. Nuclear Regulatory Commission (USNRC), where she managed the safety and safeguards regulatory program for commercial fuel cycle facilities. During her career at USNRC she also managed transportation activities and the safeguards program for nuclear power reactors. She is currently president of ETE Consulting, Inc., and is involved in consulting work for the Department of Homeland Security on vulnerability assessments through Argonne National Labs. Ms. Ten Eyck received her B.S. degree in electrical engineering from the University of Maryland.

Seth Tuler is an expert in public participation, environmental decision making, and community responses to risk communication. His research, which is funded by the National Science Foundation and private foundations, involves public and worker health risks associated with U.S. nuclear weapons production. He also participates in education and training of community members in public participation mechanisms and public health study methods in association with the Alliance for Nuclear Accountability and its member groups. He is a research fellow in the Center for Technology, Environment, and Development at the George Perkins March Institute at Clark University (Worcester, Massachusetts) and a researcher for the Social and Environmental Research Institute (Leverett, Massachusetts). He received a B.A. degree in mathematics from the University of Chicago; an M.S. degree in technology and policy from the Massachusetts Institute of Technology; and a Ph.D. degree in environmental science and policy from Clark University.

Detlof von Winterfeldt is an internationally recognized expert in applying decision and risk analysis to technology and environmental management problems—both as a researcher and as a practitioner. He is a professor of public policy and management in the School of Policy, Planning, and Development at the University of Southern California (USC). He also is director of USC's Homeland Security Center for Risk and Economic Analysis of Terrorism Events. He currently receives support from the Department of Homeland Security for work on risk and economic analysis of terrorism and from the Lawrence Livermore National Laboratory for a project on living with risk. He is the author and coauthor of several publications related to nuclear safety and nuclear waste issues, including a chapter on DOE's selection of candidate nuclear waste repository sites (Keeney, R.L., and von Winterfeldt, D. 1988. The analysis and its role for selecting nuclear repository sites. In G.K. Rand (Ed.) *Operational Research '87*. Amsterdam: Elsevier Science Publishers, pp. 686–701) and an article comparing strategies of disposing of spent fuel from power plants (Keeney, R.L., and von Winterfeldt, D. 1994. Managing nuclear waste from power plants. *Risk Analysis*, 14:107-130). He received M.A. and B.A. degrees in psychology from the University of Hamburg, Germany, and a Ph.D. degree in mathematical psychology from the University of Michigan. He is the 2000 recipient of the Ramsey Medal for distinguished contributions to decision analysis by the Decision Analysis Society.

Thomas R. Warne is known nationally for his expertise in transportation administration, public policy, and large project and program delivery. He is the founder and president of Tom Warne and Associates, LLC, a management consulting company. His previous positions include executive director of the Utah Department of Transportation and deputy director and chief operating officer of the Arizona Department of Transportation. Mr. Warne holds a bachelor's degree in civil engineering from Brigham Young University and a master's degree in civil engineering from Arizona State University; he is a registered professional engineer in Arizona and Utah.

Clive Young is an internationally recognized expert in safety standards for transport of radioactive materials. He has worked at the Department for Transport of the United Kingdom since 1978 and, since 1996, has been head of the Radioactive Materials Transport Division and transport radiological adviser to the Secretary of State for Transport. In this position he is responsible for carrying out the executive functions of the "competent authority" for the transport of radioactive material in the United Kingdom on behalf of the Secretary of State for Transport. He serves as chairman of the Transport Safety Standards Committee of the International Atomic Energy Agency and chairman of the Radioactive Material Working Group of the International Maritime Organization. He previously held the position of research engineer at the UK Atomic Energy Authority. Mr. Young earned his B.Sc. in mechanical engineering from the University of Leeds. He is a member of the Institution of Mechanical Engineers and is a chartered engineer.

APPENDIX

B

List of Presentations Received at Committee Meetings

Washington, D.C., May 16–17, 2003

- NRC's Transportation Program, E. William Bach, U.S. Nuclear Regulatory Commission (USNRC)
- Transportation of Radioactive Waste, Jeffrey Williams, Department of Energy (DOE), Office of National Transportation
- U.S. Department of Transportation's (DOT's) Role in Transportation of Radioactive Waste, Richard Boyle, DOT
- NAS Used Fuel Transportation Study: Remarks from EPRI, John H. Kessler, EPRI

Las Vegas, Nevada, July 24–26, 2003

- State of Nevada Perspectives on Spent Nuclear Fuel (SNF) and High-Level Radioactive Waste Transportation Issues, Robert Loux, Nevada Agency for Nuclear Projects
- Yucca Mountain Transportation Access Issues, Robert Halstead, Nevada Agency for Nuclear Projects
- Yucca Mountain Transportation Risk and Impact Issues, Robert Halstead, Nevada Agency for Nuclear Projects
- Full-scale Testing of Shipping Casks, Jim Hall, Hall and Associates
- Baltimore Tunnel Fire: Implications for SNF Transportation Safety, Marvin Resnikoff, Radioactive Waste Management Associates

- Security, Safeguards, and Implications for Emergency Management, James Ballard, California State University, Northridge
- Plans for Developing a National Transportation Plan, Jeffrey Williams, DOE, Office of National Transportation
- Baltimore Tunnel Fire Study and Sandia Vulnerability Studies, Jack Guttmann, USNRC
- Yucca Mountain Impacts, Irene Navis, Clark County Nuclear Waste Division
- Under the Draining End of the Transportation Funnel, Abby Johnson, Eureka County
- Lincoln County and Town of Caliente Perspectives on Spent Fuel and High-Level Radioactive Waste Transportation Issues, Kevin Phillips, City of Caliente
- Nye County's Views on Nuclear Waste Transportation, Les Bradshaw, Nye County Department of Natural Resources and Federal Facilities
- Perspectives of other Nevada-based Organizations:
  - — Kalynda Tilges, Shundahai Network
  - — Judy Treichel, Nevada Nuclear Waste Task Force
  - — Peggy Maze, Citizen Alert

Denver, Colorado, October 29–31, 2003

- Technical Aspects of DOE's Transportation Operations, Robin Sweeney, DOE, Office of National Transportation
- NRC's Role in Transportation of Spent Nuclear Fuel, Earl Easton, USNRC
- Transportation of Radioactive Waste by Highway, James Simmons, USDOT, Federal Motor Carrier Safety Administration (FMCSA)
- Federal Railroad Administration's (FRA) High-Level Radioactive Materials Program, Kevin Blackwell, USDOT, FRA
- Research and Special Programs Administration, Richard Boyle, DOT
- Western States and NWPA (Nuclear Waste Policy Act) Shipments, Doug Larson, Western Interstate Energy Board
- A State Perspective Regarding Transportation of High-Level Radioactive Waste, Bill Sinclair, Utah Department of Environmental Quality
- Transportation of Radioactive Waste Through the Western States, Bill Mackie, Western Governor's Association
- Used Nuclear Fuel Transportation Responsibilities and Operations, John Vincent (deceased), Nuclear Energy Institute (NEI)

• Trucking Companies Operations Issues, David Bennett, Tri-State Motor Transit

• Rail Perspective on Transportation of SNF, Sandy Covi, Union Pacific Railroad

Chicago, Illinois, February 2–4, 2004

• Federal Role in Responding to Emergencies, Eric Tolbert, Federal Emergency Management Agency (FEMA)

• Department of Energy Role for Emergency Response in the Yucca Mountain Transportation Program, Jozette Booth, DOE

• Corridor States Perspectives in Planning for Spent Nuclear Fuel Shipments, Thor Strong, Michigan Low-Level Radioactive Waste Authority

• Midwestern State's Views on Emergency Preparedness and Response Planning, Tim Runyon, Illinois Emergency Management Agency

• Local Roles for Planning, Training, Response, and Intergovernmental Coordination:

— Perspectives of a Large City, Chief Gene Ryan, City of Chicago

— Perspectives of a Small City/County, Ned Wright, Linn County, Iowa, Emergency Management

— Perspectives of Private Industry, Patrick Brady, Burlington Northern Santa Fe Railway

— Training Issues, Chief Gordon Vermeer, Argonne National Laboratory and International Association of Fire Chiefs

• Risk Perception, Risk Communication, and Public Response to Emergencies: What Really Happens?

— Dennis Mileti, University of Colorado

— Michael Lindell, Texas A&M University

Washington, D.C., May 5–7, 2004

• Used Nuclear Fuel Transportation: Prior Experience Is Valid, Stephen P. Kraft, NEI

• Regulator Perspective on Transportation Safety, E. William Brach, USNRC

• Irradiated Fuel Transport to Yucca Mountain: What History Tells Us, Michele Boyd, Public Citizen

• SNF Cask Crash Tests: What Can They Tell Us About the Safety of the Current Fleet of SNF Shipping Casks? Douglas Ammerman, Sandia National Laboratories

- Reexamination of Spent Fuel Shipment Risk Estimates, Douglas Ammerman, Sandia National Laboratories
- Transportation Infrastructure Acquisition Update, Ned Larson, DOE, Office of National Transportation
- SNF Transportation Cask Certification Requirements, Earl Easton, USNRC
- Update on the USNRC Package Performance Study, Mike Mayfield, USNRC
- Naval Nuclear Propulsion Program, Barry Miles, DOE, Naval Reactors
- Spent Nuclear Fuel Transportation Experience, Steven Edwards, Progress Energy
- AAR Perspective on the Dedicated Trains and Other Issues, Bob Fronczak, Association of American Railroads (AAR)
- Regional Perspectives on Radioactive Materials Transportation, Christopher Wells, Southern States Energy Board
- Northeast Council of State Governments' Perspectives on SNF Transportation, Edward L. Wilds, Council of State Governments Northeast High-Level Radioactive Waste Transportation Task Force

Albuquerque, New Mexico, July 21–23, 2004

- An Overview of Stigma and Potential Socioeconomic Impacts, James Flynn, Decision Research
- Stigma Studies on Nevada and Yucca Mountain, Bob O'Connor, National Science Foundation
- Environmental Justice Impacts, Veronica Eady, National Environmental Justice Advisory Council
- Information and Involvement for Transportation of Nuclear Waste, Christina Drew, University of Washington
- Stakeholder Involvement Strategies for Highly Technical and Controversial Issues, Janesse Brewer, The Keystone Center
- Key Problem Issues in Risk Communication, James Flynn, Decision Research
- Planning, Decision Making, and Monitoring, Veronica Eady, National Environmental Justice Advisory Council
- Planning, Decision Making, and Monitoring, Mervyn Tano, International Institute for Indigenous Resource Management
- Social Diversity in Framing Risk Problems: Implications for Risk Management and Communication Regarding the Transportation of Radioactive Waste, Elaine Vaughan, University of California, Irvine
- Office of National Transportation Update, Gary Lanthrum, DOE, Office of National Transportation

- Assessing the Role of Time in Public Perceptions of the Waste Isolation Pilot Plant (WIPP), Amy Goodin, University of New Mexico
- WIPP Transportation Information Issues, Don Hancock, Southwest Research and Information Center
- Collaboration Among States and Agencies in the WIPP Transportation Safety Program, Anne deLain Clark, New Mexico Energy, Minerals and Natural Resources Department
- State Emergency Management Perspective on Lessons Learned from Coordinating with Other Stakeholders and Communicating with the Public, W. Scott Field, New Mexico Office of Emergency Management
- Native American Perspective on Lessons Learned from Coordinating with Other Stakeholders and Communicating with the Public, Sue Loudner, WIPP Transportation Emergency Preparedness Program
- Federal Perspective on Lessons Learned from Coordinating with Other Stakeholders and Communicating with the Public, Paul Detwiler, DOE, Carlsbad Field Office
- Federal Perspective on Involving the Public in Radioactive Materials Transportation Programs, Judith Holm, DOE, Office of Civilian Radioactive Waste Management

Washington, D.C., May 26, 2005

- Research Reactor Spent Fuel Routing Study: Scope and Origin, Richard Boyle, Pipeline and Hazardous Materials Safety Administration, DOT
- Background: History, Scope, and Purpose of Foreign, University, and DOE Research Reactor Spent Fuel Programs, Alex Thrower DOE, Office of Environmental Management
- Department of Transportation Regulations, Michael Conroy, Pipeline and Hazardous Materials Safety Administration, DOT; Ryan Paquet, DOT, FMCSA
- Nuclear Regulatory Commission Regulations, Philip Brochman, Division of Nuclear Security, USNRC
- DOE Foreign Research Reactor Spent Fuel Acceptance Program, Alex Thrower, DOE; James Wade, Idaho National Laboratory, DOE; Charles E. Messick, DOE Nuclear Material Programs Division, Savannah River Site, Aiken, S.C.
- Domestic University and DOE Research Reactors, James Wade, DOE; Christopher Becker, University of Michigan, Ann Arbor
- Collaboration of States with DOE and Licensees on Routing, Chris Wells, Southern States Energy Board; Lisa Sattler, Council of State Governments—Midwest Office, Lombard, Ill.

Committee Subgroup Visit to Europe (September 2004)

Committee members Dennis Bley, Hank Jenkins-Smith, Mel Kanninen, John Poston, Seth Tuler, Detlof von Winterfeldt, and Clive Young and committee staff Kevin Crowley and Joe Morris participated in all or portions of this visit. Because it involved substantially less than half of the committee, this was not an official information-gathering meeting.

September 24 (Berlin and Environs, Germany)

- Briefing on the AREVA spent fuel transportation program, including transportation operations and communications with the public. Host: Vincent Roland, AREVA
- Tour of the German Federal Institute for Materials Research and Testing (BAM) facility near Horstwalde and attendance at the cask-drop test. Host: Thomas Böllinghaus, BAM

September 27: BNFL Sellafield Site (Seascale, Cumbria, UK, and London)

- Overview of British Nuclear Fuel Limited (BNFL) and the Sellafield Site. Briefers: Colin Boardman, Ben Children, Bob Quinn, and Stephen Stagg, BNFL
- BNFL transportation capabilities, Mark Robinson, BNFL
- Tour of the transportation cask maintenance facility and spent fuel receipt and storage areas of the Thermal Oxide Reprocessing Plant
- Lunch meeting with local government and emergency planning organizations to discuss transportation issues: David Cook (Chairman, Emergency Planning Subcommittee), David Humphreys (Cumbria County Council Emergency Planning Agency), and Mat Fox (Senior Emergency Response Manager, BNFL)

September 28: Visit to Direct Rail Service (DRS) Headquarters in Carlisle, Cumbria

- Presentation and discussion of DRS's UK rail activities, Chris Connelly, General Manager, Commercial and Business Services, DRS
- Tour of DRS Carlisle facilities
- Dinner meeting to discuss BNFL transportation program with London area stakeholders: Roger Evans (London Assembly); Linda Hayes (Cricklewood Against Nuclear Trains); Steve Robinson (Environment Business Management); Patrick van den Bulck (Campaign for Nuclear Disarmament). Dinner hosted by Rupert Wilcox-Baker, Head of Communications, Spent Fuel Services, BNFL

September 29: London and Environs

- Field trip to a BNFL rail head to observe transportation operations.

APPENDIX C

Federal Repository Transportation System

This appendix describes the system for transporting spent fuel and high-level radioactive waste to a federal repository at Yucca Mountain, Nevada. It is provided to support the discussions of this transportation system that appear in Chapter 5.

C.1 PRINCIPAL ELEMENTS OF A REPOSITORY TRANSPORTATION SYSTEM

The system for transporting spent fuel and high-level waste to a federal repository at Yucca Mountain will involve train and truck transport from sites across the continental United States (see Figure 1.1). A good deal of the complexity of this system is the unanticipated consequence of the siting process that was established in the Nuclear Waste Policy Act (NWPA).[1] When writing the NWPA, Congress did not consider the availability of transportation routes or the distance of transport from spent fuel and high-level waste storage sites to the repository. The 1987 amendments to the Nuclear Waste Policy Act fixed the end point of the transportation system at Yucca Mountain, Nevada—a state that has no commercial nuclear power plants and is geographically distant from most spent fuel and high-level

[1]The "Nuclear Waste Policy Act of 1982," P.L. 97-425 (January 7, 1983) and amendments (P.L. 100-203, Subtitle A [December 22, 1987]; P.L. 100-507 [October 18, 1988]; and P.L. 102-486 [October 24, 1992]).

waste storage sites (Figure 1.1). The Yucca Mountain site also lacks good existing transportation access, especially rail access.

The NWPA establishes that the federal government is responsible for developing and operating the transportation and disposal programs, and the owners of spent fuel and high-level waste are financially responsible for covering transport and disposal costs (Table C.1). The NWPA vests authority with the Secretary of Energy for carrying out the federal government's responsibilities and establishes the Nuclear Waste Fund to cover the costs of disposal.

TABLE C.1 Responsibilities Under the Nuclear Waste Policy Act for Storage of Spent Fuel and High-Level Waste and Their Transport to a Federal Repository

	Commercial Spent Fuel	DOE Spent Fuel and High-Level Waste[a]	Naval Spent Fuel
On-site storage	Spent fuel owners	DOE-EM	DOE-NR
Costs of transport and disposal at a federal repository	Spent fuel owners	DOE-EM	DOE-NR
Development of a federal repository	DOE-OCRWM	DOE-OCRWM	DOE-OCRWM
Transportation to a federal repository	DOE-OCRWM	DOE-OCRWM	DOE-NR
Package certification	USNRC	USNRC	DOE-NR, USNRC[b]
Advance shipping notifications of states and tribes required?	Yes	Yes	No
Emergency responder training	DOE supported	DOE supported	DOE supported

NOTE: DOE-EM = Office of Environmental Management; DOE-NR = Naval Reactors Program; DOE-OCRWM = Office of Civilian Radioactive Waste Management.

[a]Includes research reactor spent fuel described in Chapter 4.

[b]DOE seeks USNRC concurrence on package certifications under a cost-reimbursable agreement.

The NWPA also establishes the authorities of the U.S. Nuclear Regulatory Commission (USNRC) as well as the roles of states and tribal nations. It requires that the U.S. Department of Energy (DOE) transport spent fuel and high-level waste only in packages that have been certified by the USNRC (see Chapter 2). It also requires DOE to follow USNRC regulations for advance notifications of state and local governments prior to transport of these materials. Further, it requires DOE to provide technical assistance and funding to states for training of appropriate units of local government and Indian tribes through whose jurisdictions the Secretary of Energy plans to transport spent nuclear fuel or high-level radioactive waste. This training is to cover procedures for safe routine transportation and emergency response.

The NWPA explicitly recognizes the precedence of existing federal, state, and local transportation laws: "Sec. 9: Nothing in this Act shall be construed to affect Federal, State, or local laws pertaining to the transportation of spent nuclear fuel or high-level radioactive waste." Indeed, as described elsewhere in this appendix, there are many additional laws and regulations that apply to the national transportation program for Yucca Mountain.

Section 304 of the NWPA establishes the Office of Civilian Radioactive Waste Management within the Department of Energy (DOE-OCRWM) to carry out the waste disposal function. Within this office, the Office of National Transportation is taking the lead for developing a national transportation program for Yucca Mountain.[2] The Office of Civilian Radioactive Waste Management will be responsible for transporting both commercial spent fuel and DOE spent fuel (except naval spent fuel) and high-level waste to the repository. The Naval Reactors Program (DOE-NR) will transport its own spent fuel to the repository.

The DOE-OCRWM Office of National Transportation is responsible for performing the following tasks:

- Develop a national transportation plan that specifies schedules, modes, and routes for shipping spent fuel to the repository.
- Build the necessary infrastructure (e.g., a rail line in Nevada; road upgrades as necessary) to support the transportation program.
- Purchase the necessary equipment (e.g., transport packages, tractor-trailers, and railcars) to support the transportation program.
- Hire transportation contractors to operate the transportation program.

[2]DOE-OCRWM was undergoing a reorganization in early 2006. The new office structure had not been announced when the present report was completed.

- Establish a federal oversight program to ensure that transportation operations comply with all applicable federal, state, and local regulations.

DOE issued the *Standard Contract for Disposal of Spent Nuclear Fuel and/or High-Level Radioactive Waste* (Title 10, Part 961 of the Code of Federal Regulations) that describes how it expects to carry out its responsibilities under the NWPA. These regulations establish a "standard contract" to be signed by the Secretary of Energy and owners of commercial spent fuel, who are referred to as "purchasers" of DOE's services. The standard contract lays out the responsibilities of the purchasers and the DOE in carrying out the provisions under the contract. These responsibilities are summarized in Sidebar 5.1 and are discussed elsewhere in this appendix.

A flowchart for the national transportation system is illustrated schematically in Figure C.1A, and the organizations responsible for the system components are shown in Figure C.1B. Regulatory authorities are shown in Figure C.1C. The components of the system are described below.

C.1.1 System Planning

Three primary planning functions must be completed before the transportation system can be put into operation: procurement of equipment and other needed infrastructure; scheduling the acceptance of spent fuel and high-level waste from its owners; and identifying the routes to be used for shipping this material to the federal repository and undertaking any necessary upgrades to such routes. These planning functions are described in the following subsections.

Procurement

DOE plans to use the nation's existing roads and railways to ship to the repository,[3] but it may be required at its own expense to upgrade roads and rail access to plant sites or to provide heavy-haul or barge access to railheads. There is no rail spur to Yucca Mountain in Nevada, for example, so DOE will have to build one to support its decision to ship by "mostly rail." DOE has selected the location for the rail spur in Nevada (see Chapter 5) but has not yet completed a detailed route survey or the supplemental Environmental Impact Statement (EIS) that is required prior to the initiation of construction. The rail spur is expected to take several years to

[3]As described further in this appendix and in Chapter 5, rail access to the repository does not now exist, so DOE will be required to build a rail spur to the repository in Nevada. DOE may also be required to build a road to the repository from the nearest highway to support truck transport.

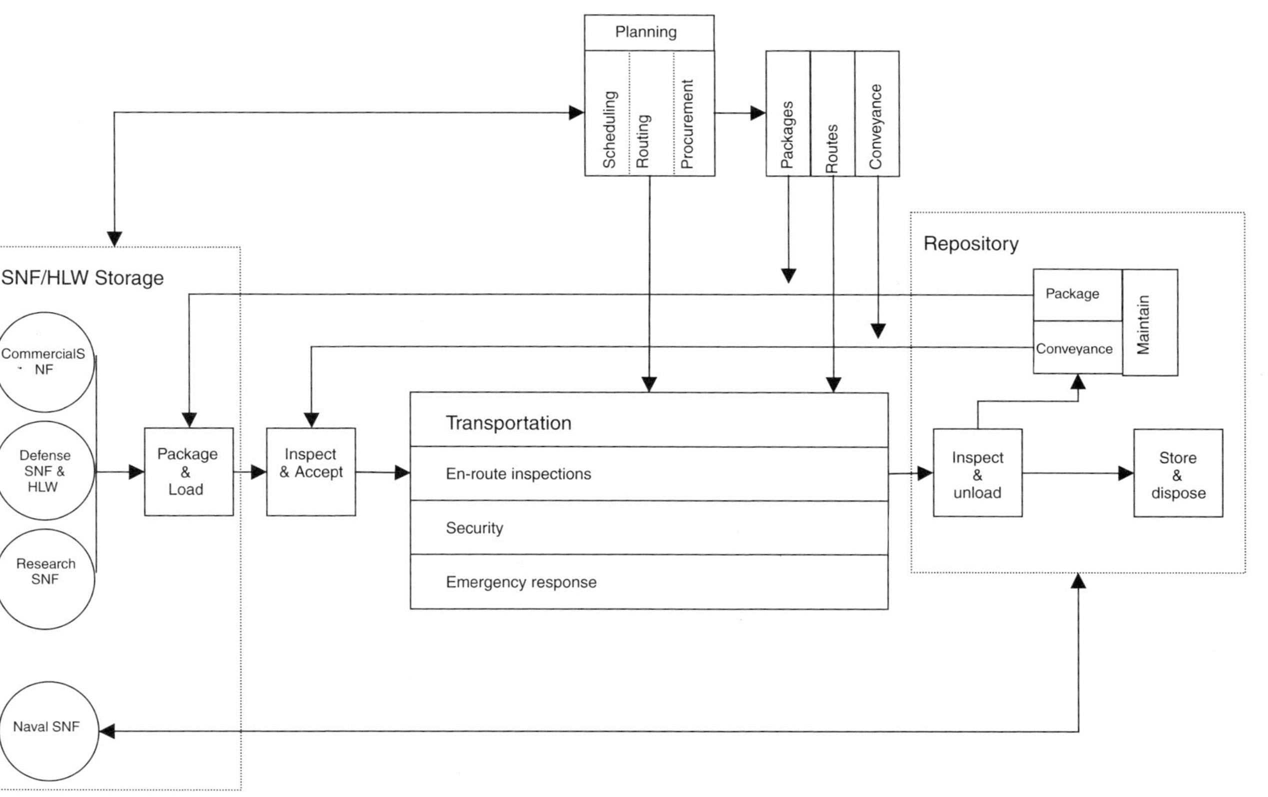

FIGURE C.1A National Transportation System

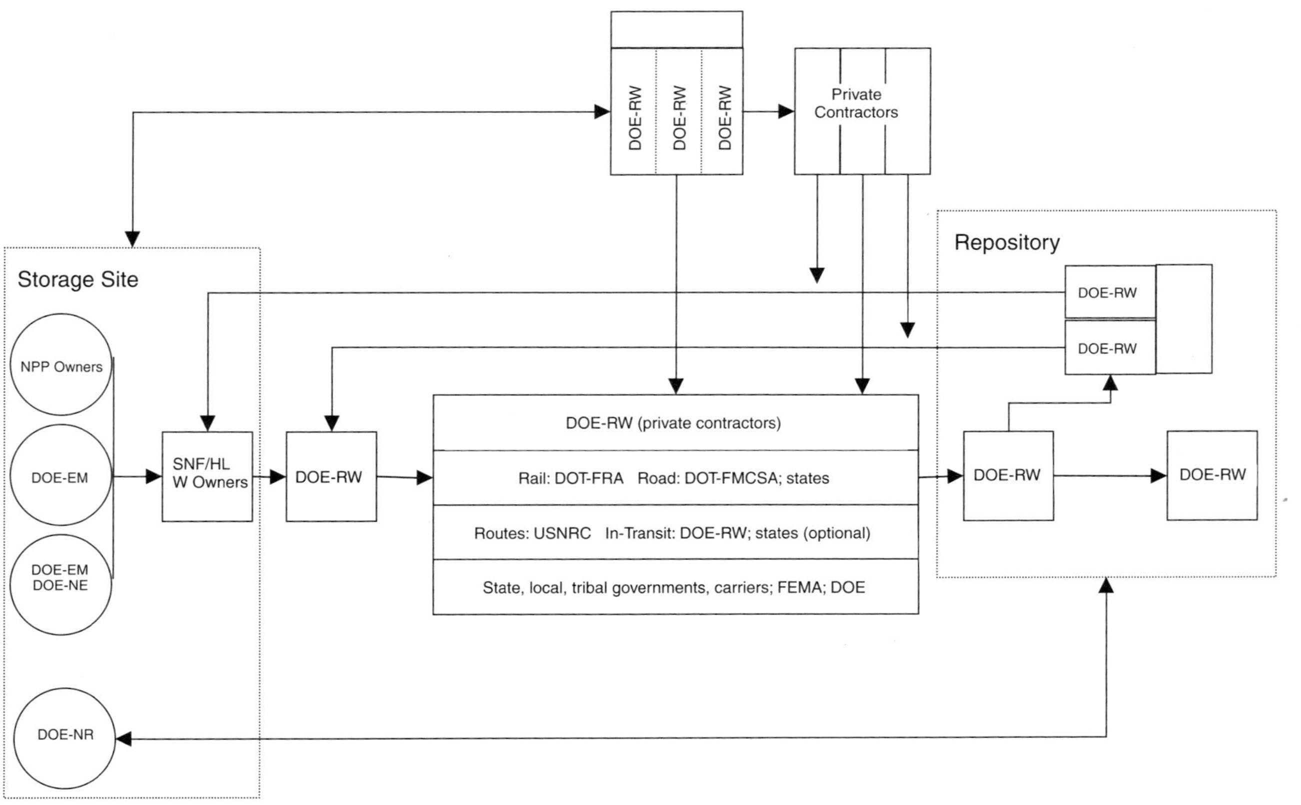

FIGURE C.1B Organizational responsiblilities.

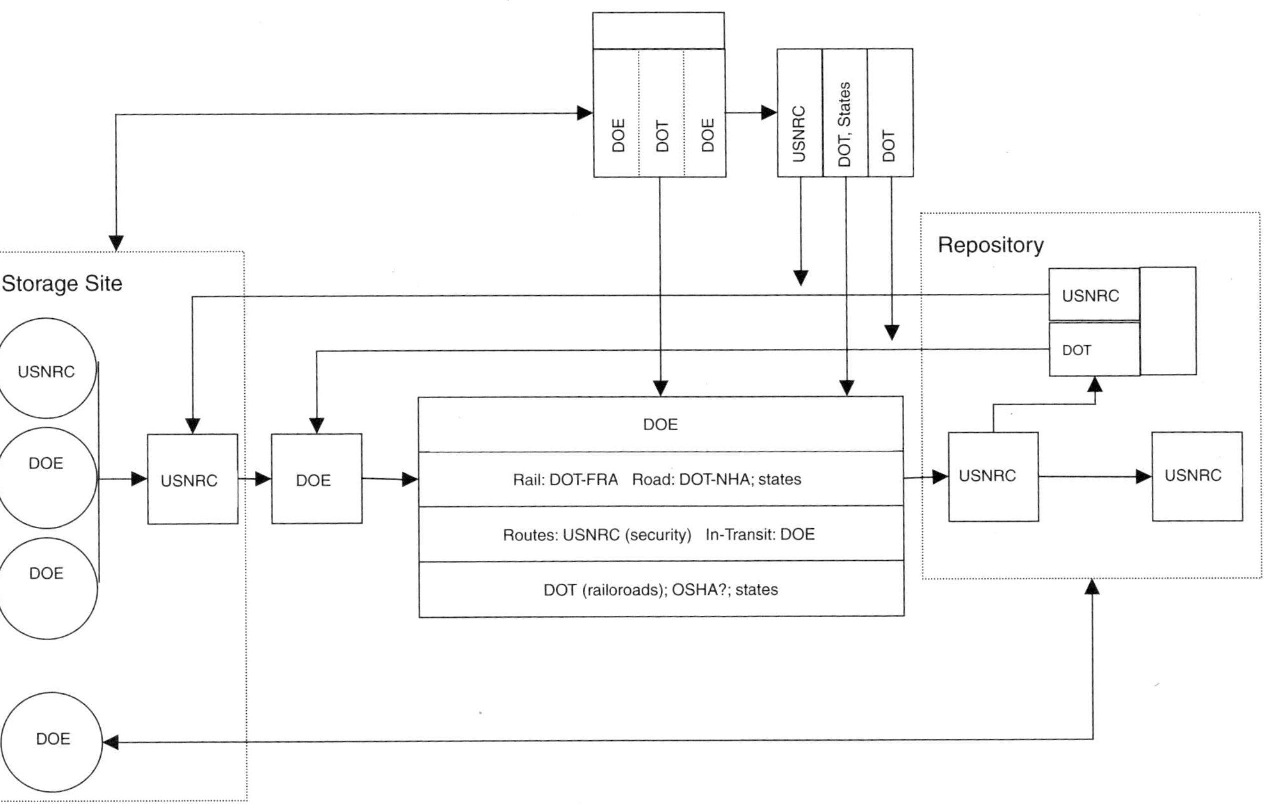

FIGURE C.1C Regulatory authorities.

construct at a DOE-estimated cost of $2 billion. Congress must appropriate the necessary construction funds to undertake this work.

DOE is also responsible for procuring the reusable transport packages that it will use to ship spent fuel and high-level waste to the repository. A variety of package designs may be needed to accommodate existing spent fuel and high-level waste types. There are certified packages available for commercial spent fuel transport (see Chapter 3) that could be purchased by DOE. In May 2004, DOE issued purchase orders to vendors possessing current USNRC certificates of compliance for transport packages to finance studies on the vendors' capabilities to take possession of spent fuel at nuclear power plant sites and deliver it to Yucca Mountain. However, in late 2005, DOE announced plans to develop a standardized package design for its federal repository transportation program.

The process for procuring transport packages will likely take several years, especially if new package designs are needed. The Nuclear Waste Policy Act requires that new package designs be certified by the USNRC, which will add additional time to the acquisition process. Procurement of waste packages, which will cost on the order of $1 million each, also will require timely appropriations from Congress. A substantial transportation fleet must be acquired to meet DOE's stated intention to ship to the repository at a steady-state rate of 3000 metric tons per year (see Table 5.1).

DOE will also have to procure purpose-built railcars and tractor-trailers (and possibly tractors) to transport these packages. These also will be designed and fabricated by private vendors. As described in Section 5.3, DOE's schedule for these activities is changing because of delays in completing the license application for the federal repository.

The equipment purchase decision is a complex systems problem within the broader transportation system and depends on several factors, for example:

- *Modal decisions.* Fundamentally different kinds and quantities of equipment will be required depending on the number of truck and train shipments to be made. This in turn will depend on the availability of rail access at each of the shipping sites and at Yucca Mountain.
- *Shipping rates.* There must be sufficient equipment to support the envisaged steady-state shipping rate of 3000 metric tons per year within a few years of the start of repository operations. This is equivalent to about 200 to 300 rail package shipments per year or 1500 truck package shipments per year.
- *Contingencies.* Spare equipment will be needed to allow for maintenance, "deadhead" shipments of packages and conveyances from the repository to spent fuel and high-level waste storage sites, and unanticipated equipment failures.

The Nuclear Waste Policy Act requires that DOE use private contractors "to the extent feasible" for the repository program. Indeed, DOE has announced its intention to use private contractors to run its national transportation program. However, DOE has not yet decided on the structure for contractor operations. In 1998, DOE issued a notice that it intended to divide the country into regions, each serviced by a "Regional Service Contractor." This plan was never implemented. In 2002, DOE issued another notice that it planned to issue a request for proposals for a single management and operating contractor to run the entire transportation program. It retracted this notice after receiving unfavorable comments. DOE has yet to announce what contracting approach it will pursue.

Once DOE decides which approach to use, it could take one or more years to hire a contractor using the usual DOE procurement processes. Once a contractor is selected, additional time will be required to develop the procedures to be used in the transportation program. The contractor will also have to hire and train workers to follow these procedures.

Scheduling

DOE must develop a schedule for accepting spent nuclear fuel and high-level waste from commercial and DOE sites for transportation to the repository. The schedule will based on two factors: (1) the order in which spent fuel or high-level waste is to be accepted from its owners (see Sidebar 5.2); and (2) the rate at which fuel can be shipped to the repository. The shipping rate will be based on several factors, but an important rate-controlling step is likely to be the capacity at the repository for receiving and unloading transportation packages. The availability of packages and other transport equipment also could become a rate-limiting step, especially if unexpected system upsets (e.g., equipment malfunctions) are encountered. There also could be site-specific shipping delays owing to equipment malfunctions and interference with reactor operations.

Routing

Two types of route planning are required to make the transportation system functional. The first is to assess whether the routes at the shipping and receiving ends of the system are adequate to accommodate shipments using trains, the preferred DOE shipping mode. If not, upgrades will be required before shipments can be made. DOE is in the process of planning a Nevada rail route, as noted previously. Road upgrades also may be needed in Nevada to support truck transport to the repository.

A second, more detailed route planning function must be carried out for each shipment of spent fuel and high-level waste. Railroads operate almost entirely on private right of way. At present, there are few regulatory restrictions on railroads' freedom to select routes for spent fuel and high-level waste shipments according to their own determinations of operational suitability and safety.[4] Such restrictions might be imposed through contractual or voluntary agreements between DOE and the railroads.

If the shipment is to be made by truck, route selection is based on Department of Transportation (DOT) regulations described in Chapter 2 and Chapter 4. DOE's current plans regarding mode and route selection, legal and regulatory restrictions on those decisions, and the states' concerns regarding routing are examined in Chapter 5.

C.1.2 Package and Load

Spent fuel and high-level waste owners are responsible for loading their waste into transportation packages at their sites. DOE-OCRWM is responsible for delivering empty packages to the sites to be loaded and can, at its option, observe the loading operations. The standard waste contracts require that DOE develop procedures for package loading and also provide the necessary training to sites for such loading and any associated package maintenance. These procedures would presumably be developed by the transportation contractor.

Under the terms of DOE's standard contracts with utilities, each shipment of spent fuel from a generator's site to the repository will involve the mass equivalent of one reactor core "offload," that is, the amount of spent fuel discharged from the reactor each time the core is reloaded with fresh nuclear fuel. Commercial power plants typically discharge about a third of a reactor core during each offload. This amounts to about 30 metric tons (33 short tons) of spent fuel for a typical 100 metric ton (110 short ton) reactor core. These offloads were commonly made on an annual basis, but they are now typically made every two years as technologies have been implemented for increasing fuel utilization.[5]

---

[4]This autonomy sometimes causes public controversy. In 2004–2005, for example, the train transport of chlorine tank cars through Washington, D.C., within blocks of the U.S. Capitol building, was the subject of controversy and led to an unsuccessful attempt by city government to restrict such shipments. The railroad announced that it had voluntarily re-routed some of these shipments around the city.

[5]The utilization of nuclear fuel is referred to as "burn-up," usually given in terms of the total amount of electricity (megawatt-days) that is generated from each metric ton of uranium in the fuel. Burn-ups on the order of 25,000 megawatt-days per metric ton were achieved during the early days of nuclear power generation. Today, technology improvements allow burn-ups approaching 60,000 megawatt days per metric ton to be achieved. Because of such increased burn-ups, the nuclear fuel in reactors requires less frequent replacement.

A single core offload will be transported to the federal repository in several packages. A typical rail package holds about 10 metric tons (11 short tons) of spent fuel, so a 30 metric ton (33 short ton) core offload would fit into two or three packages. A truck package typically holds about 2 metric tons (2.2 short tons) of spent fuel, so a single core offload would fill 15 or more packages. A site must be capable of handling multiple packages and conveyances for each shipping campaign it plans to make. There will be strong pressure from generators to ship full packages for reasons of economic efficiency, but there are likely to be many instances when the spent fuel to be shipped from a site results in less than a full package. DOE and the industry have not yet worked out arrangements for resolving this issue. The industry maintains that it wants the transportation system to run efficiently and will work with DOE to ensure that it does.

Owners are responsible for providing infrastructure for loading transportation packages for shipment to the repository. At operating commercial nuclear power plants, packages would be loaded under water in existing spent fuel pools. The packages would be moved in and out of the pools using existing overhead cranes. The overhead cranes at some plants will have to be upgraded to handle rail packages if the site intends to ship by that mode. Those packages can weigh in excess of 100 metric tons (110 short tons) when fully loaded, versus the 25 metric ton (28 short ton) weight for typical truck packages.

Several nuclear power plants, including decommissioned plants that no longer have operating spent fuel pools, have large quantities of fuel in dry-cask storage. Some of this fuel is stored in single-purpose packages that are not directly suitable for transportation. Special equipment (e.g., package over packs) or one-time variances from the USNRC will be needed to transport these packages to the repository. Otherwise, this fuel will have to be removed from the storage casks and reloaded into a package certified for transport. This will require access to pool space or heavily shielded dry-transfer facilities.

Owners are also responsible for providing access infrastructure within their plant sites so that the transport packages can be delivered and picked up. All commercial nuclear power plants have road access, and most, but not all, have rail access. Some plants may have to upgrade or add rail spurs if they choose to ship by that mode. DOE will have to assess and, where necessary, make improvements to roads and rails outside plant sites so that package deliveries and spent fuel pickups can be made.

The type of access available at each site will help determine what transport mode is selected. Sites that have adequate access to rail lines will likely ship by that mode using packages designed to be transported on railcars. Sites without rail access could ship by truck using packages designed to be transported on truck trailers. Legal-weight truck[6] shipments could be made over most of the nation's highways. Overweight truck shipments would require special permits issued by states and would be restricted to routes with appropriate load ratings. Alternatively, rail packages could be transported by heavy-haul truck[7] (again with state permits using load-rated highways) or by barge[8] over water to nearby railheads if the necessary infrastructure improvements are in place or can be made. DOE is responsible for the costs of such improvements.

DOE's Office of Environmental Management (DOE-EM) will be responsible for packaging its defense and research reactor spent fuel and high-level waste for shipment to the repository. All of the DOE sites that will ship this waste have rail access, but package-loading facilities may have to be constructed at some of these sites.

C.1.3 Inspect and Accept

Once loaded, the owner must place the spent fuel or high-level waste package onto the DOE-provided conveyance (i.e., typically railcars or trailers). This will be done at the owner's site using the owner's equipment. DOE-OCRWM will then perform an inspection to determine if the shipment is "transport ready." If so, it will accept the package for transport to the repository. Ownership of the spent fuel or high-level waste passes to DOE-OCRWM at the plant gate.

DOE told the committee that its current plans are to transport an average of three packages per rail shipment. This is approximately the number of train packages needed to ship a full core reload. DOE could, however, accept a larger or smaller number of packages in each shipment, or even partially full packages—there is no physical constraint that limits a

[6]A legal-weight truck is a truck that complies with all state and federal truck weight regulations and therefore does not require any special permits to operate. In almost all cases, legal-weight trucks would be limited to 80,000 pounds (about 36 metric tons) gross weight (i.e., the weight of the vehicle and cargo) and could carry a cargo of about 50,000 pounds (about 23 metric tons).

[7]In the context of the Yucca Mountain transportation problem, a heavy-haul truck is a highway vehicle capable of carrying a rail shipping package. Such a truck would weigh in excess of 200,000 pounds (about 90 metric tons) empty, plus the weight of the package when loaded.

[8]Even if barge access is available, DOE may not be able to ship by this mode due to state and public opposition. Such opposition appears to be especially strong in the upper Midwest, particularly for shipments that would utilize the Great Lakes. The committee was told that barge transport appears to be more acceptable in the southeastern United States, where such shipments for the disposal of low-level waste are already being made.

single train shipment to three packages. DOE could, in principle, combine packages from several sites into a single shipment by moving packages to nearby rail yards for makeup of single trains. There will likely be strong pressure to minimize time in transit, which could encourage shipments from single sites directly to the repository, and to minimize the consolidation of shipments from multiple sites into single trains. On the other hand, there might be pressure from transit states to consolidate shipments to reduce the total number of trips.

C.1.4 Transport

The transport of spent fuel and high-level waste from storage sites to the repository will be a complicated process involving DOE; its contractors; regulatory authorities; and state, tribal, and local authorities. Each shipment or shipping campaign[9] must be planned well in advance; the exact planning requirements will depend to a great extent on the transportation mode selected.

Although DOE has decided on a mostly rail transportation program, the possibility still exists that trucks will play a significant role. For truck shipments, the transportation contractor, in consultation with DOE, would select a route from DOT's approved list of preferred or alternate preferred routes (see Chapter 4). The latter are designated by states. The highway routes selected are most likely to be Interstate System highways and bypasses around cities. The transportation contractor would then pick up the shipment at the owner's site and transport it to Yucca Mountain. The only planned stops during transport would be for inspections, refueling, maintenance, and crew rests and changes.

For train shipments, the railroad will select the route to be used probably in consultation with the contractor and DOÈ. DOE could impose restrictions on the route—for example, to reduce time in transport, to minimize transport through tunnels or densely populated areas, or to avoid poor-quality tracks—but this might increase shipping costs. There is no state or local control over routing decisions. The Western Interstate Energy Board has examined the possibility of state regulation of routing of spent fuel train shipments, but no state has acted and the limits of state powers to regulate such matters have not been legally determined (WIEB, 1995, Pp. 34–36, Pp. 45–47).

---

[9]A shipping campaign consists of a number of shipments from a single site to the federal repository over a fixed period of time. Such campaigns are likely for truck transport but less likely for train transport unless a number of core offloads (each of which could be shipped to the repository using a single train) are planned for consecutive shipment.

Regardless of mode, actual transportation operations would be carried out by private trucking or railroad companies under contract to DOE. In principle, a single trucking firm could handle all truck shipments to the repository, because highways are publicly owned. This would not be the case for train shipments, however, because railroad rights of way are privately owned. DOE would have to contract with the rail operator who owns the tracks with direct access to the shipping site. That operator would pick up the shipment and move it to the end of its line. From there it would be transferred to the rail operator who owns the next segment of track. Several such transfers might take place to move a single shipment from the owner's site to the repository.

DOE has yet to develop procedures or hire contractors for shipping to the repository. These tasks would presumably be handled by the transportation contractor.

En Route Inspections

States have primary responsibility for enforcing highway safety regulations concerning federal motor carrier safety and hazardous materials transportation. Consequently, truck shipments of spent fuel and high-level waste to the repository would be subject to state inspection, and each shipment could potentially receive separate inspections by each state. State enforcement officials can stop and inspect vehicles and inspect the premises of motor carriers to check for compliance with federal and state requirements regarding equipment, documentation, and driver fitness. States can also require carriers to obtain special permits to operate these vehicles and charge fees for such permits.

However, the federal hazardous materials transportation law is explicit that federal rules preempt state rules in cases of conflict (49 USC 5125), consistent with the goal of nationally uniform regulation. DOT has administrative authority to determine when state rules are to be preempted. For example, DOT has determined that state requirements for special permits for radioactive materials highway shipments are preempted if they require documentation or prenotification in excess of federal requirements. DOT also has determined that state fees imposed on hazardous materials transport are preempted if they are excessive or if the revenue is not used for purposes related to hazardous materials transport. Additionally, DOT has determined that state routing requirements for highway radioactive materials transport are preempted if they are not identical with federal requirements (Buren, 1998).

Regulation of the safety and operation of railroads is dominated by the federal government. Railroad shipments of spent fuel and high-level waste are not subject to state regulation, but they are subject to inspection by

DOT's Federal Railroad Administration (FRA). The FRA is generally responsible for safety regulations governing the design and maintenance of track, signals, and equipment; railroad operating practices; hazardous materials handling; and rail worker safety practices and qualifications.[10] Enforcement is by railroad self-inspections, which are audited by Federal Railroad Administration inspectors with the participation of administration-certified state inspectors.

In 1998, the Federal Railroad Administration published the report *Safety Compliance Oversight Plan for Rail Transportation of High-Level Radioactive Waste and Spent Nuclear Fuel* (FRA, 1998) as a comprehensive definition of FRA activities to ensure the safety of rail transport of these materials. The plan is a statement of policy rather than a regulation. It was developed to guide the FRA's involvement in DOE's Foreign Research Reactor Spent Fuel Program, but it is also intended to be applicable to future shipments to a federal repository or to temporary storage. The plan requires coordination among the Federal Railroad Administration, DOE, utilities, railroads, and the states.

The elements of this plan are grouped in three activity areas: planning, inspections, and training. The planning provisions require the shipper to notify the Federal Railroad Administration with carrier and route information at least 90 days before the initial shipment. The FRA will inspect the track and will consult with DOE and the shipper and carrier on route selection. The plan highlights physical track quality as a route selection criterion. The Federal Railroad Administration will also assist states in estimating highway-rail grade-crossing accident risks along the route. Other provisions of this plan involve inspections by FRA personnel of all equipment, a requirement for administration personnel to be present in the railroad's dispatch center during each movement of spent fuel or high-level waste, and training for railroad personnel and emergency responders.

Security

DOE-OCRWM is responsible for providing security for its shipments of spent fuel and high-level waste. Under the Nuclear Waste Policy Act, DOE is required to follow the U.S. Nuclear Regulatory Commission's regulations in 10 CFR 73.37: *Requirements for the Physical Protection of Irradiated Reactor Fuel in Transport*. These regulations address the establishment of a physical protection system that minimizes the possibility for radiological sabotage of spent fuel shipments, especially within heavily

[10]These regulations are promulgated in Title 49, Parts 200-245 of the Code of Federal Regulations.

populated regions, and facilitates the location and recovery of spent fuel shipments that come under the control of unauthorized persons. The regulations require that the shipper take the following steps:

- Notify the USNRC in advance of each shipment. Part 73.72 requires that such notifications be received at least 10 days prior to the initiation of such shipments.
- Make arrangements with local law enforcement agencies along the route of each shipment for their response to an emergency or call for assistance.
- Obtain advance approval by the USNRC of the routes to be used. The Commission will physically survey the routes and issue an approval to use them for a fixed time period.
- Plan shipments so that scheduled intermediate stops are avoided to the extent practicable.
- Provide trained escorts for the shipments and require that these escorts make calls to a communications center at least every two hours to provide a shipment status.

Escort requirements are specified according to mode and route. Truck transport within a heavily populated region[11] is required to be occupied by a driver and at least one escort, and to be escorted by an armed member of local law enforcement in a mobile unit. Alternatively, a vehicle occupied only by a driver is required to be escorted by lead and trailing vehicles containing at least one armed escort each. A road transport vehicle not in a heavily populated region is required to have an escort on board or to be escorted by a vehicle that contains at least two escorts. Escorts must be capable of communicating with the communications center, local law enforcement, and each other. The transport vehicle must be equipped with USNRC-approved features that permit immobilization of the cab or cargo-carrying portion of the vehicle. Additional security measures were required by the USNRC after September 11, 2001. These requirements have not been made public.

Shipments by train within heavily populated regions must be accompanied by two armed escorts, at least one of whom is stationed on the train to

[11]The more stringent requirements for transporting spent fuel through heavily populated regions were developed based on the "urban studies" completed by Sandia for the USNRC (DuCharme et al., 1978; Finley et al., 1980; Sandoval et al., 1983; Sandoval, 1987). These studies suggested that sabotage of spent fuel transport packages in urban areas could have significant negative consequences. Consequently, the Commission requires tighter security for spent fuel transport through such areas. The USNRC is undertaking additional vulnerability studies in response to the September 11, 2001, attacks.

permit direct observation of the package railcar while in motion. Shipments by rail in regions that are not heavily populated must be accompanied by at least one escort stationed on the train to permit such direct observation. These escorts must have the capability to communicate with the communications center and local law enforcement.

Regulations also require that the shipper notify the governor or governor's designee of each state through which the shipment will pass. This notification must be postmarked at least seven days before the shipment reaches the state or be delivered by messenger at least four days in advance. The notification must include a description of the shipment, a listing of the routes to be used within the state, the estimated time of departure from the point of origin of the shipment,[12] and the date and time of entry into the state. Subsequent timely notifications are required if the schedule changes by more than six hours.

All spent fuel and high-level waste shipments to the repository will have DOE- or contractor-provided security escorts and, in addition to the USNRC requirements described previously, will be tracked using a global positioning system (GPS) to provide real-time monitoring of the locations of all shipments. GPS tracking is currently in use for other DOE waste shipments, including shipments to the Waste Isolation Pilot Plant in New Mexico. The information on location and status of shipments is made available to states when shipments are within their geographic boundaries.

Emergency Response

An *emergency responder* is any individual trained to provide assistance at the scene of an accident. The first emergency responder at the scene of a transportation accident might be the transport vehicle driver. The carrier (whether truck or rail) is responsible for providing the first line of emergency response should an accident or terrorist attack occur. This would include preventing the spread of contamination at the site of the accident; preventing individuals from coming into contact with spilled materials; and preventing contaminated vehicles or equipment from being used until they have been surveyed and, if necessary, decontaminated. The carrier also is responsible for notifying local authorities, the shipper, and applicable federal agencies of the accident and providing whatever additional assistance is required.

State, tribal, and local governments play a preeminent role in emergency response and preparedness. They provide the first line of government

[12]The state is to keep this information protected in accordance with 10 CFR 73.21 until at least 10 days after the shipment entered or originated within the state or, for a series of shipments, 10 days after the last shipment.

response to emergencies involving chemically hazardous and radioactive materials and can call on other levels of government for assistance as circumstances require. Their level of readiness varies nationally. Some require minimal technical assistance, whereas others will act only to secure the accident scene and evacuate endangered populations while awaiting technical assistance.

The first government responder will likely be a member of a public safety department such as sworn police officer or firefighter, or both, from the jurisdiction where the accident is located. Specialized teams (i.e., a "hazmat team") attached to the jurisdiction's emergency services agency, or from neighboring communities when mutual aid agreements are in effect, may be dispatched to the site in the event of an accident involving hazardous cargo.

Except in national defense incidents, the local government and the on-scene commander are responsible for incident command and control. The local authority (usually a government entity such as a city or county) will establish an on-scene "incident commander" to direct the response and ensure that all needed resources are made available. If the emergency needs outstrip the resources of the jurisdiction, the incident commander will work with authorities to summon additional assistance. This could include trained teams from adjacent jurisdictions that are activated through prearranged mutual assistance agreements; state and federal response teams; and private organizations.

The Department of Homeland Security (DHS) is the responsible federal agency for incident response. At the direction of the President of the United States, DHS has instituted the National Incident Management System (NIMS), which is intended to become the national command and control system for responding to incidents of any nature in the homeland. The President has issued a number of Homeland Security Presidential Directives (HSPDs)[13] to implement NIMS, and governors and state legislatures have also been requested to adopt this system. The federal government has established 2006 as the deadline for national compliance.

The NIMS establishes a National Response Plan, which tasks all federal agencies and establishes operational standards and procedures for responding to incidents.[14] Under this plan, the Secretary of Homeland Security, acting as the principal federal official, or a federal official designated by the secretary as a federal coordinating officer, can direct federal assis-

---

[13]HSPD-5 designates the Secretary of the Department of Homeland Security as the principal federal official for hazards preparedness, response, and recovery anywhere in the United States. HSPD-8 implements NIMS, defines national preparedness goals, and provides a National Uniform Task List for all hazards.

[14]The National Response Plan replaces the Federal Response Plan, which was an agreement among federal agencies on a management process for responding to incidents.

tance and resources to the on-scene incident commander. The governor of the state in which the incident occurs will appoint a state coordinating officer to direct and control state assets and request federal assistance on behalf of the on-scene commander.

The federal government also is the primary provider of training to the nation's firefighters and emergency management personnel through the National Emergency Training Center in Emmitsburg, Maryland. This center houses the National Fire Academy, which provides training to the nation's firefighters, and the Emergency Management Institute, which serves as the national focal point for executive-level emergency management training. These organizations provide resident and nonresident training; the latter includes Web-based training and other distance training using video tapes and other instructional materials.

The Department of Transportation, in cooperation with Canadian and Mexican transportation organizations, has developed the *Emergency Response Guidebook* (DOT, 2004) for use by emergency responders during the initial phase of a dangerous goods or hazardous materials incident. It was developed to help responders quickly identify actual or potential hazards in an incident and take steps to protect themselves and the public during the initial phases of a response. Federal regulations[15] require that first responders be trained in the use of this guidebook.

The Department of Energy provides technical assistance and training to emergency responders through its Transportation Emergency Preparedness Program (TEPP). Under the Nuclear Waste Policy Act, DOE is responsible for providing emergency responder training assistance along routes that will be used to ship spent fuel and high-level waste to the repository. This program is described in Chapter 5. DOE plans to initiate this training assistance once it identifies transport routes.

The Institute for Nuclear Power Operations (INPO[16]) has established voluntary agreements among the nation's nuclear utilities to provide mutual assistance in the event of a radioactive materials transportation accident involving commercial spent fuel. Through these agreements, nuclear utilities located near the scene of an accident would provide technical advice and assistance to federal, state, and local emergency responders regardless of who owns the spent fuel involved in the accident.

---

[15]These are provided in 29 CFR 1910.120 and 40 CFR Part 311.

[16]INPO was formed by the nuclear power industry in 1979 as a result of the Three Mile Island Unit 2 accident near Harrisburg, Pennsylvania. Its purpose is to promote the safety and reliability of nuclear power plant operations. INPO sets performance guidelines and objectives for nuclear power plant operations; conducts evaluations of plant performance; and accredits utility training programs for plant operators and supervisors. All U.S. nuclear power plant operators are members of this organization, and nuclear plant operators in other countries are participants.

C.1.5 Inspect and Unload

Once the shipment reaches the repository, the package will be removed from its conveyance and unloaded. A surface facility at the repository site will be constructed for this purpose. The package unloading function will be handled by the repository contractor as part of disposal system operations. The rate of unloading will be governed by the availability of surface storage and the rate at which spent fuel and high-level waste can be moved to storage or emplaced underground. This rate will help determine the transport package procurement strategy.

DOE-OCRWM is responsible for constructing facilities for receiving and unloading waste packages at the repository and maintaining the packages and conveyance equipment. DOE will include the design of these facilities in its application to the USNRC for a repository construction license. Construction of these facilities could take several years.

C.1.6 Maintenance

Once the package has been unloaded, it and its conveyance would be inspected for wear and damage, and maintenance would be performed as required. They would then be returned to service. Empty packages and conveyances would be sent to the next site in the waste acceptance queue. While DOE-OCRWM has not yet determined where it will carry out this maintenance activity, it sees advantages in locating this facility at or near the Yucca Mountain site. DOE-OCRWM noted that it also does not yet have a design for such a maintenance facility. The design requirements will depend on the mix of transport modes employed (i.e., train versus truck), because this will affect the size and throughput of transport packages.

DOE-OCRWM may find that it needs more than one facility to service a national transportation program, given the amount of equipment involved and the geographic extent of its planned operations. The transportation program in the United Kingdom, for example, has one maintenance facility for locomotives, two geographically separated facilities for maintaining railcars, and yet another facility for maintaining its transportation packages.

C.2 DISCUSSION

As described in the foregoing sections, the system to transport spent fuel and high-level waste to a federal repository will involve a large number of operations, responsible parties, and interdependencies. Many of the major decisions that have to be made to establish this system are under DOE-OCRWM's direct control, most notably modal and routing decisions, equipment purchases, and contractor selections. Of course, establishment of this system is dependent on DOE's ability to obtain a license to construct and operate the repository, including surface facilities to handle spent fuel and high-level waste shipments, and to obtain adequate appropriations from Congress for procuring equipment and hiring contractors.

One of the most important decisions in the transportation program—the order for accepting spent fuel from generators—is established by the Nuclear Waste Policy Act (see Sidebar 5.2). It has no geographic coherence, and neither will the transport system developed to accommodate it. Spent fuel is likely to be transported to the repository simultaneously from several parts of the country along several different routes. It could be difficult for DOE to achieve efficiencies with respect to many of its transport operations, most notably route planning and emergency responder training, if it has no legal control of the acceptance order.

The national transportation program was in the early stages of development when this report was being written. Most of the planning and a great deal of the infrastructure will have to be developed from scratch, most notably the following:

- The back-end facilities at the repository for handling the packages and conveyances have not been licensed or constructed.
- A rail route has not been constructed in Nevada.
- The packages and conveyances to move the spent fuel and high-level waste have not yet been ordered. New designs may have to be developed for some of this equipment. New package designs will require certification by the USNRC. DOE will not be able to fully determine its package needs until it has updated data on rail and heavy-haul access to owner's sites (see Chapter 5).
- DOE has not yet decided on contractor roles in the program, nor has it advertised for a transportation contractor.
- Procedures for loading and moving spent fuel and high-level waste have not yet been established.
- DOE has not yet selected routes so that emergency responder training can begin.
- Many states, including Nevada, have yet to specify alternate preferred routes for truck shipments.
- Procedures, facilities, and criteria for inspection and maintenance of packages and conveyances have not been developed.

In short, the national transportation program is very much a work in progress. The committee makes several recommendations in Chapter 5 for improving this transportation program.

APPENDIX D
Glossary

10 CFR Part 71: Title 10, Part 71 of the Code of Federal Regulations, Packaging and Transportation of Radioactive Material. These regulations were promulgated by the U.S. Nuclear Regulatory Commission.

10 CFR Part 72: Title 10, Part 72 of the Code of Federal Regulations, Licensing Requirements for the Independent Storage of Spent Nuclear Fuel, High-Level Radioactive Waste, and Reactor-Related Greater than Class C Waste. These regulations were promulgated by the U.S. Nuclear Regulatory Commission.

49 CFR Part 173: Title 49, Part 173 of the Code of Federal Regulations, Shippers—General Requirements for Shipments and Packagings. These regulations were promulgated by the U.S. Department of Transportation.

A_1**:** The maximum activity of special form radioactive material permitted in a Type A package as provided in Appendix A of 10 CFR Part 71.

A_2**:** The maximum activity of normal form radioactive material permitted in a Type A package as provided in Appendix A of 10 CFR Part 71.

Absorbed dose: The quantity of ionizing radiation deposited into an organ or tissue, expressed in terms of the energy absorbed per unit mass of tissue. The basic unit of absorbed dose is the rad or its SI equivalent the gray (Gy).

Acceleration: The rate of change of velocity of an object.

Accelerometers: Devices that measure the acceleration of an object.

Accident conditions of transport: Severe conditions that well exceed normal conditions of transport. Such conditions could result in the application of thermal and mechanical forces that have the potential to damage the vital containment functions of the package.

Accident consequence value: The estimated collective dose that would be received by a population as a result of an accident scenario.

Accident scenario: A postulated sequence of events during a transportation accident that result in the application of elevated thermal and mechanical loads to a transportation package.

Accident source term: The amount of radioactive material released from a loaded transportation package in an accident.

Actinides: Any of a series of chemically similar radioactive elements with atomic numbers ranging from 89 (actinium) through 103 (lawrencium). This group includes uranium (atomic number 92) and plutonium (atomic number 94).

Activity: The rate of decay of a radioactive isotope.

Acute radiation exposure: A radiation exposure that occurs over a relatively short period of time (e.g., seconds to hours). A chest X-ray is an acute radiation exposure.

Affected communities: Communities that are impacted by a transportation program, for example, communities along a route used to ship spent fuel and high-level waste.

Agreement State: States that have assumed authority under Section 274b of the Atomic Energy Act to license and regulate by-product materials (radioisotopes), source materials (uranium and thorium), and certain quantities of special nuclear materials.

Atoms for Peace Program: A U.S. program begun under the Eisenhower administration to supply research reactor technology and nuclear fuel to foreign nations that agreed to forgo the development of nuclear weapons.

Bare-fuel packages: See *Package*.

Becquerel: A unit of radioactive decay equal to 1 disintegration per second.

Bounding accident scenarios: Physically realistic accident scenarios that would be expected to produce large thermal and mechanical loading conditions.

Burn-up: A measure of the degree to which the uranium-235 in nuclear fuel has been used up (fissioned), which determines the amount of radioactivity and heat generation in the fuel after it has been removed from the reactor.

BWR: Boiling water reactor, a type of nuclear reactor in which the reactor's water coolant is allowed to boil to produce steam. The steam is used to drive a turbine and electrical generator to produce electricity.

Byproduct material: Defined by the Atomic Energy Act as radioactive material (except special nuclear material) yielded in or made radioactive by exposure to the radiation incident to the process of producing or using special nuclear material; and tailings or wastes produced by the extraction or concentration of uranium or thorium from any ore processed primarily for its source material content.

Caliente corridor: A corridor in Nevada that has been selected by the U.S. Department of Energy for the construction of a rail line to Yucca Mountain. The corridor begins near Caliente, Nevada; passes north of the Nevada Test and Training Range; and then runs south to Yucca Mountain.

Cancer incidence: Also known as the *incidence rate*. The rate of occurrence of cancer within a specified period of time per unit of population; for example, the number of cancers per year per 100,000 people.

Cancer mortality: Also known as the *mortality rate*. The rate of death from cancer within a specified period of time per unit of population; for example, number of cancer deaths per year per 100,000 people.

Canister-based packages: See *Package*.

Cask: See *Package*.

Centralized interim storage: See *Interim storage*.

Certification tests: See *Full-scale testing*.

Chronic radiation exposures: Radiation exposures that occur over extended periods of time (e.g., months to years). Exposure to natural background is a chronic radiation exposure.

Closed fuel cycle: A nuclear fuel cycle in which spent fuel is processed to recover its usable contents.

Cloud shine: Radiation exposure from airborne radioactive material.

Collective dose: The sum of all radiation exposures received by all members of a specified population.

Commercial spent fuel: Spent fuel produced from commercial nuclear power plants.

Community: A group of people having similar characteristics, interests, or interactions; for example, people living in a particular geographic region, people with common cultural backgrounds, or people with common professional interests.

Comparative risk: See *Risk*.

Complementary cumulative distribution function (CCDF): Also known as the *risk curve*. A graphical representation of estimates of risk to a population from a particular accident scenario involving spent fuel and high-level waste. The estimate is developed through computer modeling studies.

Compound CCDF: A graphical representation of the statistical properties (e.g., mean, median) of a group of complementary cumulative distribution functions.

Containment effectiveness: The ability of a transportation package to contain its radioactive contents and maintain its radiation shielding effectiveness during routine use and under severe accident conditions.

Conventional vehicular impact: Health and safety impacts that result from both normal transportation and transportation accidents involving spent fuel and high-level waste. These include the health impacts of transport vehicle exhaust emissions as well as traffic-related fatalities, injuries, and property damage.

Conveyance: Any transport vehicle or vessel used to move spent fuel and high-level waste packages; for example, a truck and trailer or a locomotive and railcar.

Criticality events: Self-sustaining nuclear reactions in spent fuel like those that occur when the fuel is in the reactor.

Crud: Non-fixed radioactive contamination on the external surfaces of spent fuel assemblies.

Curie: A unit of radioactive decay equal to 3.7×10^{10} (37 billion) disintegrations per second.

Dedicated train: A train that transports only spent fuel or high-level waste and no other cargo.

Demonstration tests: See *Full-scale testing.*

Depleted uranium: Uranium from which much of the uranium-235 has been removed.

Direct inhalation: See *Inhalation.*

Direct socioeconomic impact: Loss of economic or social well-being as a direct result of transportation program operations; for example, economic losses from a transportation accident.

DOE spent fuel: Spent fuel that is being managed by the U.S. Department of Energy, including spent fuel from defense reactors, research reactors, naval reactors, and some commercial power plants.

Dose: See *Radiation dose.*

Dose rate: See *Radiation dose rate.*

Dry-cask storage: Packages used to store spent fuel in a dry state.

Effective dose: The equivalent dose averaged over all organs that accounts for the varying sensitivity of different organs and tissues to the biological effects of ionizing radiation. The effective dose has the same units as the equivalent dose.

Elastic limit: The maximum stress that can be applied to an object without causing permanent deformation.

Emergency responder: A individual trained to provide assistance at the scene of an accident.

Environmental justice: The fair treatment and meaningful involvement of people regardless of race, gender, national origin, or level of attained education in the development of laws, regulations, and policies that affect them.

Equivalent dose: The absorbed dose averaged over the organ or tissue of interest multiplied by a weighting factor that accounts for the differences in the biological effects per unit of absorbed dose for different types of radiation. The basic unit of equivalent dose is the rem or its SI equivalent the sievert (Sv).

Essentially unyielding surface: A surface that, because of its large mass and stiffness, absorbs minimal energy when impacted with other objects.

Event trees: A graphical illustration of the sequence of events leading to an accident along with the probability of occurrence of each event. Each branch of the tree depicts the sequence of events that leads to the accident outcome depicted at the end of the branch. The probability of that accident is equal to the product of the probabilities of each segment along the branch.

Exclusive use: Defined in 10 CFR 71.4 as "sole use by a single consignor of a conveyance for which all initial, intermediate, and final loading and unloading are carried out in accordance with the direction of the consignor or consignee. The loading and unloading must be carried out by personnel having radiological training and resources appropriate for the safe handling of the consignment."

Extraregulatory conditions: Accidents that impose thermal or mechanical loads on transportation packages that exceed those generated in the hypothetical accident conditions specified in 10 CFR Part 71.

Federal repository: A federally operated underground facility for the permanent disposal of spent fuel and high-level waste.

Fissile: Capable of being fissioned with thermal (low-energy) neutrons, the primary process in most nuclear reactors. The two most important fissile materials in spent fuel are uranium-235 and plutonium-239.

Fission: The splitting of a nucleus into at least two fragments accompanied by the release of neutrons and energy.

Free-drop test: One of the hypothetical accident conditions in 10 CFR Part 71 involving the free fall a transportation package from a height of 9 meters (about 30 feet) onto a flat, essentially unyielding horizontal surface, with the package striking the surface in the position expected to produce maximum damage.

Fuel assembly: A bundle of fuel rods arranged in a square array.

Fuel basket: A latticed container that is inserted into a transportation package or dry-storage cask and is designed to hold spent fuel assemblies in a fixed configuration.

Fuel cladding: The thin-walled metal tube, usually fabricated from a zirconium alloy, that forms the outer jacket of a nuclear fuel rod.

Fuel pellets: Small cylinders of uranium, usually in an oxide form, that provide the fuel for a nuclear reactor.

Fuel rod: Sometimes referred to as a *fuel element* or *fuel pin*. A long, slender, sealed tube that holds uranium fuel pellets. Fuel rods are bundled in square-arrays bundles called *fuel assemblies*.

Full-scale testing: Tests carried out on full-scale transportation packages or package components. Tests carried out to simulate the hypothetical accident conditions specified in 10 CFR Part 71 are referred to as *certification tests*. Tests carried out to simulate other severe accident conditions are often referred to as *demonstration tests*.

Fully engulfing fire: An optically dense, hydrocarbon-fuel fire that is sufficiently large to completely engulf a transportation package.

g: The acceleration imparted by Earth's gravity field; 1 *g* is equal to 9.8 meters per second squared or 32 feet per second squared.

Galactic cosmic radiation: Radiation originating from distant stars and galaxies.

General trains: Also known as *general service trains*. Trains that carry other freight in addition to spent fuel or high-level waste.

Generic package: A conceptual package design that incorporates the salient features of packages that have been certified for actual service. Generic package designs have been constructed for use in package performance modeling studies.

Gray literature: Foreign or domestic open-source material that usually is available through specialized channels and may not enter normal channels or systems of publication, distribution, bibliographic control, or acquisition by booksellers or subscription agents.

Greater-than-Class-C waste: Radioactive waste that contains concentrations of certain radionuclides above the Class C limits in 10 CFR 61.55.

Ground shine: Radiation exposure from material deposited on the ground.

Hazardous Materials Incident Reporting System: A computerized database maintained by the U.S. Department of Transportation that contains information on incidents involving the interstate transportation of hazardous (including radioactive) materials by air, highway, rail, and water.

Health and safety risks: In the context of this report, risks that arise from exposures of people to radiation as a direct result of spent fuel and high-level waste transport.

Heavy-haul truck: As used by the U.S. Department of Energy, an overweight truck that is capable of transporting a full-size rail package.

High-income economy countries: Defined by the U.S. Department of Energy to include Australia, Austria, Belgium, Canada, Denmark, Finland, France, Germany, Israel, Italy, Japan, Netherlands, Spain, Sweden, Switzerland, Taiwan, and the United Kingdom.

High-level radioactive waste (HLW): Defined in Title 42 of the U.S. Code as the highly radioactive waste material resulting from the reprocessing of spent nuclear fuel, including liquid waste produced directly in reprocessing and any solid material derived from such liquid waste that contains fission products in sufficient concentrations; and other highly radioactive material that the U.S. Nuclear Regulatory Commission, consistent with existing law, determines by rule to require permanent isolation.

Highly enriched uranium (HEU): Uranium enriched in the isotope uranium-235 to concentrations greater than or equal to 20 percent.

HIGHWAY: A U.S. Department of Energy computer model used to route highway shipments of radioactive materials.

Highway route controlled quantity: Defined in 49 CFR 173.403 as a quantity within a single package that exceeds 3000 times the A_1 or A_2 values for special form and normal form radioactive material, respectively, or that contains 1000 terabecquerels (27,000 curies) of radioactivity, whichever is least.

HTGR: High-temperature gas-cooled reactor, a type of nuclear reactor in which helium is used as the coolant instead of water. The helium is used to heat water to generate steam, which in turn drives a turbine and electrical generator to produce electricity.

Hypothetical accident conditions: A series of package tests described in 10 CFR Part 71 that includes a free-drop test, puncture test, thermal test, and immersion test.

Immersion test: One of the hypothetical accident conditions in 10 CFR Part 71 in which an undamaged package is subjected to a pressure head equivalent to immersion in 15 meters (about 50 feet) of water. Additionally, 10 CFR 71.61 specifies that for packages designed for transport of spent fuel with activity exceeding 37 petabecquerels (1 million curies), the undamaged containment system must be able to withstand an external water pressure of 2 megapascals (290 pounds per square inch) for one hour without collapse, buckling, or in-leakage of water.

Impact limiters: Protective coverings attached to the ends of transportation packages that are designed to absorb mechanical forces and provide thermal protection for the package closure system.

Incidents: (1) Any event that involves the actual or suspected release of radioactive material from or surface contamination of a transportation package. (2) An intentional criminal act intended to damage a transportation package or disrupt a shipment.

Independent Spent Fuel Storage Installations (ISFSIs): See *Interim storage.*

Ingestion: Radiation exposures resulting from the uptake of radioactive material into the body via the digestive tract.

Inhalation: Exposures resulting from the uptake of material into the body via the respiratory tract. The exposures of interest in this report are from inhalation of radioactive materials.

Interim storage: The temporary storage of spent fuel and high-level waste, either in pools or in dry casks. Facilities designed to store spent fuel or high-level waste from several different sites are also informally referred to as *centralized interim storage facilities*.

INTERLINE: A Department of Energy computer model used to route rail shipments of radioactive materials.

Intermodal transportation: Movement of spent fuel or high-level waste using a combination of rail, highway, and barge.

INTERTRAN 2: An implementation of the RADTRAN code for international applications.

Ionizing radiation: Radiation that is sufficiently energetic to ionize the matter (i.e., remove electrons from the atoms) through which it moves.

Irradiated nuclear fuel: See *Spent nuclear fuel*.

Large-quantity shipping programs: Transportation programs that ship on the order of hundreds to thousands of metric tons of spent fuel or high-level waste.

Latent cancer: Cancerous lesions in a living organism that have not progressed to a stage to be detectable.

Legal-weight truck: A truck having a total gross weight (i.e., including cargo) of 80,000 pounds (about 36 metric tons) or less.

Low enriched uranium (LEU): Uranium enriched in the isotope uranium-235 to concentrations less than 20 percent.

Magnox package: A transportation package used in the United Kingdom to transport spent fuel from Magnox reactors. Also referred to as a *Magnox flask*.

Margin of safety: Also known as a *safety margin*. A philosophy for designing engineering structures that relies on conservative assumptions about the mechanical properties of the materials used in the structure and the applied loads that the structure must resist.

Maximally exposed individual: The individual in a population who is expected to receive the largest radiation dose.

Maximum reasonably foreseeable accident: Defined by the U.S. Department of Energy as an accident characterized by extremes of mechanical and thermal forces, and other conditions not specified, that leads to the "highest reasonably foreseeable consequences." DOE defines any accident that has the chance of occurring more than 1 in 10 million times per year as being reasonably foreseeable.

Mean collective dose risk: The collective dose received by the population from an accident times the probability of occurrence of that accident.

Mixed oxide fuel: Nuclear fuel that contains uranium and plutonium.

Mode: See *Transportation mode*.

Modular Emergency Response Radiological Transportation Training (MERRTT): U.S. Department of Energy-developed training materials that provide information on fundamental concepts and procedures for responding to radioactive materials transport incidents.

Mostly rail scenario: U.S. Department of Energy's preferred scenario for using trains to ship spent fuel and high-level waste to a federal repository at Yucca Mountain, Nevada.

Mostly truck scenario: U.S. Department of Energy's alternate scenario for using trucks to ship spent fuel and high-level waste to a federal repository at Yucca Mountain, Nevada.

MTHM: Metric tons of heavy metal, where the heavy metal is usually uranium. This is a commonly used measure of fuel quantity.

Natural background radiation: Radiation that exists naturally in the environment. It includes cosmic and solar radiation, radiation from radioactive materials present in rocks and soil, and radioactivity that is inhaled or ingested.

Naval spent fuel: Spent fuel from nuclear submarines, ships, and training reactors belonging to the U.S. Navy.

Nominal probability coefficient: The slope of the linear no-threshold relationship between radiation dose and effect, the latter usually expressed in terms of fatal cancer.

Non-fixed surface contamination: Contamination that adheres to the outer surfaces of a package and can be detected by wiping.

Normal conditions of transport: Transport conditions in which the package is subjected to minor mishaps due to rough handling or exposure to weather. Such minor mishaps would not be expected to compromise the vital containment functions of the package.

Normal form radioactive material: Defined in 10 CFR Part 71 as radioactive material that is not special form radioactive material.

Nuclear fuel cycle: The cradle-to-grave processes for obtaining, using, recycling, and disposing of uranium for nuclear applications such as electric power generation.

Nuclear Waste Fund: A fund established by the U.S. government and funded through a 1.0 mil (0.1 cents) per kilowatt-hour fee on nuclear-generated electricity to pay for the costs of disposing of commercial spent fuel.

Nuclear Waste Policy Act: A federal law passed in 1982 and amended thereafter that provides for the development of a federal program to develop and operate a federal repository for the disposal of spent fuel and high-level waste.

Open fuel cycle: Also known as the *once-through fuel cycle*. A type of nuclear fuel cycle in which fuel is disposed of after it becomes spent, with no effort made to recover its usable components.

Operational controls: Restrictions placed on individual shipments of spent fuel and high-level waste to reduce the likelihood and/or consequences of accidents.

Other than high-income economy countries: Defined by the U.S. Department of Energy to include Argentina, Bangladesh, Brazil, Chile, Colombia, Greece, Indonesia, Iran, Jamaica, Malaysia, Mexico, Pakistan, Peru, Philippines, Portugal, Romania, Slovenia, South Africa, South Korea, Thailand, Turkey, Uruguay, Venezuela, and Zaire.

Overweight truck: A truck exceeding the legal-weight limit of 80,000 pounds (approximately 36 metric tons) but within the range of weights that states approve for operation on public roads after issuance of special permits.

Package: See *Transportation package.*

Package closure system: The system of seals and lids on a transportation package that provides for an airtight closure to prevent the release of radioactive materials to the environment.

Package dose rates: The dose rate on the external surface of the package.

Package performance: The ability of a transportation package to maintain a high level of containment effectiveness in long-term routine use and under extreme mechanical and thermal loading conditions.

Package Performance Study: A U.S. Nuclear Regulatory Commission study of package response to extreme thermal and mechanical loading conditions.

PATRAN/PThermal: A commercial computer code used to model heat convection, conduction, and radiation transport processes.

Perceptions: Defined by *Webster's Third New International Dictionary* as the "integration of sensory impressions of events in the external world . . . as a function of nonconscious expectations derived from past experience."

Physical protection: Security measures to protect spent fuel and high-level waste shipments from malevolent acts.

Plutonium (Pu): A naturally radioactive actinide element with atomic number 94.

Private Fuel Storage, LLC: A private company that is developing a surface facility in Utah for the interim storage of commercial spent fuel.

PRONTO 3D: A computer code developed by Sandia National Laboratories that has been used to model transportation package response to thermal and mechanical loading conditions.

Publics: Groups of people differentiated by both demographic (ethnicity, income) and interest-based criteria.

Puncture test: One of the hypothetical accident conditions in 10 CFR Part 71 in which a package is dropped through a distance of 1 meter (about 40 inches) onto the upper end of a 6-inch (15.2-centimeter) diameter solid, vertical, cylindrical mild steel bar mounted on an essentially unyielding horizontal surface. The package is dropped onto the bar in a position that is expected to produce maximum damage.

PWR: Pressurized water reactor, a type of nuclear reactor in which the reactor's water coolant is kept at high pressure to prevent it from boiling. The heat from this water is transferred to a secondary water system that produces steam to drive a turbine and electrical generator to produce electricity.

Quality assurance program: Defined in 10 CFR Part 71 as those planned and systematic actions necessary to provide adequate confidence that a system or component will perform satisfactorily in service.

Radiation dose: The quantity of radiation deposited in an object divided by the mass of the object. The radiation dose of interest in this report is ionizing radiation. Ionizing radiation doses can be expressed as an absorbed dose, equivalent dose, or effective dose.

Radiation dose rate: The quantity of ionizing radiation deposited into an object per mass of the object per unit time.

Radiation exposure: The act of being exposed to radiation. Also referred to as *irradiation.*

Radiation shine: Radiation emitted from a transportation package containing spent fuel or high-level waste.

Radioactive: Elements that are unstable and transform spontaneously (i.e., decay) through the emission of ionizing radiation, a process known as *Radioactive decay*.

Radioactive decay: See *Radioactive*.

Radioactive Material Incident Report Database: A database maintained by Sandia National Laboratories that contains information about radioactive materials transportation incidents that have occurred in the United States since 1971.

RADTRAN: A computer code developed by Sandia National Laboratories that has been used to estimate radiological exposures and consequences under both normal transport conditions and accidents.

Railhead: The terminus of a rail line, usually within easy access of a public roadway or water port.

Rail package: A transportation package designed to be transported by rail.

Repository: See *Federal repository*.

Research reactor: Small nuclear reactor used primarily to conduct research, to develop theoretical practices, and for education or medical purposes. These serve as sources of neutrons for spectrographic and radiographic applications and for the manufacture of radionuclides for medical and other uses.

Research reactor spent fuel: Spent fuel produced from research reactors.

Resuspension inhalation: Inhalation of radioactive materials that were deposited onto the ground and later resuspended in air.

Risk: As used in this report, the potential for an adverse effect from the transport of spent fuel or high-level waste. This potential can be estimated quantitatively if answers to the following three questions can be obtained: (1) What can go wrong? (2) How likely is it? (3) What are the consequences? Risk can be expressed in absolute terms or in comparison to other types of risks.

Risk curves: See *Complementary cumulative distribution function.*

RISKIND: A computer code developed by Argonne National Laboratory that has been used to estimate local, scenario-specific radiological doses to maximally exposed individuals.

Routine conditions of transport: Transport conditions that are free of minor mishaps (see *Normal conditions of transport*) or accidents (see *Accident conditions of transport*).

Safeguards: Steps taken to ensure that special nuclear material is not stolen or otherwise diverted for possible use in nuclear explosives or for radiological sabotage.

Safe havens: Preplanned locations where a transport vehicle can stop in case of an emergency and receive protection by police or other security forces.

Safety: Measures taken to protect spent fuel and high-level waste during handling and transport from failure, damage, human error, and other inadvertent acts.

Scale model: A dimensionally accurate representation of an object, usually at a reduced size, that is used for design and testing purposes.

Security: In the context of this report, measures taken to protect spent fuel and high-level waste during handling and transport from sabotage, attacks, and theft.

SI: International System of Units (from the French Système International d'Unités), also sometimes referred to as the metric system.

Small-quantity shipping programs: Transportation programs that ship on the order of tens of metric tons of spent fuel or high-level waste.

Social amplification of risk: Social processes that increase the consequences of a particular risk.

Social process: A characteristic mode of social interaction. Such interactions shape the communities in which people live by influencing choices about where to purchase or rent a home, where to work, and where to send children to school.

Social risk: Risks that arise from social processes and human perceptions.

Societal risk: All risks that affect society, including the health and safety risks and social risks discussed in this report.

Source materials: Defined in the Atomic Energy Act to include any combination of uranium and thorium, in any physical or chemical form, or ores that contain at least 0.05 percent by weight of uranium, thorium, or any combination thereof. Depleted uranium is considered to be a source material.

Special form radioactive material: Defined in 10 CFR Part 71 as radioactive material that exists as a single solid piece or is encapsulated material that meets certain other requirements.

Special nuclear material: Defined in the Atomic Energy Act as plutonium, uranium-233, or uranium enriched in the isotopes uranium-233 or uranium-235. These isotopes are fissile and can be used in nuclear explosive devices.

Spent fuel: See *Spent nuclear fuel.*

Spent fuel basket: See *Fuel basket.*

Spent fuel pool: A water-filled pool that is used for storage of spent fuel assemblies after their removal from a nuclear reactor.

Spent nuclear fuel: Fuel whose fissile radionuclides have been consumed by fission to a point where it is no longer efficient for its intended purpose (e.g., for producing heat, electricity, or neutrons).

Standard contract: Created by the Nuclear Waste Policy Act, it specifies the responsibilities of the U.S. Department of Energy and owners of commercial spent fuel for handling, receipt, and transportation of spent fuel to a federal repository for permanent disposal.

Stigma: An event or condition that marks a person, place, product, or technology as deviant, flawed, or undesirable. When the particular stigmatizing characteristic is observed, the person, place, product, or technology may be denigrated or avoided.

Strain: The deformation of an object under an applied force.

Thermal creep: Deformation of an object over an extended period of time when subjected to elevated temperatures.

Thermal test: One of the hypothetical accident conditions in 10 CFR Part 71 in which a package is fully engulfed in a hydrocarbon-fuel fire with an average flame temperature of at least 800°C (1472°F) for a period of 30 minutes.

Thermally induced creep: See *Thermal creep.*

Thermomechanical conditions: Describes the mechanical forces and heat loads applied to a transportation package.

Transportation Emergency Preparedness Program: A program developed to support training of federal, state, tribal, and local authorities in emergency preparedness and response to transportation incidents involving DOE radioactive materials shipments.

Transportation mode: Describes the means by which spent fuel and high-level waste is transported, for example, by truck, train, or barge.

Transportation operations: The spectrum of activities associated with the shipment of spent fuel and high-level waste.

Transportation package: In the context of this report, containers used for the transport of spent fuel or high-level waste, whether loaded or empty. Loaded packages are referred to as *packages* in international standards, whereas the containers themselves without their contents are referred to as *packagings*. The terms *casks* and *flasks* (the latter term is commonly used in the United Kingdom) are sometimes used synonymously with packages. This report distinguished between two types of transportation packages: *bare-fuel packages* in which the spent fuel and the fuel basket are placed directly into the package, and *canister-based packages* in which the spent fuel and fuel basket are placed into a welded steel canister that in turn is placed in the package.

Transuranic waste: Radioactive waste containing long-lived radioactive transuranic elements (elements with atomic numbers greater than 92) such as plutonium in concentrations greater than 100 nanocuries per gram.

Truck package: A transportation package designed to be transported by road.

Type A package: A package designed for the transport of materials of limited radioactivity—for example, uranium hexafluoride and fresh nuclear fuel.

Type B package: A package designed for the transport of larger quantities of material including spent fuel, high-level waste, and mixed oxide fuel.

Type C package: A package designed for the air transport of quantities of radioactive material exceeding a defined threshold including, for example, plutonium and mixed oxide fuel.

Uranium (U): A naturally radioactive actinide element with atomic number 92.

Very long duration fires: Fires that burn for periods of hours (or longer), which can produce thermal loading conditions that exceed those for the regulatory thermal test specified in 10 CFR 71.73.

Vitrification: A process for immobilizing radioactive waste, particularly high-level waste, in glass matrices.

Vulnerable communities: Communities of people who, because of disproportionate exposures to other health-affecting substances, or because of ethnic, linguistic, or socioeconomic issues, may be less able to read or understand information from the authorities, to act in a first-responder role, to exit the area in a timely manner in an emergency, or to otherwise cope with an emergency.

APPENDIX

E

Acronyms

AAR	Association of American Railroads
AEA	Atomic Energy Act
AEC	Atomic Energy Commission
ALARA	as low as reasonably achievable
ASME	American Society for Mechanical Engineers
BAM	German Federal Institute for Materials Research and Testing
BLM	Bureau of Land Management
BNFL	British Nuclear Fuels Limited
BWR	boiling water reactor
CCDF	complementary cumulative distribution function
CEGB	Central Electricity Generating Board
CFR	Code of Federal Regulations
CT	computed tomography
CVTR	Carolinas-Virginia Tube Reactor
DHS	U.S. Department of Homeland Security
DNA	deoxyribonucleic acid
DOE	U.S. Department of Energy
DOE-EM	U.S. Department of Energy, Office of Environmental Management

DOE-NE U.S. Department of Energy, Office of Nuclear Energy
DOE-NR U.S. Department of Energy, Naval Reactors Program
DOE-RW U.S. Department of Energy. Office of Civilian Radioactive Waste Management
DOT U.S. Department of Transportation
DOT-FMCSA U.S. Department of Transportation, Federal Motor Carrier Safety Administration
DOT-FRA U.S. Department of Transportation, Federal Railroad Administration
DOT-NHA U.S. Department of Transportation, National Highway Administration
DU depleted uranium

EIS Environmental Impact Statement
EJ Environmental Justice
EPA U.S. Environmental Protection Agency
EPRI Electric Power Research Institute
ERDA Energy Research and Development Administration

FEMA Federal Emergency Management Agency
FHWA Federal Highway Administration
FICA Facility Interface Capabilities Assessment
FRA Federal Railroad Administration
FRR foreign research reactor

GAO U.S. Government Accountability Office
GPS global positioning system

HEU highly enriched uranium
HFIR High Flux Isotope Reactor
HLW high-level waste
HMIS Hazardous Materials Incident Reporting System
HSPD Homeland Security Presidential Directive
HTGR high-temperature gas reactor

IAEA International Atomic Energy Agency
ICRP International Commission on Radiological Protection
IME Institution of Mechanical Engineers
INEL Idaho National Engineering Laboratory
INL Idaho National Laboratory
INPO Institute for Nuclear Power Operations
ISO International Organization for Standardization

LCF	latent cancer fatality
LEU	low enriched uranium
MERRTT	Modular Emergency Response Radiological Transportation Training
MN	meganewton
MOU	memorandum of understanding
MOX	mixed oxide fuel
MRFA	Maximally Reasonably Foreseeable Accident
MTHM	metric tons of heavy metal
NCRP	National Council on Radiation Protection and Measurements
NEPA	National Environmental Policy Act
NIMS	National Incident Management System
NIST	National Institute of Standards and Technology
NPP	nuclear power plant
NRC	National Research Council
NSTI	Near Site Transportation Interface
NWPA	Nuclear Waste Policy Act
NWS	Naval Weapons Station
OCRWM	Office of Civilian Radioactive Waste Management (DOE)
ORNL	Oak Ridge National Laboratory
OSHA	Occupational Safety and Health Administration
PATRAM	International Symposium on Packaging and Transportation of Radioactive Materials
PFS	Private Fuel Storage, LLC
PWR	pressurized water reactor
RITA	Research and Innovative Technology Administration
RMIR	Radioactive Material Incident Report
ROD	Record of Decision
RSPA	Research and Special Programs Administration
SI	International System of Units as defined by the General Conference of Weights and Measures in 1960
SNF	spent nuclear fuel
SRS	Savannah River Site
SSEB	Southern States Energy Board

TEC	Transportation External Coordination (Working Group)
TEPP	Transportation Emergency Preparedness Program
TRANSSC	Transport Safety Standards Committee
TRUPAC II	Transuranic Waste Transport Package
UER	Urban Environmental Institute, LLC
USNRC	U.S. Nuclear Regulatory Commission
WGA	Western Governors' Association
WIPP	Waste Isolation Pilot Plant
W/MTU	watts per metric ton of heavy metal